Energy simulation in building design

Energy simulation in building design

J A Clarke
BSc, PhD
Advanced Research Fellow
ABACUS
Faculty of Engineering
University of Strathclyde
Glasgow, Scotland

Adam Hilger Ltd
Bristol and Boston

British Library Cataloguing in Publication Data

Clarke, J. A.
  Energy simulation in building design.
  1. Buildings—Energy conservation—
  Mathematical models
  I. Title
  696   TJ163.5.B84

  ISBN 0-85274-797-7

Published by Adam Hilger Ltd
Techno House, Redcliffe Way, Bristol BS1 6NX
PO Box 230, Accord, MA 02018, USA

Typeset by Mathematical Composition Setters Ltd, Salisbury
Printed in Great Britain by J W Arrowsmith Ltd, Bristol

To
Kathryn,
Fiona, Karen and Andrew

Contents

Preface

Internationally the energy picture is a complex and changing one. In the UK, for example, there has been a steady increase in the total consumption of primary fuels (except coal) up to 1973, the year of the oil embargo. Since then, escalating costs have promoted the view that pre-1973 consumption levels were profligate and so post-1973 years have seen the UK and other major consumers become more energy conscious: in the UK the 1980 consumption level was not significantly different from the 1970 level.

Consideration of energy in relation to the built environment reveals that, in Europe and the US for example, in excess of 50% of all delivered energy can be associated with buildings and of this some considerable portion (more than 60% in the UK, for example) is consumed to moderate spatial conditions. Thus retrofit and innovatory design measures are concerned with the management and potential conservation of significant amounts of energy. By effective implementation of such measures, the most commonly agreed and achievable target reduction is around 30%, with more optimistic expectations—up to 70% relative to 1973 figures—for buildings incorporating new technology features.

Is a prosperous future feasible then, at low energy cost? One prerequisite of this is, perhaps, the existence of modelling systems which can be used to promote effective but frugal use of available energy resources. Traditionally, designers have relied on indicative calculation methods based on empirical considerations or simplifying assumptions such as steady state behaviour or perfect control. Such methods have many deficiencies which, until the energy 'crisis', were latent because energy matters had low priority. Since 1973 a demand for explicit and accurate energy evaluation has exposed these deficiencies and, in response, a new breed of model has emerged which seeks to replace—as the next generation—the traditional methods.

A number of factors serve to distinguish between these models and the previous generation: they utilise simulation techniques rather than simple calculations, and so they deal by *explicit* means with the complex dynamic behaviour of buildings as they respond to real climatic influences; they attempt to adhere to the reality rather than being 'lumped parameter' and therefore merely indicative; they are *dynamic* rather than *static*; and they are highly flexible in use.

This book describes the elements and tasks inherent in the development of such an energy modelling system. Whilst the main issues covered—theoretical basis, validation, software development and use in practice—are not model specific, the material covered draws heavily on a major model research and development project undertaken at the ABACUS unit within the Faculty of Engineering at the University of Strathclyde and funded by the UK Science and Engineering Research Council. Although the book is technical in nature—addressing the theory of the various heat transfer processes occurring within buildings and describing a numerical solution scheme which allows exact and efficient solution of large equation sets representing combined building/plant systems—an attempt has been made to mix matters of fact and experience as appropriate. The objectives of the book are therefore threefold: to present material of use to those developing advanced modelling systems, to demonstrate the building and testing of one such system and to detail the current state of the art for those wishing to apply the new technology.

I am deeply indebted to many people who, over the years, have given guidance and encouragement. To Professor Tom Maver and colleagues at ABACUS I give my thanks for creating a social and intellectual environment within which it was a pleasure to work, and to Dr Don McLean I give my thanks for his help and support throughout manuscript preparation. I am also grateful to the many colleagues—too numerous to mention—who gave helpful comments, trivial and comprehensive, on the draft manuscript.

Finally, a very special thanks to my supportive and tolerant wife and children to whom I dedicate this book.

Joe Clarke

1

Introduction

Since the advent of the digital computer the potential to predict future reality has taken a quantum jump. In each application area—weather prediction, satellite control, economic forecasting, process control and so on—the availability of increasing computing power has allowed a large number of possibilities to be examined and assessed in the context of some underlying model, established in software form, which embodies the causal relationships observed in reality. In almost every field the potential is considerable and such computer-based modelling systems, if properly integrated within the decision-making process, hold much promise for the improvement in efficiency and quality of the related process or product.

In the field of building design many subsystems can be identified which can benefit from a computer-aided building design (CABD) approach:

Movement simulation can be used to check building population build-up at critical points during anticipated events such as evacuation under emergency conditions.

Spatial synthesis can be aided by software which makes allowance for area requirements, inter-space dependencies (associations) and journey time criteria.

Capital and running cost predictions can be made against any (variable) background inflation rate, separate fuel inflation, general interest rate and building life. And, since cost estimation is an imprecise process, it is possible to vary the estimation technique from probability-based at an early design stage to quantity-based as more detailed design information becomes available.

Computer generated line perspective images can be used to check sightlines for complex obstruction geometries, photomontage techniques aid meaningful visual impact assessment for environmental planning, and the introduction of advanced colour generation can aid final finish and lighting fitting selection to promote aesthetic aspects by providing direct experiential information.

Energy analysis systems can be used to provide insight, during the design phase or post occupancy, into issues of comfort, energy saving or management strategies, the optimum configuration of the building components, and the impact of various operational regimes.

It is against this general CABD background that any particular application can and must be meaningfully placed. This book is concerned with the last application: dynamic energy modelling of the energy flowpaths within buildings and their associated climatic control systems.

With the growing implications of the terms 'conservation' and 'management' those concerned with the design and management of buildings are becoming increasingly aware of the need for valid design appraisal tools—tools which would permit an accurate prediction of energy consumption and performance characteristics at an early design stage where the consequence of design intervention is maximised or, alternatively, which can be used to appraise operational options throughout the life of a building to maintain optimum fuel consumption. Unfortunately the building and testing of quality appraisal tools is a non-trivial task requiring detailed knowledge of the mechanisms of the contribution (to overall energy flow) of a number of subsystems: the building form and fabric; the layout and operational characteristics of plant; the leakage and pressure distribution which affect air movement; and the actions of building users. Thus the energy 'signature' of a building is a complex function of many interrelating energy flowpaths each of which can be substantially influenced (in a positive or negative sense) by design intervention or occupancy behaviour.

Several models have emerged over the years following the oil embargo of 1973 and in response to the ensuing energy cost escalation. Based on advanced theories, they seek to replace, as the new generation, the many traditional techniques based for the most part on simplifying assumptions or empirical paradigms which (it can be argued) do not offer acceptable accuracy or flexibility when applied to contemporary design problems.

The objectives of this book are threefold: to demonstrate the theories underlying advanced energy modelling techniques; to construct, from first principles, a simulation model capable of processing a building and its plant system simultaneously and in a mathematically exacting manner; and to introduce the reader to the problems surrounding model validation and implementation of the technology in practice.

Chapter 1 discusses the various energy flowpaths occurring inside and outside buildings and considers prevailing interactions. The case for advanced modelling is argued and the available modelling techniques are summarised.

Chapter 2 derives, in detail, the two main analytical formulations for advanced building energy modelling—time-domain and frequency-domain response function methods—and sets out the elements for an alternative numerical method based on finite difference heat balance considerations applied to control volumes.

Chapter 3 demonstrates the step-by-step formulation of such a numerical model by deriving time-dependent energy balance equations—one for each characteristic building region—for use in the structuring of equation sets which relate to some discrete nodal equivalent of a distributed building system.

Chapter 4 demonstrates equation-set (or matrix) formulation and describes various techniques, based on matrix partitioning protocols, by which fast simultaneous matrix solution can be achieved.

Chapter 5 addresses the various sub-problems of the energy modelling task. A basic geometrical description technique is defined which allows the specification of non-orthogonal geometries and from which quantities such as planar angles, contained volumes and areas can be directly extracted. The theoretical and algorithmic basis is also given for each of the following time-dependent processes: external and internal surface shading and insolation; shortwave and longwave radiation; natural and forced surface convection; infiltration and zone-coupled air movement; radiant and convective casual heat sources; climatic data availability and severity assessment.

Chapter 6 addresses plant simulation by applying the numerical technique of chapter 3 to selected plant systems. The interlocking of the energy flow matrices to emerge with the previously derived building matrices is then demonstrated.

Chapter 7 considers the transformation of the developed numerical model into some software equivalent. Software logic structure and data handling are discussed and recommendations given on development languages and target operating systems. Indicative performance data are also given to enable the assessment of new systems during the development phase.

Lastly, chapter 8 considers the issues surrounding model validation and implementation in practice. In the former case, a validation methodology is discussed and current and planned work described. In the latter case, the barriers to implementation are outlined and, by reference to a series of short case studies of a developed model in use, the potential of an advanced modelling approach is demonstrated.

The book is concerned to demonstrate the development and testing of advanced building/plant energy simulation systems which achieve, as their prime objective, the conservation of energy, mass and system integrity by processing combined building/plant configurations rigorously and simultaneously.

The format of the book adheres to the software development process as expounded by Maver and Ellis (1982), in which development is viewed as a sequence of stages progressing from initial research to commercial exploitation as follows:

1   Research into model needs, methods, algorithms and organisation. This leads to a research prototype embodying the fundamental laws governing heat flow.
2   Development of a pilot applications program based on the research findings and which offers a reasonable user interface.
3   Validation of the model to test the physical assumptions and the selected numerical scheme.
4   Implementation trials to test the robustness, relevance and efficacy of the software in the real-world, real-time context of design practice.
5   Improvement of the software and documentation with respect to commercial standards by the incorporation of the lessons learned through the validation and trial implementation studies.

6 Commercial exploitation.

Thus chapters 1–6 are concerned with step 1 and chapters 7 and 8 with steps 2–6.

1.1 Energy flowpaths and causal effects

Figure 1.1 shows the various flowpaths commonly encountered inside and outside buildings and which interact, in a dynamic manner, to dictate comfort levels and energy demands. Underlying these flowpaths is the concept of energy, mass and momentum balance which requires, in turn, a knowledge of the fundamental processes of conduction, convection and radiation exchange. Many excellent texts exist which cover the fundamentals of heat transfer (for example, Kreith 1973, Ozisik 1977, Incropera and DeWitt 1981) and no attempt is made here to commence this book with an elementary treatment of the subject. Instead, specific mathematical models are introduced for each flowpath and, as the book develops, these are combined to form a single, unified mathematical structure which is the statement of whole system balance. It is necessary therefore to commence with an explanation of the flowpaths encountered in building systems and so candidates for inclusion in the simulation model.

Transient conduction
This is the process by which a fluctuation of heat flux at one boundary of a solid material finds its way to another boundary, being diminished in magnitude due to material storage, and shifted in time. Within the building fabric, transient conduction is a function of the temperature and energy excitations at exposed surfaces, the temperature-dependent (and therefore time-dependent) thermophysical properties of the individual homogeneous materials, and their relative position. For modelling and simulation purposes it is usual to declare external climatic excitations as known time-series data with the objective to determine internal transient energy flow and hence the dynamic variation of heat flux at internal surfaces.

The thermophysical properties of interest include conductivity (W m^{-1} °C^{-1}), density (kg m^{-3}) and specific heat capacity (J kg^{-1} °C^{-1}) as well as physical dimensions. These properties are themselves time-dependent because of intra-material temperature and/or moisture fluctuations. However, in many applications, such dependencies are ignored and properties are assumed to be invariant in the time dimension. Materials with high thermal diffusivity values (conductivity divided by the product of density and specific heat; units m^2 s^{-1}) transmit boundary heat flux fluctuations more rapidly than do materials with correspondingly low values. Appendix B lists the basic thermophysical properties of a range of building materials. Listed properties include conductivity, density, specific heat, surface emissivity and solar absorptivity.

The relative position of the different materials within composite constructions can also greatly influence transient behaviour. Traditionally designers have relied

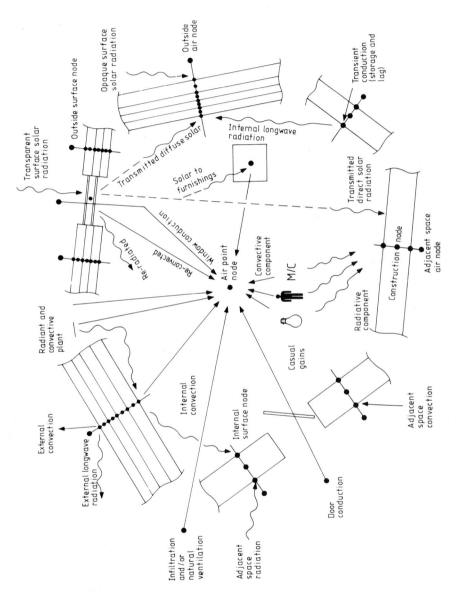

Figure 1.1 Building energy flowpaths.

on the simple steady-state U-value concept to assess the heat loss characteristics of the building fabric. In addition to ignoring the dynamic aspects of fabric behaviour, this approach does not preserve the spatial integrity of a multi-material construction since in reality different constructions can perform differently although each may have the same U-value.

As an example, if insulation is located at the innermost position of a wall then any shortwave solar radiation penetrating windows and striking that internal surface cannot be readily stored in the construction since the insulation will act as a barrier. Instead, the solar energy will cause a surface temperature rise which, in turn, will increase the rate of energy release to the adjacent air by the process of natural convection. A space experiencing high solar energy penetration is therefore likely to overheat if cooling is not introduced in some way. Conversely, if the insulation is relocated externally with capacity elements exposed to the inside then internal surface shortwave gain can access capacity to be stored. By proper design this stored energy can later be harnessed (passively rather than by mechanical means) to minimise heating requirements rather than cause overheating. On the other hand internal capacity may give rise to increased peak plant demand due to the initial rush of energy to capacity at plant start-up in an intermittent scheme. With continuous operation, capacity can help to minimise the peaks and maximise the troughs of plant demand and so promote good load levelling. This in turn will give a stable environment and encourage efficient plant operation by allowing plant to operate consistently at or near full load. The risk of interstitial condensation is, however, greater in the case of internally located insulation, since a substantial portion of the construction may fall below the dew point temperature of moist air permeating through the construction in the absence of an effective vapour barrier.

In summary, transient conduction will affect energy requirements, load diversity and peak plant demand, load levelling and plant operating efficiency, and condensation potential.

Unfortunately there is no simple design paradigm which can be used to select an optimum construction. Table 1.1, for example, shows the effect on cooling energy requirements and load levelling for a number of combinations of construction and plant operation. If these results are accepted, then it is clear that there is a need to utilise dynamic models to determine the performance of alternative constructions when each are combined with the many other combinatorial design features: whilst a U-value can be used as a simple selection index, it has little if no role as a sophisticated indicator of construction energy performance.

Surface convection
This is the process by which the heat flux, emanating at some opaque or transparent surface, is transmitted to an adjacent fluid layer. For building surfaces it is usual to differentiate between external and internal exposures. In the former case convection is usually considered as forced whereas, with internal surfaces, natural and/or forced air movement can be assumed, depending on the available information on mechanical equipment and the convective field to result.

Table 1.1 Effect of construction and plant operation on building performance (from Clarke 1980).

	Construction†	Energy (kW h)	Load levelling ($Q_{max} - Q_{min}$; kW)	Notes
continuous operation	40I /150C/ 40I‡	410	22	Load levelling (LL)
	5I /150C/ 75I	396	18	best when more
	75I /150C/ 5I	413	25	insulation to outside of capacity
	75C/ 80I / 75C	398	17	LL best when more
	25C/ 80I /125C	411	20	capacity to inside of
	125C/ 80I / 25C	391	16	insulation
intermittent operation	40I /150C/ 40I	268	37	LL best when insulation
	5I /150C/ 75I	305	39	split equally either side
	75I /150C/ 5I	275	43	of capacity
	75C/ 80I / 75C	330	44	LL best when more
	25C/ 80I /125C	336	53	capacity to inside of
	125C/ 80I / 25C	328	42	insulation

†Each construction comprises 3 homogeneous elements each of which can be insulation (I) or capacity (C). The specifications are all inside to outside and all constructions have the same U-value.
‡Dimensions in mm.

It is normal practice in simulation modelling to make use of dimensioned convection coefficients (W m^{-2} $^\circ$C^{-1}) which represent some average value for a particular finite surface area but can change with time. In the fluid flow literature it is common to express fluid-to-surface heat transfer in dimensionless terms but, in recent years, some workers (Alamdari and Hammond 1982) have recognised the needs of the modelling community and produced correlation expressions to allow estimation of time-dependent but surface-averaged convection coefficients of much use in simulation applications.

Forced convection is a function of the prevailing fluid flow vector. Typically, for external building surfaces, wind speed and direction data are available for some reference height and simple techniques exist to estimate non-reference height values in terms of characteristic vertical velocity profiles. Forced convection estimation for internal surfaces is more problematic, requiring knowledge of the distribution and operation of air handling equipment and heat emitters and the nature of the boundary layer at each surface position (laminar, turbulent or transitional).

Natural convection is an easier problem to study and many formulations have emerged which express coefficients as simple functions of the surface-to-fluid temperature difference; surface aspect, roughness and dimension; and direction of heat flow.

Inter-surface longwave radiation

In many calculation methods surface heat transfer coefficients are treated as combinations of convection and longwave radiation although the values used are often dubious. In reality the two processes are related by the fact that they both conspire to raise or lower surface temperatures and so influence each other.

Inter-surface longwave radiation is a function of the prevailing surface temperatures; the emissivity of each surface; the extent to which the surface pair are in visual contact, often referred to as the view factor; and the nature of the surface reflection, specular or diffuse. The flowpath will tend to establish surface temperature equilibrium by cooling hot, and heating cold, surfaces. It is most important under conditions of asymmetric heating often found within passive solar applications in which an attempt is made to capture shortwave solar energy at some elected surface.

The mathematical representation of the flowpath is non-linear in the temperature term and this will introduce complications in simulation modelling where the condition of linearity is desirable.

External surface longwave radiation

The exchange of energy by longwave radiation between external (opaque and transparent) surfaces and the sky vault, surrounding buildings and ground can result in a substantial lowering of surface temperatures especially under clear sky conditions at night. This process alone can result in sub-zero surface temperatures, especially in exposed roofs, and can become critical in cases of low insulation level. Conversely, the flowpath can result in a net gain of energy, although under most conditions this would be negligible.

From a modelling viewpoint an adequate treatment of this flowpath will require the ability to estimate the effective sky temperature as a function of prevailing cloud cover and type; the temperature of surrounding buildings; the temperature of the ground as a function of terrain conditions, air temperature and incident shortwave energy; and the relevant view factor information which visually couples the surface with the three portions of its scene.

Shortwave radiation

In most buildings the gain of energy from shortwave penetration constitutes a significant portion of the total loading and therefore the method of treatment of shortwave flowpaths can largely determine the accuracy of the overall predictions.

Some portion of the shortwave energy (arriving directly from the sun or diffusely after atmospheric scatter and terrain reflections) impinging on an exposed surface may—depending on subsequent temperature and energy variations affecting transient conduction—eventually find its way through the structure where it will contribute to the inside surface heat flux at some time later. It is not uncommon for exposed surfaces to be as much as 10–15 °C above ambient temperatures. Many existing techniques utilise the concept of a 'sol–air' temperature which represents some suitably elevated ambient temperature for use as the index in subsequent wall conduction calculations. This is clearly inadequate on two points:

1 Unless the solar contribution to the sol–air temperature is determined on the basis of time-dependent surface properties such as shading, film coefficients and incidence angles, then a difference will prevail between actual solar absorption and that predicted.
2 Insulation/capacity structures are often a mix of opaque, transparent and translucent materials and so it is in some cases important to permit deep structure shortwave penetration.

In the case of completely transparent structures, the shortwave energy impinging on the outermost surface is partially reflected and partially transmitted. Within the layers and substrates of the system many further reflections take place and some portion of the energy is absorbed within the material to raise its temperature. This temperature rise will augment the normal transient conduction processes and by this mechanism help to establish extreme innerside and outerside surface temperatures which then, in turn, drive the surface convective and longwave radiative flowpaths. Thus, in effect, absorbed shortwave radiation penetrates the building via convection and longwave radiation.

The component of the incident beam which is transmitted will eventually (with no perceptible time lag) strike some internal exposed surface or surfaces where it behaves as did the external surface impingement: opaque surface absorption and reflection, transparent surface reflection and transmission (to outside or another zone), and behind-the-surface transient conduction where it is stored and lagged.

Accurate solar modelling therefore requires a number of algorithmic methods for the prediction of surface position relative to the solar beam as well as exposed surface shading and the moving pattern of insolation of internal and external surfaces. Surface/solar position is a function of site latitude and longitude, time of day and year, and surface geometry. Accurate shading/insolation estimation requires the existence of ray tracing techniques of the kind derived in chapter 5.

The thermophysical properties of interest include shortwave absorptivity for opaque elements and absorptivity, transmissivity and reflectivity for transparent elements. The magnitude of these properties is dependent on the angle of incidence of the shortwave beam and on its spectral composition. With regard to the latter, it is common practice to accept properties which are averaged for the spectral portion under consideration.

Shading and insolation

These processes control the magnitude and point of application of solar energy and so dictate the overall accuracy of any solar processing. Both time-series are usually expressed as proportions of one or as percentages and will require sophisticated point projection or hidden line/surface techniques for their estimation, as well as access to a data structure which contains obstruction features.

It is usual to assume that facade shading caused by remote obstructions (such as buildings, trees etc) will reduce the magnitude of direct insolation leaving the diffuse beam undiminished. Conversely, shading caused by facade obstructions (such as overhangs, window recesses etc) should also be applied to the diffuse beam since the effective solid angle of the external scene, as subtended at the surface in question, may be reduced.

At any point in time the shortwave radiation directly penetrating an exposed window will be associated with one or more internal surfaces, depending on the prevailing solar angle relative to the window and internal building geometries. Thus the receiving surface(s) may be an opaque surface, a window in another wall (connecting the zone to another zone or to ambient conditions), items of furniture, and so on, depending on the established data structure. While it is true that disregarding the proportioning of window transmitted shortwave energy between the associated receiving planes can have a significant effect on both the quantitative and qualitative aspects of thermal predictions, the smearing of the portion received by one surface over its entirety will have little quantitative effect if the surface can be regarded as uniform, in the sense that it is the same composition with no spatially-dependent boundary conditions (Robinson 1979). Likewise, there is no appreciable qualitative effect for the case of uni-directional conduction heat flow representations.

Fluid flow
In building simulation modelling two fluid flowpaths predominate: infiltration and zone-coupled air flow. And these flowpaths give rise to advective (fluid-to-fluid) heat exchanges. Both are vector quantities in that only air flow into a region is considered to cause thermal loading, any air loss merely being the driving force for a corresponding replacement to maintain a mass balance.

Infiltration is the name given to the leakage of air from outside and can be considered as comprising two components: the unavoidable movement of air through distributed leakage paths such as the small cracks around windows and doors, through the fabric itself and at material junctions; and the ingress of air through intentional openings (windows, vents etc), often referred to as *natural ventilation*.

Zone coupling, like infiltration, is caused by pressure variations and by buoyancy forces caused in turn by density variations due to the temperature difference between the coupled volumes of air.

Thus random occurrences such as window and door opening and changes in the prevailing wind conditions or the intermittent use of mechanical ventilation will have some effect on infiltration, natural ventilation and zone-coupled air flow. Although the effect (of these occurrences) on air movement is difficult to determine, models of varying complexity can nevertheless be constructed. Such models will span the spectrum from simple whole building predictors based, say, on linear regression methods to complex simulation systems involving a numerical solution of some governing partial differential flow equation.

At a level appropriate to building energy modelling, air movement is often represented by a simplified nodal network in which nodes represent volumes and nodal connections represent the distributed leakage connecting the volumes and through which air movement may occur. Numerical techniques can then be applied to this network to establish the mass balance corresponding to any particular nodal pressure and temperature field.

Casual gains

In many buildings the effects of heat gains from lighting installations, occupants and miscellaneous equipment can be considerable, and it is therefore important to process these heat sources in a realistic manner. This will entail the separate processing of the radiant and convective components and the ability to work with time-dependent profiles to allow any casual source to change on an hourly, daily, weekly and/or seasonal basis. It is usual to assume that the convective component is experienced instantaneously as an air load but that the radiant portion, behaving in a similar manner to shortwave radiation penetrating the building envelope, is apportioned among internal opaque and transparent surfaces according to some angular distribution strategy, and so has a relationship with system capacity and will be lagged.

Plant

The problem of predicting the energy consumption of a building is usually divided into two distinct stages. As shown in figure 1.2, the first stage is concerned to predict the energy requirements to satisfy the demands of the building activity. This is found by in some way modifying the various instantaneous heat gains and losses by the ever present thermal storage and lag effects. In the second stage these energy requirements are modified by the operating characteristics of the selected

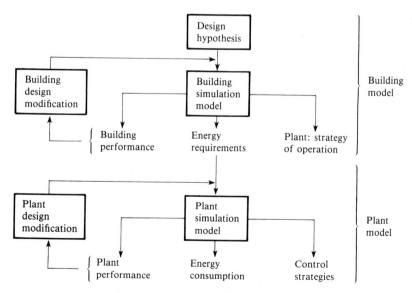

Figure 1.2 The role of building and plant models.

plant to give the energy actually consumed. Thus the first stage is concerned with the design of the building to reduce the energy requirements (and so, perhaps, consumption) whilst the second stage is concerned with the design of the installed plant to best match these requirements and minimise consumption. This is the strategy underlying many of the simulation systems presently available.

It is possible, however, to develop systems which can model the combined building/plant configuration in a simultaneous manner. This is demonstrated in chapter 6 where selected plant systems are combined with the building model developed in chapter 3. Irrespective of the level of plant modelling, the building model will be required to handle control statements which superimpose complex time and thermostatic constraints on the availability of plant capacity to the building system.

Also, it has been the general practice to control simulations on the basis of simple tests applied to zone air temperatures or some equivalent single index and to declare plant in a conceptual manner such as convective (air volume interaction) or mixed (air and surface interaction). It is now possible (and often desirable) to operate in terms of actual controller-sensed temperature determined on the basis of air movement and radiant effects assessed from subtended solid angle considerations. It is also possible to allow plant to interact with building regions in a manner which more closely adheres to the reality.

Moisture

Fluctuations in moisture level will obviously affect cooling loads in systems which permit humidity control. It can also affect the thermophysical properties so often assumed constant for building modelling purposes; an assumption difficult to justify for plant modelling and in certain building applications such as passive solar designs where temperature variations within some components may be considerable.

Other factors

In recent years many designers have come to favour the use of the so-called passive solar features. These act to capture and process solar radiation passively and without recourse to mechanical systems. Consider figure 1.3 which summarises the main passive solar elements. In each case certain factors can be identified which, in particular, will impose technical complexity on any modelling exercise. These are:

(*a*) *Non-diffusing direct gain* systems will require adequate treatment of the mapping of the solar beam on to the internal receiving surfaces or obstruction objects such as furniture.

(*b*) *Diffusing direct gain* systems will require the accurate determination of the spectral behaviour of the window system.

(*c*) *Earth banking* will introduce complexity in the modelling of ground exchange processes.

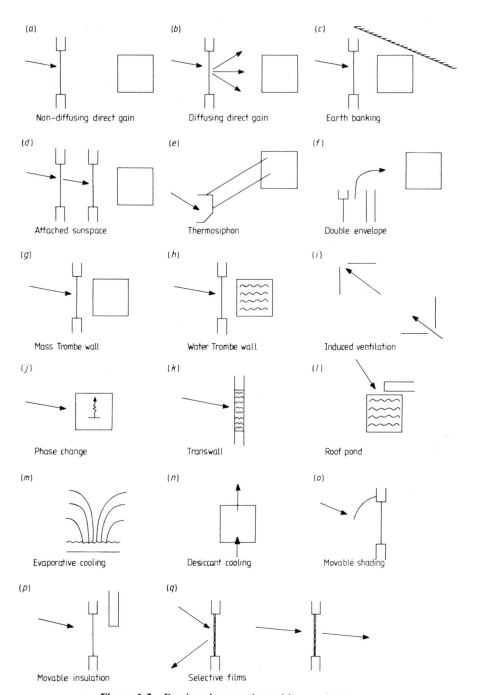

Figure 1.3 Passive elements in architectural design.

(*d*) *Attached sunspace* systems will require that the model be able to establish the level of penetration of solar radiation to interior contained zones.

(*e*) *Thermosiphon* systems require buoyancy driven air flow modelling.

(*f*) *Double envelopes* require sophistication on the part of solar algorithms since radiation may penetrate the external skin to cause 'deep' construction absorption.

(*g*) *Mass Trombe walls* will require the existence of a quality convection model to handle adequately the convection current effects caused by asymmetric heating.

(*h*) *Water Trombe walls* will experience internal convection currents and will allow the transmission of shortwave energy.

(*i*) *Induced ventilation* schemes will require accurate modelling of stack and zone-coupled air flow effects.

(*j*) *Phase change* materials will necessitate the switching from sensible heating behaviour to constant temperature behaviour in the transient conduction schemes.

(*k*) *Transwalls* will impose demands on the solar and conduction algorithms since direct transmission and fluid motion will occur.

(*l*) *Roof ponds* will require accurate external longwave radiation assessment.

(*m*) *Evaporative cooling* systems may require an approach capable of modelling combined heat and mass transfer.

(*n*) *Desiccant cooling* material will exhibit a change of cooling power with time.

(*o*) *Movable shading* will require device movement prediction and sophistication on the part of shading/insolation software.

(*p*) *Movable insulation* implies a time-dependent system definition which, in turn, will place demands on the numerical methods underlying the model.

(*q*) *Selective thin films* act to enhance or reduce solar pick-up and will require detailed spectral analysis facilities for the prediction of angle-dependent short-wave response.

Advanced modelling systems seek to include each of the energy flowpaths outlined in this section, whilst respecting the inevitable interactions and underlying complexities.

1.2 The need for accuracy and flexibility

It is impossible to establish *a priori* the optimum level of model accuracy and flexibility in the field of energy systems appraisal. Indeed, the trade-off between accuracy and flexibility is itself a dynamic concept which will vary according to the modelling and design objectives. Nevertheless, it is important to differentiate between simplified models and comprehensive models which are capable of simple model emulation.

In the former case a number of simplifying assumptions are applied to the underlying thermal network and/or solution scheme so that some flowpaths are omitted entirely or approximated. The model to result is then valid only when applied to problems which embody the same or near-same simplifications.

In the latter case a comprehensive model is designed to operate on input data ranging from 'simplified' to 'detailed' depending on the application in hand. This is achieved by incorporating 'dynamic defaults' to allow the inclusion of any flowpath not explicitly addressed in the input data set.

The latter model is obviously more flexible, with the accuracy level changing as a function of the quality of the design information supplied. The following example is included to demonstrate the consequences of the two approaches.

Following on the introduction of the 1978 Building Regulations for England and Wales, a study was undertaken (ABACUS and VALTOS 1979) to examine the consequences of compliance or non-compliance with the 'deemed-to-satisfy' provisions. Regulation FF3 addressed the 'Conservation of fuel and power' and stated:

> A building or part of a building to which this part applies shall be so designed and constructed that the enclosing structure provides adequate resistance to the passage of heat the loss of which from the building or part would entail the consumption of fuel or power to enable temperature conditions normal for the proposed use of the building or part to be maintained.

Two approaches to complying with FF3 are set out in the provisions of FF4:

1 Walls, floors and roofs of a building must be designed and constructed to meet prescribed U-values and the total percentage areas of the openings provided for windows and rooflights in these walls and roofs must not exceed prescribed limits.
2 A wall, floor or roof may have a higher U-value provided the total rate of heat loss through all the walls, floors and roofs does not exceed that which would have resulted if the first approach had been adopted. Similarly the limits on openings for windows and rooflights may be exceeded provided that the total rate of heat loss through the glazed areas does not exceed that which would have resulted had the limits been observed, for example through the use of double or triple glazing.

The study team felt that it was important to draw a distinction between prescriptive and performance requirements. The deemed-to-satisfy provisions of FF4, by focusing on heat loss rate per square metre of fabric, prescribed allowable construction in large measure, thus precluding innovatory facade treatment. More worryingly, it is entirely possible to satisfy the provisions with a design which, in terms of geometry, thermal mass, insulation, orientation, plant control, etc, may be profligate in energy consumption. Had the provisions dealt directly with performance—maximum annual energy consumption based on typical occupancy and operational statistics—the onus would be on the designer to present a design solution, however innovatory, together with appropriate predictive evidence of its energy behaviour. The issue then, if such a performance concept is accepted, is one of modelling accuracy and flexibility.

As part of the study a hypothetical but entirely typical test case, a multi-storey hotel complex, was subjected to rigorous simulation to determine annual energy requirements against alternative glazing scenarios. Figure 1.4 gives the results (the full curve) and demonstrates that areas of glazing greater than the deemed-to-satisfy limit (25% single glazing in this case) can offer a significant saving in cost-in-use terms (point C4 compared with point R1, the latter obtained by subjecting the regulation limit scheme to the same simulation). In other words, the provisions by their apparent exclusion of building geometry, orientation, thermal inertia, shading, climatic variability, etc, may limit a designer to a solution which is somewhat removed from the optimum in terms of energy consumption and comfort performance.

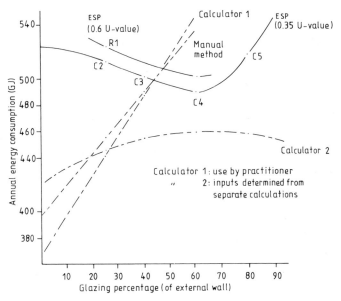

Figure 1.4   Energy consumption predictions by alternative methods.

Also shown on figure 1.4 (the chain curves) are the results obtained from two of the more commonly employed techniques: the RIBA calculator method and manual calculation methods taken from the CIBS Guide. These results raise an additional point: if in certain cases effective energy management can only be achieved by going beyond the constraints of the regulations then the designer cannot rely on the available simplified methods to provide the necessary evidence of performance since the results so obtained may be inaccurate and therefore misleading. And, of course, this problem is further compounded if additional technical complexity is introduced such as with passive solar elements or advanced control systems.

As a general strategy it would seem reasonable to aim for a high level of accuracy combined with a model structure which is capable of adapting to the information available at any design stage. It is likely that a truly 'simple' model, as perceived by a user, will be internally comprehensive in its treatment of the energy flowpaths, relying on the proper design of the software/hardware combination for its operational flexibility. This is the philosophy underlying the modelling approach developed in this book. The contention throughout is that accurate and flexible appraisal models can only be achieved by an approach which:

Achieves conservation of energy whilst including *all* participating energy flowpaths.

Ensures integrity of the mathematical model vis-à-vis the reality.

And wins acceptability through proper interface design in conjunction with rigorous validity and application testing.

The three-pole axiom of conservation of energy, conservation of integrity and conservation of flexibility is the essential target of the modelling system derived in chapters 3–7.

1.3 Energy modelling techniques

The undoubted importance, then, of accurately assessing building energy performance, coupled with the increasing availability of powerful but low-cost computing power, has resulted in much activity in the field of computer-based energy simulation modelling. As detailed in §1.1 any advanced system must be capable of accurately and dynamically handling the following processes:

The transient conduction of heat through the enclosure envelope and therefore the associated lag and thermal storage effects.

The time-dependent sensible and latent gains from occupants, lights, processing equipment etc and the relative split of these gains into radiant and convective portions which will dictate how they are delayed in time by the system.

Infiltration, natural and controlled ventilation, and inter-zone air movement.

The effects of shortwave solar radiation impinging on exposed external and internal surfaces.

The longwave radiation exchange between exposed external surfaces and the sky vault and the surroundings.

The corresponding longwave radiation exchange between internal surfaces.

The shading of external opaque and transparent surfaces as caused by surrounding buildings as well as a variety of facade obstructions.

The mapping of moving insolation patches from windows to internal receiving surfaces.

Time varying convection and, perhaps, other system thermophysical properties.

The essential link between controller location and type, plant characteristics and interaction point(s), and properties and operation of the building system.

Effects of moisture.

Many modelling systems exist which purport to address these processes at different levels of accuracy. Table 1.2, for example, summarises the main facilities of a number of UK programs at the time of a 1979 survey; table 1.3 gives summary details of the programs participating in the 1978–83 validation activities of the International Energy Agency; and table 1.4 lists the attributes of several models when assessed from the viewpoint of passive solar architecture. These models represent a wide range of accuracy and flexibility levels and indeed some will be grossly deficient on both counts.

From a thermal point of view, a building is a complex network of thermal resistances and capacitances linking different regions and representing conductive, convective, advective, radiative and heat storage processes. The manner in which this network is treated mathematically—some portion may be neglected, fixed values may be assigned or simplifying boundary condition assumptions might be made—will determine the flexibility of the modelling technique to emerge. In broad terms most existing and planned building models will fall into one of five 'catch-all' categories: steady-state; simple dynamic; response function; numerical; electrical analogue.

Each method is concerned, at its own level, to satisfy the first and second laws of thermodynamics but, as the level of sophistication of the method falls, so many of the active flowpaths are ignored and the method becomes indicative rather than deterministic and subject to the inaccuracies touched on in the previous section.

Steady state

These methods have no mechanism for the accurate inclusion of the effects of solar gains, casual gains, longwave radiation exchanges, plant operational strategies etc, and so such models typically address only fabric heat flow (under very special boundary conditions) and not building energy. Typical inadequacies include the omission of any consideration of the dynamic response of buildings, an inability to deal realistically with many of the energy flows occurring within buildings, and an inability to effect the correct relationship between building fabric and installed plant operation. In consequence these methods are being subsumed by the dynamic theories and will play a diminishing role even at an early design stage where, as well as accuracy problems, their ability to provide even indicative results can be seriously questioned (ABACUS and VALTOS 1979).

Simple dynamic

In recent years a number of simplified methods of energy assessment have been

Table 1.2 Facilities of some UK programs (from Burgess 1979).

ENERGY PROGRAMS	HEAT LOSSES		NET HEAT GAINS				SIMULATION PROGRAMS		ENERGY USE PREDICTION	
	U-VALUES	HEAT LOSS	HEAT GAIN	SPACE LOAD	SUMMERTIME TEMPERATURES	ENERGY REQUIREMENTS	DYNAMIC BLDG SIMULATION	PLANT & SYSTEM SIMULATION	ENERGY CONSUMPTION	TOTAL RUNNING COSTS
1 E/1F - U-values	●									
2 BP111 - Thermal performance of elements	●									
3 CPA1 - Thermal properties	●									
4 E/1A - G-values		●								
5 HEAT LOSSES		●								
6 LOSS - Heat losses		●								
7 HTLOSS		●								
8 HEATLOSS		●								
9 HEAT - Heat losses		●								
10 BP102 - Solar cooling load			●	●						
11 COOL		●	●	●						
12 E/1B,E/1C - Space heating requirements			●							
13 CLOADS - Cooling loads			●	●						
14 HG2-7 - Heat gain & ventilation suite			●	●						
15 BS1 - Heat gain & ventilation		●	●	●						
16 BUFLA - Built form load analysis			●	●						
17 CD24 - Built form & energy needs			●	●						
18 CARRIER - Air conditioning loads		●	●	●						
19 ARTEMP*GEN - Summertime temperatures					●					
20 SOLARS - Summertime temperatures					●					
21 STEMP - Summertime temperatures					●					
22 BP103 - Summertime temperatures					●					
23 ANT1 - Part of the ANNEX suite				●	●					
24 HEATGAIN		●	●	●	●					
25 SUMTEMP		●	●	●	●					
26 SOLCOOL - Summertime temperatures		●	●	●		●				
27 ARGAIN*GEN - Cooling loads			●	●		●				
28 ENDSOP - Design of internal environment		●	●	●		●				
29 COLO - Temperature prediction & A/C load			●	●	●	●				
30 ARCO - Air conditioning		●	●	●		●				
31 CEDAR3 - Energy module		●	●	●		●				
32 GAINS - Cooling loads		●	●	●		●				
33 CD11 - Heat gains		●	●	●		●				
34 SPEED - Environmental envelope design		●	●	●	●	●				
35 THERM - Model of thermal behaviour		●	●	●	●	●	●			
36 SYSTEM - Central heating performance								●		
37 ENPRO - Building energy analysis		●	●	●	●	●			●	●
38 ATKOOL - Energy analysis suite	●	●	●	●	●	●			●	●
39 ATKOOL365 - Energy analysis suite	●	●	●	●	●	●			●	●
40 NBSLD - Load determination	●	●	●	●	●	●	●		●	
41 ANT5 - Part of the ANNEX suite	●	●	●	●	●	●			●	●
42 TOTAL BEEP - Building energy estimating	●	●	●	●		●			●	●
43 ESP - Environmental systems performance	●	●	●	●	●	●	●		●	
44 ENERGY		●	●	●	●		●	●	●	
45 BP101 - Heat loss,fuel use & annual cost		●								●
46 TRACE - Air conditioning economics		●	●	●	●	●	●	●	●	●

Table 1.3 International Energy Agency (IEA) Annex 1 and 4 participant programs (1978–83).

Program Name	Organisation
Annex 1	
ATKOOL	W S Atkins Group, UK
OFFICE	Electricity Council Research Centre, UK
THERM	British Gas Corporation, UK
HTB	UWIST, UK
ENPRO	Faber Computer Operations, UK
ESP	University of Strathclyde, UK
ANTS	Pilkington Flat Glass Ltd, UK
ECUBE '75	American Gas Association, USA
SCOUT	Gard Inc, USA
DOE2	Lawrence Berkeley Laboratory, USA
ECUBE3	American Gas Association, USA
MS	Reid Crowther & Partners, USA
JULOTTA	Swedish Council for Building Research, Sweden
VENTAC	AB Svenska Flaktfabriken, Sweden
DOE2	EMPA, Switzerland
WTEO1	TNO, Holland
LPB1	University of Liège, Belgium
?	Methiwether & Associates, USA
Annex 4	
AMBER	Faber Computer Operations, UK
ATKOOL	W S Atkins Group, UK
DOE2	EMPA, Switzerland
ENCO2	Pilkington Flat Glass Ltd, UK
ESP	University of Strathclyde, UK
LPB1	University of Liège, Belgium
TEMPER	CSIRO, Australia
THERM	British Gas Corporation, UK
WTEO1	TNO, Holland

produced which address dynamic performance. These methods are mostly based on regression techniques applied to the results of multiple parametric runs of more powerful modelling systems. The results to emerge can often be reduced to simple relationships or presented in tabular or graphical form.

Response function

It is possible, by so specifying system boundary conditions, to solve the partial differential heat equation, which governs the flow of heat within the building fabric, to provide a means of modelling the dynamic response of a building. Two main branches of this method exist—time-domain response function and

Table 1.4 Passive solar modelling features and status of four energy programs (basic data from Littler 1982).

Topic		ESP	BLAST	DEROB	SUNCODE
Designs	Direct gain	Y	Y	Y	Y
	Attached sun space	Y	Y	Y	Y
	Thermosiphon	Y	Y	Y	Y
	Roof space collector	Y	Y	Y	Y
	Double envelope	Y	N	1	N
	Mass wall (vented)	Y	Y	Y	Y
	Mass wall (unvented)	Y	Y	Y	Y
	Under floor rock bed	Y	Y	Y	Y
Physical problems					
	Air/heat movement by convection or by fans	Y	Y	Y	Y
	Infiltration	Y(2)	2	2	2
	Solar radiation mapping around spaces	Y	Y	Y	N
	Variable glass emissivity	Y	N	Y	Y
	Variable room colour	Y	Y	Y	Y
	Effect of furnishings	1	1	1	N
	Air/temperature stratification	Y	N	Y	N
	Movable window insulation	Y	Y	Y	Y
	Isothermal and non-isothermal storage	Y	Y	Y	Y
	Phase change walls	Y	Y	N	Y
	Isolated storage	Y	3	Y	Y
	Adequate handling of beam radiation	Y	Y	Y	Y
	Adequate treatment of sky temperature	N	N	N	N
Weather input Complete 'set' supplied		Y	Y	Y	Y
Daylighting		Y	Y	Y	N
Surface temperatures for comfort assessment		Y	1	Y	Y
Validation		4	4	4	4
Documentation		V	P	G	V
Graphics input		V	N	V	N
Building input via digitising tablet		Y	N	N	N
Building input format		5	5	6	3
Continuing improvement		Y	Y	Y	Y

1: with difficulty   2: by schedule   3: see reference   4: tested in a number of validation exercises   5: Cartesian   6: standard shapes   Y: yes   N: no   V: very good   P: poor   G: good.

frequency-domain response function methods—and many workers are actively pursuing technique refinement.

Numerical
With the advent of powerful computing systems many problems of varying complexity can be solved by numerical means. Two main numerical techniques

exist—finite difference and finite element—although the former is the technique
most commonly applied to the problem of building energy modelling.

Electrical analogue

The analogy that exists between electrical flow and heat flow has led to the
construction of electrical analogue devices useful in the study of complex heat
flow phenomena. The technique is extremely useful as a research tool, allowing
long-term simulations to be completed in a short elapsed time, but has little
application in a design context.

The extant systems for building energy simulation are based either on the
response function method or on numerical techniques in finite difference form
and, for this reason, only these methods are described in detail in the following
chapters. For the interested reader other texts exist which describe steady state
formulations (IHVE(CIBS) 1971, Clarke 1977), the technique of regression applied to the results of multiple parametric runs of large modelling systems (Claux
et al 1982), Markus *et al* 1982), the finite element method (Zienkiewicz 1977) and
the components of analogue modelling (Forrest 1979).

1.4   The issue of integrity in building energy modelling

Underlying the flowpaths discussed in §1.1 is the concept of heat and mass
balance at every point within a building/plant network. Dynamic modelling
methods will be required to integrate a large number of equation types whilst
respecting the integrity of the total system: parabolic partial differential equations
are used to represent solid materials requiring detailed conduction modelling;
ordinary differential equations can be used to represent lumped thermal property
regions of lesser significance, or surface interfaces undergoing simultaneous conduction, convection and radiation exchanges; non-linear empirical equations
(perhaps complex), or, perhaps, partial differential equations, will represent
fluid flow; and a number of algebraic relations will define control loops, time-
dependent thermophysical properties and regional heat injection (from solar
radiation for example). The necessary equation types are explored further as the
book develops and an integration scheme is developed which is capable of
preserving the integrity of real world building systems.

   The implicit thesis of the book is that the issue of integrity is of singular
importance and if disregarded will greatly influence simulation predictions and,
through these, the related design decisions. If this is true, then it follows that
great care must be taken with simulation models which degrade integrity by:
omitting flowpaths or modelling the overall system in a piecemeal manner, thus
destroying temporal or spatial relationships; ignoring important features such as
time-varying thermophysical properties, the actual location of control system
sensing elements or non-linear processes; not allowing full and dynamic coupling
between building zones (air movement, longwave radiation, solar penetration

etc); or not permitting the simultaneous treatment of energy and mass flow or building and plant processes.

Some indication of the effect of such factors on model predictions is reported elsewhere (Clarke 1983, 1984). The need then is for a modelling infrastructure which can conserve integrity by numerically processing *all* elements of the problem as a single system, fully connected in space and time, and under the influence of dynamic influences and constraints. Chapters 3–7 describe such a modelling infrastructure.

References

ABACUS and VALTOS 1979 Deemed to Satisfy? *Architects Journal* October 345–55

Alamdari F and Hammond G P 1982 Time-Dependent Convective Heat Transfer in Warm-Air Heated Rooms *Rep. SME/J/82/01* (Cranfield: Applied Energy Group, Cranfield Institute of Technology)

Burgess K S 1979 Computer Programs for Energy in Buildings *Evaluation Rep. No 5* (Cambridge: Design Office Consortium (now CICCA))

Clarke J A 1977 Environmental Systems Performance *PhD Thesis* University of Strathclyde

—— 1980 The Impact of Alternative Constructions *ABACUS Occasional Paper* (Glasgow: University of Strathclyde)

—— 1983 Collins Building Sensitivity Analysis *Topic Paper 9, IEA Annex 4 Final Report, Oct 1983*

—— 1984 The Issue of Integrity in Building Energy Modelling *Proc. Seminar on Dynamic Thermal Behaviour of Buildings, Saint-Rémy-les-Chevreuse, December 1984*

Claux P, Franca J P, Gilles R, Pesso A, Pouget A and Raoust M 1982 *Method 5000* (Paris: Claux Pesso Raoust)

Forrest I D 1979 Electrical Analogues: Theory, Development and Application *MSc Thesis* University of Strathclyde

IHVE (Now CIBS) 1971 *Guide Book A*

Incropera F P and DeWitt D P 1981 *Fundamentals of Heat Transfer* (New York: Wiley)

Krieth F 1973 *Principles of Heat Transfer* (3rd edn) (New York: Harper & Row)

Littler J G F 1982 Overview of Some Available Models for Passive Solar Design *CAD Journal* 14(1)

Markus T A, Clarke J A and Morris E N 1982 Climatic Severity *Occasional Paper 82/5* (Glasgow: Department of Architecture, University of Strathclyde)

Maver T W and Ellis J 1982 Implementation of an Energy Model within a Multidisciplinary Practice *Proc. CAD82, March 1982*

Ozisik M N 1977 *Basic Heat Transfer* (New York: McGraw-Hill)

Robinson F 1979 Investigation of the Common Assumptions Applied to Internal Surface Insolation in Buildings *BSc Thesis* University of Strathclyde

Zienkiewicz O C 1977 *The Finite Element Method* (New York: McGraw-Hill)

2

Advanced modelling techniques

This chapter describes the theoretical basis and development background of the much popularised response function approach to building energy modelling. Both branches of the method—time and frequency responses—are derived for the case of transient conduction and intra-zone energy balance modelling and, in each case, use in practice is described.

The elements of an alternative modelling approach, based on a finite difference representation method, are then presented as the essential introduction to chapters 3 and 4 where a numerical scheme is formulated and solved respectively.

Finally §2.5 discusses the criteria on the basis of which an answer might be obtained to the question, 'Which method?'

Each of the methods provides a solution to the differential equations which govern the flow of heat in solids, heat transfer at surface layers and heat exchange between connected fluid volumes. The response function approach is usually applied to differential problems of low order with time-invariant significance whereas the numerical method is more suited to time varying problems of high order.

2.1 Response function methods

To begin the description of the principles underlying a response function approach consider figure 2.1 which shows a simple homogeneous element of thickness defined by $0 < x < l$ and, at time t, temperature $\theta(x, t)$, heat flux $q(x, t)$. Two relationships are of interest: the change of temperature with distance and the change of flux with distance. That is

$$\frac{\partial\theta(x,t)}{\partial x} = -\frac{1}{k}\, q(x,t) \tag{2.1}$$

$$\frac{\partial q(x,t)}{\partial x} = -\varrho C\, \frac{\partial\theta(x,t)}{\partial t}. \tag{2.2}$$

Combination of these equations gives the usual governing partial differential heat equation, the Fourier equation (as derived in appendix A):

$$\frac{\partial^2 \theta(x,t)}{\partial x^2} = \frac{1}{\alpha} \frac{\partial \theta(x,t)}{\partial t}. \tag{2.3}$$

An analytical approach to the solution of these equations involves the use of the Laplace transformation (Carslaw and Jaeger 1959, Churchill 1958, Davies 1978). This is essentially a three stage procedure:

The given equation in the time domain is transformed into a subsidiary equation in an imaginary space.

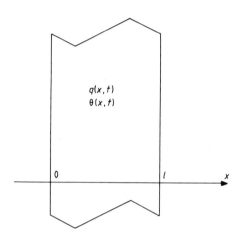

$q(x,t)$
$\theta(x,t)$

0   l x

Figure 2.1 A simple homogeneous element.

This subsidiary equation is then solved by purely algebraic manipulations.

An inverse transformation is then applied to this solution to obtain the solution in the time domain of the initial problem.

The interesting feature of the method is that in many cases ordinary differential equations are transformed into purely algebraic equations and partial differential equations are transformed to ordinary differential equations. In practice transforms and inverse transforms are often obtained from tables of Laplace transforms such as that shown in table 2.1 (see Healey 1967 for a more comprehensive list).

With respect to the temperature variable $\theta(x,t)$, the Laplace transform is given by:

$$L[\theta(x,t)] = \theta(x,p) = \int_0^\infty e^{-pt}\theta(x,t)\ dt \tag{2.4}$$

where p is a complex number whose real part is positive and large enough to cause the integral to converge.

A number of theorems accompany the transform and two in particular are

Table 2.1　Some common Laplace transform pairs.

$f(t)$		$f(p)$
Unit impulse	$\delta(t)$	1
Unit step	$H(t)$	$1/p$
Unit ramp	t	$1/p^2$
	t^n	$n!/p^{n+1}$　　n + ve integer
Delayed unit impulse	$\delta(t-\Delta)$	$e^{-p\Delta}$
Delayed unit step	$H(t-\Delta)$	$e^{-p\Delta}/p$
	e^{-at}	$1/(p+a)$
	$e^{-a(t-\Delta)}H(t-\Delta)$	$e^{-p\Delta}/(p+a)$
	te^{-at}	$1/(p+a)^2$
	$t^n e^{-at}$	$n!/(p+a)^{n+1}$
	$\sin bt$	$b/(p^2+b^2)$
	$\cos bt$	$p/(p^2+b^2)$

given here:

$$L\left[\frac{\partial\theta(x,t)}{\partial t}\right] = pL[\theta(x,t)] - \theta(x,0).$$

$$L\left[\frac{\partial^n\theta(x,t)}{\partial x^n}\right] = \frac{\partial^n L[\theta(x,t)]}{\partial x^n}.$$

The Laplacian of equations (2.1)–(2.3) can now be written as:

$$\frac{\partial\theta(x,p)}{\partial x} = -\frac{1}{k}\,q(x,p) \tag{2.5}$$

$$\frac{\partial q(x,p)}{\partial x} = -\varrho Cp\theta(x,p) + \varrho C\theta(x,0) \tag{2.6}$$

$$\frac{\partial^2\theta(x,p)}{\partial x^2} = \frac{p}{\alpha}\,\theta(x,p) - \frac{1}{\alpha}\,\theta(x,0). \tag{2.7}$$

These are the subsidiary equations which when solved give the Laplace transform $\theta(x,p)$ of the solution of the original equations. If $\theta(x,p)$ is found in a table of transforms then the solution for $\theta(x,t)$ is determined immediately. If no such transform exists then $\theta(x,t)$ is determined from $\theta(x,p)$ by the inversion theorem which states:

$$\theta(x,t) = L^{-1}[\theta(x,p)] = \frac{1}{2\pi i}\int_{\gamma-i\infty}^{\gamma+i\infty} e^{pt}\theta(x,p)\,dp \tag{2.8}$$

where γ is a large number such that all the singularities of $\theta(x,p)$ lie to the left of the line $(\gamma-i\infty,\ \gamma+i\infty)$.

The solution of the subsidiary equations (Carslaw and Jaeger 1959) is given by

$$\theta(x,p) = \cosh\left[(p/\alpha)^{\frac{1}{2}}x\right]\theta(0,p) - \frac{\sinh\left[(p/\alpha)^{\frac{1}{2}}x\right]q(0,p)}{k(p/\alpha)^{\frac{1}{2}}}$$

$$q(x,p) = -k(p/\alpha)^{\frac{1}{2}}\sinh\left[(p/\alpha)^{\frac{1}{2}}x\right]\theta(0,p) + \cosh\left[(p/\alpha)^{\frac{1}{2}}x\right]q(0,p).$$

It is convenient to represent these temperature and heat flux relationships in matrix notation (Pipes 1957) so that:

$$\begin{bmatrix} \theta(l,p) \\ q(l,p) \end{bmatrix} = \begin{bmatrix} m_{11}(p) & m_{12}(p) \\ m_{21}(p) & m_{22}(p) \end{bmatrix} \times \begin{bmatrix} \theta(0,p) \\ q(0,p) \end{bmatrix} \tag{2.9}$$

where, by inspection, the elements of matrix **M** are given by

$$m_{11}(p) = m_{22}(p) = \cosh\left[(p/\alpha)^{\frac{1}{2}}l\right]$$

$$m_{12}(p) = -\sinh\left[(p/\alpha)^{\frac{1}{2}}l\right]/k(p/\alpha)^{\frac{1}{2}}$$

$$m_{21}(p) = -k(p/\alpha)^{\frac{1}{2}}\sinh\left[(p/\alpha)^{\frac{1}{2}}l\right]$$

and the matrix **M** has unit determinant, that is $m_{11}m_{22} - m_{12}m_{21} = 1$.

In the terminology the matrix **M** is the transmission matrix and its entries are the transfer functions. For composite constructions $0 < x < L$ comprised of a number of layers in intimate contact, the formulation can be directly extended by simple matrix manipulation techniques to give

$$\begin{bmatrix} \theta(L,p) \\ q(L,p) \end{bmatrix} = \begin{bmatrix} A(p) & B(p) \\ C(p) & D(p) \end{bmatrix} \times \begin{bmatrix} \theta(0,p) \\ q(0,p) \end{bmatrix} \tag{2.10}$$

where the value of the elements $A(p)$, $B(p)$, $C(p)$ and $D(p)$ of the overall transmission matrix will depend on the properties of the component elements of the multi-layered construction and the order in which the individual element transmission matrices are combined. Thus for a multi-layered construction with homogeneous elements e1,e2,e3,...,en specified outside ($x = 0$) to inside ($x = L$) the overall transmission matrix is given by:

$$\begin{bmatrix} A(p) & B(p) \\ C(p) & D(p) \end{bmatrix} = \mathbf{M}_{e1} \times \mathbf{M}_{e2} \times \mathbf{M}_{e3} \times \ldots \times \mathbf{M}_{en}$$

where in general $m_{11} = m_{22}$ but $A(p) \neq D(p)$.

Equation (2.10) is the fundamental relationship underlying the time-domain and frequency-domain response function methods described in the following sections. For the former method a rearrangement of equation (2.10) is necessary which relates the flux at both surfaces to surface temperatures:

$$\begin{bmatrix} q(0,p) \\ q(L,p) \end{bmatrix} = \begin{bmatrix} D(p)/B(p) & -1/B(p) \\ 1/B(p) & -A(p)/B(p) \end{bmatrix} \times \begin{bmatrix} \theta(0,p) \\ \theta(L,p) \end{bmatrix}. \tag{2.11}$$

If two- or three-dimensional transient heat conduction is to be considered then the partial differential equation $\nabla^2\theta(x,t) = \alpha^{-1}\,\partial\theta(x,t)/\partial t$ can still be treated by the Laplace transform technique where the subsidiary equation to result will still be a partial differential equation but in three (space) variables instead of four (space plus time).

It is at this point—the application of the inverse transform—that the time- and frequency-domain methods take on separate identities; one concerned with the response of multi-layered constructions to time-series temperature or flux pulses, the other with the response to periodic excitations of differing frequencies.

2.2   Time-domain response functions

This solution technique—concerned with the solution of equation (2.10) in the time domain—is most commonly referred to as the response function (or factor) method. In application the technique can closely resemble numerical techniques although, unlike finite differencing or control volume heat balance techniques, it can usually only be applied to a system of equations which are both linear and invariable (especially if response functions are predetermined time-series). Some workers (for example Mitalas 1965) have stated that such a requirement need not impose severe restrictions in a building design context. The method is capable of handling both periodic and non-periodic flux and temperature time-series and for this reason, perhaps, has enjoyed wider application, especially in North America, than the frequency-domain (or harmonic) method of §2.3.

The basic strategy of the method is to predetermine the response of a system to some unit excitation applied under the same boundary conditions as are anticipated will prevail under actual operation. A unit excitation function has a value of unity at the start of its excitation and zero thereafter and is typically designated by (1,0,0,0,...). The response of a linear, invariant equation system to this unit time-series excitation function is termed the unit response function (URF) and the time-series representation of this URF, that is the individual terms of the series, are the response factors.

The number of URFs considered in any problem will depend on the number of combinations of excitation function (solar radiation, dry bulb temperature etc) and responses of interest (cooling loads, internal temperatures etc). Figure 2.2 shows one such combination in which a unit change in external air temperature (the unit excitation function) produces the URF of the heat flux at the internal surface of some multi-layered construction.

In general terms there are three steps inherent in the method after the various URFs have been determined:

The various excitation functions are resolved into an equivalent time-series. This can be achieved by triangular, rectangular or some equivalent representation technique, contemporary systems favouring the former method.

The second step involves the combination of URFs and some corresponding excitation function to determine the system response to each excitation in turn. This is achieved by application of the convolution theorem which states that the response of a linear, invariant system is given by the products of the response of the same system to a unit excitation (the URF) and the actual excitation given that the appropriate time adjustments are made during the multiplication process. Stated mathematically:

$$R(t) = \sum_{m=0}^{\infty} RF(m\Delta) \, E(t - m\Delta) \qquad (2.12)$$

where $R(t)$ is the system response at time $t = m\Delta$ (m is an integer), $RF(m\Delta)$ is a response factor at time $m\Delta$, $E(t - m\Delta)$ is the excitation at time $(t - m\Delta)$ and Δ is the URF time-step.

The last step involves the superimposition of the individual responses to give the overall response.

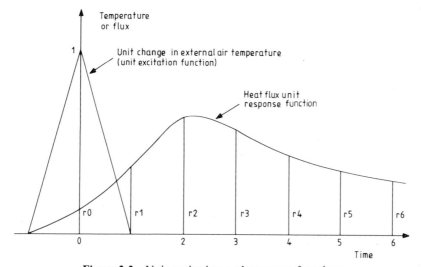

Figure 2.2 Unit excitation and response functions.

URFs are only dependent on design parameters and assumptions regarding thermophysical properties and therefore, if assumptions of invariability are acceptable, need only be determined once for any given design application. This is one of the main attractions of the method over the more generalised numerical methods where a computational exercise, equivalent to URF computation, must be implemented at each time-step as a simulation proceeds. However, should system properties vary with time, requiring that the URF be computed anew, then the computational distinction between response function and numerical schemes will become less obvious.

Pratt and Ball (1963) were among the earlier workers in the field of response function modelling. They developed a method for the calculation of room loads and temperatures using URFs derived for multi-layered constructions of up to three homogeneous elements. Stephenson and Mitalas (1967a) are largely responsible for the present day form of the popularised response function method. Their formulation builds on the earlier work of Brisken and Reque (1956) who were among the first to consider response factors as a set of numbers denoting the time-series values of a URF at equally spaced intervals of time. A triangular pulse representation technique was developed in which each term in some continuous excitation function is considered as the magnitude of a triangular pulse centred at the particular time in question and with a base equal to twice the selected time-step. The summation of such overlapping triangles is equivalent to a trapezoidal approximation and represents a continuous function comprised of straight line segments. Figure 2.3 demonstrates the triangular pulse technique and, for comparative purposes, shows the rectangular pulse representation common in earlier formulations. Although for most applications triangular approximation gives a good fit, Gupta *et al* (1974) have stated that switched inputs such as lighting loads would be better treated by the rectangular method.

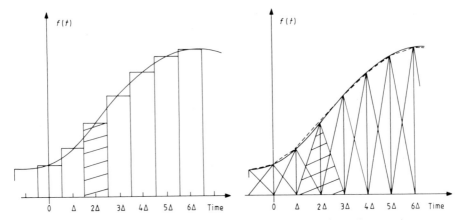

Figure 2.3 Rectangular and triangular pulse representation of a continuous function.

Many subsequent enhancements have been made to the technique including: the derivation of additional equations for the evaluation of interfacial temperatures and heat fluxes within multi-layered constructions (Kusuda 1969); the concept of whole building response functions (Muncey 1979); and an approach to the calculation of wall response based on an eigenfunction representation which is computationally attractive (Gough 1982).

The unit pulse response function method, like its analytical counterpart, the frequency response method, can be used to estimate the internal air temperature prevailing in an unconditioned building and the heating/cooling requirements for constant or varying internal climatic conditions. Figure 2.4 gives the sequence of steps involved in the computation of the overall response of a building zone to variations in external and internal climatic conditions. In most response factor implementations the overall response is conveniently considered in two stages:

The load profiles are determined relative to some fictitious and constant internal reference temperature.

Any deviation of internal temperatures from the chosen reference is determined as a function of known plant operational characteristics and capacities. In other words, the first stage is concerned with the effect of heat flow across the system boundary and subsequent internal flowpath interaction to produce a design condition plant demand. The second stage addresses the operational strategy of the installed plant.

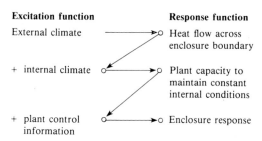

Figure 2.4 Overall zone response: sequence of steps.

2.2.1 Multi-layered constructions

Section 2.3.1 describes the derivation of the thermal response factors relating to the solution of equation (2.10) in the frequency domain. An alternative approach is to operate in the time domain by considering the boundary condition as a train of discrete pulses. Figure 2.5 shows the variation of heat flux at one surface of a homogeneous element due to a unit temperature pulse at either surface. The

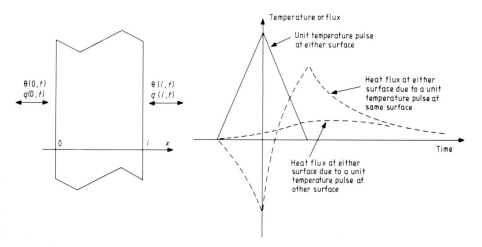

Figure 2.5 Heat flux due to a unit temperature pulse at either surface of a homogeneous element.

URF for such an element represents the heat flux at innermost or outermost surfaces caused by a unit triangular excitation applied to the other surface whilst the surface in question is held at constant temperature.

Recalling equation (2.11), which is the rearranged form of equation (2.10),

$$\begin{bmatrix} q(0,p) \\ q(L,p) \end{bmatrix} = \begin{bmatrix} D(p)/B(p) & -1/B(p) \\ 1/B(p) & -A(p)/B(p) \end{bmatrix} \times \begin{bmatrix} \theta(0,p) \\ \theta(L,p) \end{bmatrix}. \tag{2.13}$$

Flux unit response functions are the heat fluxes which result when first $\theta(0,p)$ then $\theta(L,p)$ is set to the Laplacian of a unit triangular pulse whilst the opposite surface $\theta(L,p)$ or $\theta(0,p)$ respectively is held at zero. In this case three URFs will result:

$X(n\Delta)$; the heat flux URF into construction at $x = 0$ for unit pulse at $x = 0$.

$Y(n\Delta)$; the heat flux URF out of construction at $x = 0$ or $x = L$ for a unit pulse at $x = L$ or $x = 0$ respectively.

$Z(n\Delta)$; the heat flux URF into construction at $x = L$ for unit pulse at $x = L$ where $n = 0,1,2,3,4,...$

It follows then from equation (2.12) that at time $t = n\Delta$:

$$q(0,n\Delta) = \sum_{m=0}^{\infty} \theta[0,(n-m)\Delta]X(m\Delta) - \sum_{m=0}^{\infty} \theta[L,(n-m)\Delta]Y(m\Delta) \tag{2.14}$$

$$q(L,n\Delta) = \sum_{m=0}^{\infty} \theta[0,(n-m)\Delta]Y(m\Delta) - \sum_{m=0}^{\infty} \theta[L,(n-m)\Delta]Z(m\Delta) \tag{2.15}$$

where $q(0,n\Delta)$ signifies heat flow *into* construction at $x = 0$ and $t = n\Delta$; $q(L,n\Delta)$ signifies heat flow *out of* construction at $x = L$ and $t = n\Delta$.

In the time domain a unit pulse can be represented by the superimposition of three ramp functions $r(t)$ defined as

$$r(t) = \begin{cases} 0 & t < 0 \\ t & t \geq 0. \end{cases}$$

Thus a unit triangular pulse is given by

$$f(t) = r(t + \Delta) - 2r(t) + r(t - \Delta) \tag{2.16}$$

so that for $\Delta = 1$: $f(-1) = 0$; $f(0) = 1$; $f(1) = 0$.

The Laplacian of the ramp function $f(t) = t$ is $1/p^2$ and so from equation (2.13) the URF due to a ramp function at $x = 0$ with $\theta(L,p) = 0$ gives

$$q(0,p) = D(p)/p^2 B(p)$$
$$q(L,p) = 1/p^2 B(p)$$

and establishing the ramp function at $x = L$ with $\theta(0,p) = 0$ gives

$$q(0,p) = -1/p^2 B(p)$$
$$q(L,p) = -A(p)/p^2 B(p).$$

Application of the inverse transform of equation (2.8) gives the heat flux time-series in the time domain: these (with appropriate signs) are the URFs $X(m\Delta)$,

$Y(m\Delta)$ and $Z(m\Delta)$. That is

$$X(m\Delta) = L^{-1}[D(p)/p^2 B(p)]$$
$$Y(m\Delta) = L^{-1}[1/p^2 B(p)]$$
$$Z(m\Delta) = L^{-1}[A(p)/p^2 B(p)].$$

The method of residues can now be used to achieve the integration of the inverse transform. Noticing that the expressions for the URFs can each be represented by $\xi(p) = R(p)/p^2 S(p)$ then the residue theorem gives:

$$\xi(t) = L^{-1}[\xi(p)]$$

$$= \frac{R(0)}{S(0)} t + \frac{d}{dp}\left(\frac{R(p)}{S(p)}\right)\bigg|_{p=0} + \sum_{j=1}^{\infty} \frac{R(a_j)}{a_j^2 S'(a_j)} e^{a_j t}$$

where $R(a_j)/S'(a_j)$ is the residue of the transfer function R/S at its jth singularity (or pole), a_j.

The poles of R/S are the zeros of S. And so the task in hand is to determine the roots of $S(p) = 0$ where, in this case, $S(p) = B(p)$ as determined from the matrix analysis applied to the multi-layered construction in question (equation (2.10)). Thus for $X(m\Delta)$:

$$\xi(t) = \frac{D(0)}{B(0)} t + \frac{d}{dp}\left(\frac{D(p)}{B(p)}\right)\bigg|_{p=0} + \sum_{j=1}^{\infty} \frac{D(a_j)}{a_j^2 B'(a_j)} e^{a_j t} \tag{2.17}$$

with the response factor terms given (from equation (2.16)) by:

$$\begin{aligned} X(0) &= \xi(\Delta) \\ X(\Delta) &= \xi(2\Delta) - 2\xi(\Delta) \\ X(m\Delta) &= \xi[(m+1)\Delta] - 2\xi(m\Delta) + \xi[(m-1)\Delta] \qquad m = 2,3,4,\ldots \end{aligned} \tag{2.18}$$

with similar expressions emerging for $Y(m\Delta)$ and $Z(m\Delta)$.

With most multi-layered constructions the elements of the overall transmission matrix are complex hyperbolic functions and determination of the roots of $B(p) = 0$ is achieved by numerical search procedures which attempt to locate a sign change in $B(p)$ as p is incremented in small steps. Hittle (1981) has proposed an improved root finding procedure which, it is claimed, allows larger search increments whilst still ensuring root location. This method makes use of a discovery that the roots of $B(p) = 0$ are bracketed by the roots of $A(p) = 0$. Gough (1982) has also proposed an efficient root-finding algorithm which makes use of the fact that $B(p)$ can be expressed as a product expansion in terms of its zeros a_j.

For a homogeneous element (from equation (2.9))

$$p^2 B(p) = p^2 \sinh[(p/\alpha)^{\frac{1}{2}} l]/k(p/\alpha)^{\frac{1}{2}}$$

and the roots of $\sinh[(p/\alpha)^{\frac{1}{2}} l]$ are given by

$$a_j = -j^2 \pi^2 \alpha/l^2.$$

Stephenson *et al* (1967a) have shown that (from equation (2.18))

$$X(0) = -kl/\alpha\Delta\left[-1/3 - \alpha\Delta/l^2 - 2/\pi^2 \sum_{j=1}^{\infty} \gamma_j/j^2\right]$$

$$X(1) = -kl/\alpha\Delta\left[1/3 + 2/\pi^2 \sum_{j=1}^{\infty} (\gamma_j^2 - 2\gamma_j)/j^2\right]$$

$$X(n) = -2kl/\alpha\Delta\pi^2 \sum_{j=1}^{\infty} [(\gamma_{j(n+1)} - 2\gamma_{j(n)} + \gamma_{j(n-1)})/j^2] \quad \text{for } n \geqslant 2$$

$$Y(0) = -kl/\alpha\Delta\left[1/6 - \alpha\Delta/l^2 + 2/\pi^2 \sum_{j=1}^{\infty} (-1)^j\gamma_j/j^2\right]$$

$$Y(1) = -kl/\alpha\Delta\left[-1/6 + 2/\pi^2 \sum_{j=1}^{\infty} (-1)^j(\gamma_j^2 - 2\gamma_j)/j^2\right]$$

$$Y(n) = -2kl/\alpha\Delta\pi^2 \sum_{j=1}^{\infty} [(-1)^j(\gamma_{j(n+1)} - 2\gamma_{j(n)} + \gamma_{j(n-1)})/j^2] \quad \text{for } n \geqslant 2$$

(2.19)

where $\gamma_{j(\xi)} = \exp(-j^2\pi^2\alpha\Delta/l^2)$ evaluated at time-row ξ.

Note that for a homogeneous element $A(p) = D(p)$ and so $X(m\Delta) = Z(m\Delta)$.

These unit response functions relate to some time-step Δ and can be converted to the values corresponding to a time-step of 2Δ by:

$$r_{2\Delta(n)} = 0.5r_{\Delta(2n-1)} + r_{\Delta(2n)} + 0.5r_{\Delta(2n-1)}$$

where

$r_{2\Delta(n)}$ = response factor for 2Δ time-step and at time n
$r_{\Delta(\xi)}$ = response factor at Δ time-step and at time ξ.

The URFs for a multi-layered construction are found by computation of the overall transmission matrix to determine the transfer functions for use in equation (2.18). Alternatively, the factors given by equation (2.19) can be combined for each homogeneous element in turn. As proof of this, and with reference to figure 2.6, the heat fluxes at the interfaces 1, 2 and 3 are:

$$q_1 = X_A\theta_1 - Y_A\theta_2 \tag{2.20}$$

$$q_2 = Y_A\theta_1 - Z_A\theta_2 \tag{2.21}$$

$$= X_B\theta_2 - Y_B\theta_3 \tag{2.22}$$

$$q_3 = Y_B\theta_2 - Z_B\theta_3 \tag{2.23}$$

and from equations (2.21) and (2.22):

$$\theta_2 = (Y_A\theta_1 - Y_B\theta_3)/(Z_A + X_B)$$

so that substitution in equations (2.20) and (2.23) gives

$$q_1 = \left(Y_A - \frac{Y_A^2}{Z_A + X_B}\right)\theta_1 - \left(\frac{Y_A Y_B}{Z_A + X_B}\right)\theta_3 = U\theta_1 - V\theta_3$$

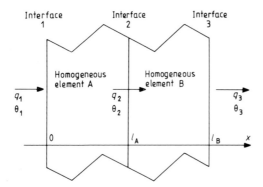

Figure 2.6 Combination of response factors.

$$q_3 = \left(\frac{Y_A Y_B}{Z_A + X_B}\right)\theta_1 - \left(Z_B - \frac{Y_B^2}{Z_A + X_B}\right)\theta_3 = V\theta_1 - W\theta_3$$

where U, V and W are the response factor time-series for the two-layer composite. Continued combination in the same manner for the remaining layers gives the overall URF.

Gough (1984) has stated that this procedure is slightly suspect as it embodies an assumption of linear interpolation between sample values at every layer boundary. The errors so introduced may not be negligible for a wall composed of several layers.

Concerning the infinite series of equations (2.17) and (2.19), on evaluation it is observed that after a finite number of terms all subsequent terms decay with a common ratio and that each successive term is given by

$$X(m\Delta) = X(m\Delta - 1)\exp(a_j).$$

In the early seventies Stephenson and Mitalas (1971) introduced a further transform, known as the z transform from a method by Jury (1964), which allows the boundary conditions to be represented by a train of samples with linear (or some other form of) interpolation used to convert the samples to a continuous function as with the response factor method. The results from the z transform method are identical to those from the response factor method—for identical interpolation assumptions—but substantially quicker to obtain. In this method the heat flux at the inside surface of a multi-layered construction is given as a weighted finite sum of both input and output excitation histories. That is:

$$q(l,t) = \sum_{m=0}^{N} E(m\Delta)\theta(0,t - m\Delta) - \sum_{m=1}^{M} H(m\Delta)\, q(l,t - m\Delta)$$

where E and H are z transfer functions.

The z transfer functions are derived from consideration of the ratio of input and output time-series. They are related to the p transfer functions $X(m\Delta)$,

$Y(m\Delta)$ and $Z(m\Delta)$ by

$$E(m\Delta) = \sum_{j=0}^{N} X(j\Delta)H(m\Delta - j\Delta)$$

$$F(m\Delta) = \sum_{j=0}^{N} Y(j\Delta)H(m\Delta - j\Delta)$$

$$G(m\Delta) = \sum_{j=0}^{N} Z(j\Delta)H(m\Delta - j\Delta)$$

and $H(m\Delta)$ $(m = 1,2,3,...)$ is obtained as the coefficient of Z^{-m} in the expansion:

$$\sum_{m=0}^{\infty} H(m\Delta)Z^{-m} = \prod_{m=1}^{\infty} (1 - Z^{-m} e^{a_j\Delta})$$

where a_j are the poles of $B(p)$ as in the formulation of equation (2.17).

2.2.2 Zone energy balance

For any given zone, in order to determine the overall URF under any unit excitation it is necessary, as a result of the interrelationship of the various heat flowpaths, to consider the entire energy field as an integrated and single system. This can be done by formulating energy balance equations for each major region within the zone, effectively to link all regions over space and time dimensions. The subsequent simultaneous solution of this equation set, when subjected to some unit excitation function with all other excitations set at zero, gives the corresponding URF. Repetition of this procedure for each unit excitation gives the complete URF set from which the total zone response is determined.

Consider figure 2.7 which shows a simple zone comprised of six multi-layered opaque constructions, one window and one fluid volume representing the thoroughly mixed internal air. Surface 1 is an external construction with internal surface denoted by 1i and external surface by 1e. Since the thermal storage effects of the glazing (surface 7) can be assumed negligible, and since the glass resistance contributes little to the overall resistance (glass + boundary layers), then the inside and outside surface temperatures are here considered to be equal and so only a one-surface representation is required. This restriction is easy to remove as demonstrated in chapter 3. The remaining external surfaces 2e–6e are assumed to make contact with adjacent zones of known temperature θ_{A2} through θ_{A6} respectively. Explicit matrix representations for other systems such as furnishings, multiple glazing for which the foregoing assumption is unacceptable, variable temperatures, multi-zone geometries etc are given in chapters 3 and 4.

Assuming that internal surfaces have no associated thermal capacity, since all structural capacity effects are included in the response functions of the preceding section, then the surface heat balance is given by

$$q_c(t) = q_R(t) + q_K(t) + q_S(t) + q_M(t) \tag{2.24}$$

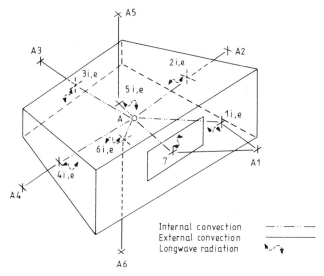

Figure 2.7 Example for zone energy balance.

where $q_c(t)$ is the surface convection at time t, $q_R(t)$ is the longwave radiation gain from surrounding surfaces at time t, $q_K(t)$ is the uni-directional conduction flux at surface at time t, $q_S(t)$ is the shortwave radiation excitation at time t, and $q_M(t)$ is the other miscellaneous radiant excitations from lights, appliances, etc. All terms are measured in W m^{-2}.

The convective gain is given by

$$q_c(t) = h_{cs}(\theta_x(t) - \theta_s(t)) \tag{2.25}$$

where h_{cs} is the time-invariant (or time-averaged) convective heat transfer coefficient for surface s under consideration (W m^{-2} $^{\circ}$C^{-1}), $\theta_x(t)$ is the adjacent air temperature at time t ($^{\circ}$C) and equals $\theta_A(t)$ for an inside surface or $\theta_{Aj}(t)$ for any outermost surface j, and $\theta_s(t)$ is the surface temperature at time t ($^{\circ}$C).

The longwave radiation gain is given by:

$$q_R(t) = \sum_{j=1}^{N} h_{rj,s} \, (\theta_j(t) - \theta_s(t)) \tag{2.26}$$

where N is the total number of surfaces in thermal communication, $h_{rj,s}$ is the time-invariant linearised radiation coefficient between surface j and surface under consideration (W m^{-2} $^{\circ}$C^{-1}), and $\theta_j(t)$ is the temperature of the communicating surface at time t ($^{\circ}$C).

For outermost surfaces it is convenient here to combine radiative and convective exchanges into a single term relating to the adjacent air layer. This is necessary since, in this formulation, no external or adjacent zone surface temperature information is available. The heat flux for outermost surfaces is given by

$$q_c(t) + q_R(t) = h_{Ts}(\theta_{Aj}(t) - \theta_s(t)) \tag{2.27}$$

where h_{Ts} is the time-invariant combined heat transfer coefficient related to the air temperature (W m^{-2} °C^{-1}).

Sections 5.4 and 5.5 cover the evaluation of time-dependent h_c and h_r values for use in the numerical method formulated in chapter 3, and chapter 4 describes a matrix organisation, partitioning and processing technique which allows treatment of combined multi-zone problems and allows the removal of the assumption of linearity applied to the non-linear longwave radiation term.

The conduction heat flux at an inside surface is dependent on the history of temperature and heat flux variations at the exposed surfaces of any multi-layered construction as well as the thermophysical properties of the individual layers. Section 2.2.1 demonstrated URF formulation for transient conduction within multi-layered constructions which, when combined with the surface temperature time-series, gives the conduction heat flux. Thus from equation (2.15):

$$q_K(t) = - \sum_{m=0}^{\infty} \theta_{si}(t - m\Delta)z(m\Delta) + \sum_{m=0}^{\infty} \theta_{se}(t - m\Delta)y(m\Delta) \qquad (2.28)$$

where m is the time-step index measured from commencement of the application of some unit excitation ($m = 0$), $z(m\Delta)$ represents the response factor terms of the inside surface URF due to a unit temperature excitation at the inside surface (W m^{-2} °C^{-1}), $y(m\Delta)$ the response factor terms of the inside surface URF due to a unit temperature excitation at the outside surface (W m^{-2} °C^{-1}), and si, se denote inside and outermost surfaces respectively.

Because of the dependence of heat flux on the relative position of component elements within composite constructions, the heat flux at an outermost surface is given by:

$$q_K(t) = \sum_{m=0}^{\infty} \theta_{se}(t - m\Delta)x(m\Delta) - \sum_{m=0}^{\infty} \theta_{si}(t - m\Delta)y(m\Delta) \qquad (2.29)$$

where $x(m\Delta)$ represents the response factor terms of the outermost surface URF due to a unit temperature excitation at the outermost surface and $y(m\Delta)$ the response factor terms of the outermost surface URF due to a unit temperature excitation at the inside surface (both W m^{-2} °C^{-1}).

Section 5.3 describes a technique for the estimation of internal shortwave radiation distribution ($q_S(t)$) and §5.7 discusses the treatment of the convective and radiant portions of casual heat sources ($q_M(t)$).

Substitution of equations (2.25)–(2.28) in equation (2.24) and collecting all unknown terms on the equation left-hand side gives, for an internal surface:

$$-\left(h_{cs} + \sum_{j=1}^{N} h_{rj,s} + z(0)\right)\theta_{si}(t) + y(0)\theta_{se}(t) + \sum_{j=1}^{N} h_{rj,s}\theta_j(t)$$

$$= - h_{cs}\theta_A(t) - q_S(t) - q_M(t) + \sum_{m=1}^{\infty} \theta_{si}(t - m\Delta)z(m\Delta)$$

$$+ \sum_{m=1}^{\infty} \theta_{si}(t - m\Delta)y(m\Delta) \qquad (2.30)$$

and for an outermost surface:

$$- [h_{Ts} - x(0)]\theta_{se}(t) - y(0)\theta_{si}(t)$$

$$= -h_{Ts}\theta_{As}(t) - q_S(t) - q_M(t) - \sum_{m=1}^{\infty} \theta_{se}(t - m\Delta)x(m\Delta)$$

$$+ \sum_{m=1}^{\infty} \theta_{si}(t - m\Delta)y(m\Delta). \tag{2.31}$$

In forming these relationships the assumption is made that shortwave distribution and miscellaneous radiant gains can be independently assessed for all time-rows. Also, at this stage in the procedure, the internal air temperature is held constant at some design value and so is prescribed for all time.

Heat balance applied to the internal air volume yields:

$$\sum_{j=1}^{N} A_j h_{cj}(\theta_{sj}(t) - \theta_A(t)) + \frac{\varrho C V}{\delta t}(\theta_A(t) - \theta_A(t - \delta t)) + q_p(t) = 0 \tag{2.32}$$

where δt is the calculation time-step (s), ϱ the air density (kg m^{-3}), C the air specific heat (J kg^{-1} °C^{-1}), V the enclosed volume (m^3) and q_p the convective plant requirement to maintain the design temperature (W).

The heat balance relationships of equations (2.30) and (2.31) can now be applied to each internal and external surface in turn and this equation set (including the window heat balance equation) represented in standard matrix notation as

$$\mathbf{A} \times \boldsymbol{\theta} = \mathbf{B} \tag{2.33}$$

where $\mathbf{A}$ is a non-homogeneous matrix of future time-row coefficients of the surface temperature terms as represented by the left-hand side of equations (2.30) and (2.31), $\boldsymbol{\theta}$ is a column matrix (vector) of future time-row (t) surface temperatures, and $\mathbf{B}$ is a column matrix formulated from the terms which are dependent on the excitation components at time t and the entire surface temperature history from the start of the solution stream to the previous time-step. Note that in practice only a finite number of terms need be considered (see §2.2.3).

The contents of the $\mathbf{B}$ matrix are defined by the right-hand side of equations (2.30) and (2.31).

The solution of equation (2.33) is now obtained as:

$$\boldsymbol{\theta} = \mathbf{A}^{-1} \times \mathbf{B} \tag{2.34}$$

where the inverted $\mathbf{A}$ matrix is unchanged with time and so only the $\mathbf{B}$ matrix need be reformulated at each new time-step using the latest values determined at the previous time-row. The objective then is to obtain the surface temperature URFs from equation (2.34) and hence the cooling and/or heating URFs from equation (2.32).

Returning to the single zone problem, figure 2.8 gives the contents of the $\mathbf{A}$,

$$
A = \begin{bmatrix}
-h_{c1} - \displaystyle\sum_{j=1}^{7} h_{rj,1} - z_1(0) & h_{r2,1} & h_{r3,1} & h_{r4,1} & h_{r5,1} & h_{r6,1} & h_{r7,1} \\[2ex]
h_{r1,2} & -h_{c2} - \displaystyle\sum_{j=1}^{7} h_{rj,2} - z_2(0) & h_{r3,2} & h_{r4,2} & h_{r5,2} & h_{r6,2} & h_{r7,2} \\[2ex]
h_{r1,3} & h_{r2,3} & -h_{c3} - \displaystyle\sum_{j=1}^{7} h_{rj,3} - z_3(0) & h_{r4,3} & h_{r5,3} & h_{r6,3} & h_{r7,3} \\[2ex]
h_{r1,4} & h_{r2,4} & h_{r3,4} & -h_{c4} - \displaystyle\sum_{j=1}^{7} h_{rj,4} - z_4(0) & h_{r5,4} & h_{r6,4} & h_{r7,4} \\[2ex]
h_{r1,5} & h_{r2,5} & h_{r3,5} & h_{r4,5} & -h_{c5} - \displaystyle\sum_{j=1}^{7} h_{rj,5} - z_5(0) & h_{r6,5} & h_{r7,5} \\[2ex]
h_{r1,6} & h_{r2,6} & h_{r3,6} & h_{r4,6} & h_{r5,6} & -h_{c6} - \displaystyle\sum_{j=1}^{7} h_{rj,6} - z_6(0) & h_{r7,6} \\[2ex]
h_{r1,7} & h_{r2,7} & h_{r3,7} & h_{r4,7} & h_{r5,7} & h_{r6,7} & -h_{c7} - \displaystyle\sum_{j=1}^{7} h_{rj,7} - z_7(0) \\[2ex]
-y_1(0) & -y_2(0) & -y_3(0) & -y_4(0) & -y_5(0) & -y_6(0) &
\end{bmatrix}
$$

$$\mathbf{A}\,\boldsymbol{\theta} = \mathbf{B}$$

$$
\begin{bmatrix}
-y_1(0) & & & & & & & & & & & & \\
& -y_2(0) & & & & & & & & & & & \\
& & -y_3(0) & & & & & & & & & & \\
& & & -y_4(0) & & & & & & & & & \\
& & & & -y_5(0) & & & & & & & & \\
& & & & & -y_6(0) & & & & & & & \\
& & & & & & & & & & & & \\
-h_{\mathrm{T}1}+x_1(0) & & & & & & & & & & & & \\
& -h_{\mathrm{T}2}+x_2(0) & & & & & & & & & & & \\
& & -h_{\mathrm{T}3}+x_3(0) & & & & & & & & & & \\
& & & -h_{\mathrm{T}4}+x_4(0) & & & & & & & & & \\
& & & & -h_{\mathrm{T}5}+x_5(0) & & & & & & & & \\
& & & & & -h_{\mathrm{T}6}+x_6(0) & & & & & & &
\end{bmatrix}
\times
\begin{bmatrix}
\theta_{1i} \\ \theta_{2i} \\ \theta_{3i} \\ \theta_{4i} \\ \theta_{5i} \\ \theta_{6i} \\ \theta_{7} \\ \theta_{1e} \\ \theta_{2e} \\ \theta_{3e} \\ \theta_{4e} \\ \theta_{5e} \\ \theta_{6e}
\end{bmatrix}
=
$$

$$
\begin{bmatrix}
-h_{c1}\theta_{\mathrm{A}}(t) + \displaystyle\sum_{m=1}^{\infty}\theta_{1i}(t-m)z_1(m) - \sum_{m=1}^{\infty}\theta_{1e}(t-m)y_1(m) - q_{\mathrm{E},1i}(t) \\[2ex]
-h_{c2}\theta_{\mathrm{A}}(t) + \displaystyle\sum_{m=1}^{\infty}\theta_{2i}(t-m)z_2(m) - \sum_{m=1}^{\infty}\theta_{2e}(t-m)y_2(m) - q_{\mathrm{E},2i}(t) \\[2ex]
-h_{c3}\theta_{\mathrm{A}}(t) + \displaystyle\sum_{m=1}^{\infty}\theta_{3i}(t-m)z_3(m) - \sum_{m=1}^{\infty}\theta_{3e}(t-m)y_3(m) - q_{\mathrm{E},3i}(t) \\[2ex]
-h_{c4}\theta_{\mathrm{A}}(t) + \displaystyle\sum_{m=1}^{\infty}\theta_{4i}(t-m)z_4(m) - \sum_{m=1}^{\infty}\theta_{4e}(t-m)y_4(m) - q_{\mathrm{E},4i}(t) \\[2ex]
-h_{c5}\theta_{\mathrm{A}}(t) + \displaystyle\sum_{m=1}^{\infty}\theta_{5i}(t-m)z_5(m) - \sum_{m=1}^{\infty}\theta_{5e}(t-m)y_5(m) - q_{\mathrm{E},5i}(t) \\[2ex]
-h_{c6}\theta_{\mathrm{A}}(t) + \displaystyle\sum_{m=1}^{\infty}\theta_{6i}(t-m)z_6(m) - \sum_{m=1}^{\infty}\theta_{6e}(t-m)y_6(m) - q_{\mathrm{E},6i}(t) \\[2ex]
-h_{c7}\theta_{\mathrm{A}}(t) - q_{\mathrm{E},7}(t) \\[2ex]
-\displaystyle\sum_{m=1}^{\infty}\theta_{1e}(t-m)x_1(m) + \sum_{m=1}^{\infty}\theta_{1i}(t-m)y_1(m) - q_{\mathrm{E},1e}(t) \\[2ex]
-\displaystyle\sum_{m=1}^{\infty}\theta_{2e}(t-m)x_2(m) + \sum_{m=1}^{\infty}\theta_{2i}(t-m)y_2(m) - q_{\mathrm{E},2e}(t) \\[2ex]
-\displaystyle\sum_{m=1}^{\infty}\theta_{3e}(t-m)x_3(m) + \sum_{m=1}^{\infty}\theta_{3i}(t-m)y_3(m) - q_{\mathrm{E},3e}(t) \\[2ex]
-\displaystyle\sum_{m=1}^{\infty}\theta_{4e}(t-m)x_4(m) + \sum_{m=1}^{\infty}\theta_{4i}(t-m)y_4(m) - q_{\mathrm{E},4e}(t) \\[2ex]
-\displaystyle\sum_{m=1}^{\infty}\theta_{5e}(t-m)x_5(m) + \sum_{m=1}^{\infty}\theta_{5i}(t-m)y_5(m) - q_{\mathrm{E},5e}(t) \\[2ex]
-\displaystyle\sum_{m=1}^{\infty}\theta_{6e}(t-m)x_6(m) + \sum_{m=1}^{\infty}\theta_{6i}(t-m)y_6(m) - q_{\mathrm{E},6e}(t)
\end{bmatrix}
$$

Figure 2.8 The matrix equation $\mathbf{A}\boldsymbol{\theta} = \mathbf{B}$.

θ and **B** matrices where

$$q_{E,si}(t) = \text{internal surface excitation at time } t \text{ (W m}^{-2})$$
$$= q_s(t) + q_M(t)$$
$$q_{E,7}(t) = \text{window surface excitation at time } t \text{ (W m}^{-2})$$
$$= q_s(t) + q_M(t) + h_{T7}\theta_{A1}$$
$$q_{E,se}(t) = \text{outermost surface excitation at time } t \text{ (W m}^{-2})$$
$$= q_s(t) + q_M(t) + h_{Ts}\theta_{As}.$$

Once established equation (2.34) is solved at each successive time-step by re-establishing the **B** matrix from previously computed temperature data. This gives the solution for $\theta_s(t)$, the surface temperature response factors for any given unit excitation function. The strategy is to let one excitation component assume a unit time-series variation (1,0,0,0,...) whilst all other excitations are zero (0,0,0,0,...) and the zone air temperature is held constant at the reference temperature. The contents of the **B** matrix are then evaluated. Subsequent post-multiplication on the (constant) inverted **A** matrix by the **B** matrix gives the surface temperatures prevailing at time $t = 0$. The surface temperature history—at this stage for the first time-step only—is then used to re-evaluate the **B** matrix at time $t = 1$ and the surface temperatures prevailing at the next time-step found. The process then continues until the ratio of successive terms in the respective surface temperature time-series becomes constant. Continuation of this geometric progression, for as many terms as desired, gives the surface temperature URF. The entire process is then repeated for all other unit excitations of interest. Plant capacity response factors corresponding to any unit excitation function are then given by equation (2.32):

$$P_r(t) = - \sum_{j=1}^{N} A_j h_{cj} [S_{rj}(t) - \theta_A(t)]$$

where $P_r(t)$ is the plant capacity response factor at time t and corresponding to some unit excitation function (W), and $S_{rj}(t)$ is the surface temperature response factor at time t for surface j and corresponding to some unit excitation ($^\circ$C).

Thus there are as many plant capacity URFs as unit excitations of interest; external air temperature, solar radiation etc. When combined with the appropriate surface excitation time-series, the plant capacity response factors yield the heat transfer to or from the zone. This is the amount of heat which must be removed or added to the air to maintain constant temperature levels.

In practice, zone air temperatures will deviate from the design value due to intermittent plant operation and the unavoidable temporal mismatch between load and extraction (or addition) rates caused by control system and equipment inertia.

For an internal air volume, equation (2.32) can be rewritten in the following form:

$$\sum_{j=1}^{N} A_j h_{cj} \theta_{sj}(t) - \sum_{j=1}^{N} A_j h_{cj} \theta_A(t) + \frac{\varrho CV}{\delta t} \theta_A(t)$$
$$= \frac{\varrho CV}{\delta t} \theta_A(t - \delta t) - q_p(t) - q_L(t) - q_I(t)$$

where $q_L(t)$ is the convective heat gain from casual sources such as lights (W), and $q_I(t)$ is the advective load due to infiltration, zone-coupled air flow etc (W).

This equation can now be incorporated in the matrix of figure 2.8 so that for each unit excitation considered (infiltration, casual gains, plant etc) a corresponding air temperature URF will emerge. Alternatively, the equation system can be subjected to a unit time-series variation in internal air temperature with all other excitations held at zero. In each case the surface temperature URFs give the plant capacity response factors. These are the zone weighting factors $W(m\Delta)$.

2.2.3 Response function application

The total heating or cooling load is determined by modifying the individual heat loss and gain flowpaths by the corresponding weighting factors so that:

$$Q(t) = \sum_{m=0}^{\infty} \sum_{j=1}^{N} W_j(m\Delta)H_j(t - m\Delta) \tag{2.35}$$

where $W_j(m\Delta)$ is the weighting factor for flowpath j and $H_j(t - m\Delta)$ is the heat gain or loss due to flowpath j (W).

Kimura (1977) gives a detailed description of the algorithmic counterpart of this process as embodied in the SHASE program package (Saito and Kimura 1974).

In applying construction response or zone weighting factors, it is necessary to operate with finite summations rather than with the infinite summations of §§2.2.1 and 2.2.2. Equation (2.35) can be rewritten (for any flowpath j) to give

$$Q(t) = \sum_{m=0}^{k} W(m\Delta)H(t - m\Delta) + \sum_{m=1}^{\infty} C^{m\Delta}W(k)H(t - k - m\Delta) \tag{2.36}$$

where C is the common ratio of $W(m\Delta)$ so that $C^{m\Delta}W(k)$ defines the time-series after the kth term where the geometric progression is assumed to commence.

Also from equation (2.36) the expression for $Q(t - 1)$ is given by

$$Q(t - 1) = \sum_{m=0}^{k} W(m\Delta)H(t - 1 - m\Delta) + \sum_{m=1}^{\infty} C^{m\Delta}W(k)H(t - 1 - k - m\Delta). \tag{2.37}$$

Multiplying equation (2.37) by C and subtracting from equation (2.36) gives the expression containing only a finite summation:

$$Q(t) = CQ(t - 1) + W(0)H(t) + \sum_{m=1}^{k} W(m\Delta - 1)H(t - \Delta). \tag{2.38}$$

The final zone air temperature will depend on the design temperature, the cooling or heating load corresponding to this design temperature and the actual heat extraction or addition rate. For a cooling unit with a simple proportional control system characterised by

$$E = C_d + D\delta(t)$$

where E is the actual rate of heat extraction (W), C_d is the plant capacity when operating at the design temperature (W) and D is the change in plant potential caused by a one degree rise in air temperature (W $^\circ$C^{-1}), Stephenson and Mitalas (1967b) have shown that the deviation of the actual air temperature from the design value is given by

$$\delta(t) = \frac{L_d(t) - C_d + \sum_{m=1}^{k} W_1(m\Delta)\delta(t - m\Delta) - \sum_{m=1}^{k} W_2(m\Delta)[E(t - m\Delta) - L_d(t - m\Delta)]}{(D - W_1(0))}$$

where $\delta(t)$ is the deviation of the air temperature from the assumed design value ($^\circ$C), $L_d(t)$ is the capacity demand relative to the design temperature (W), and W_1, W_2 are appropriate weighting factors.

Kimura (1977) gives an alternative formulation for the calculation of zone temperature during plant OFF times:

$$\theta(t) = \frac{-1}{W_T(0) + K(t)} \left(\sum_{m=1}^{\infty} W_T(m\Delta)\theta(t - m\Delta) + L_d(t) \right)$$

where W_T is the total weighting function ($= \sum_{j=1}^{N} W_j$), and $K(t)$ is the infiltration conductance coefficient.

Many computer-based modelling systems exist based on the conduction transfer and zone weighting factor approach and the interested reader is directed to other texts for details of technique implementation (for example ASHRAE 1971, Mitalas and Arseneault 1970, Hittle 1979, York and Tucker 1980).

2.3 Frequency-domain response functions

The fundamental assumption underlying the frequency-domain (or harmonic) method is that climatological time-series can be represented by a series of periodic cycles. In this way the climatic influence can be represented by a steady-state term accompanied by a number of pure sine wave harmonics with, in general, increasing frequency and decreasing amplitude. The division of real climatic time-series into component sinusoidal variations about some mean condition can be readily achieved by Fourier series representation by which a given function can be approximated by a series of sine (and/or cosine) functions such that for some continuous function $f(x)$

$$f(x) = a_0 + \sum_{m=1}^{k} a_m \cos(2\pi mx/L) + \sum_{m=1}^{k} a_m \sin(2\pi mx/L)$$

where $1/L$ is the frequency.

Each selected harmonic (frequency) of such a series can then be processed separately and modified by thermal response factors appropriate to its frequency; factors which can be mathematically determined from equation (2.10) only because the boundary condition has been prescribed as a sine wave in the first instance. The principle of superimposition (Stephenson 1967b) is then invoked to allow the system response to be obtained by summation of the individual effects of the separate harmonics with respect to the mean condition.

In its most rigorous form the harmonic response method relies on representing actual climate time-series by a Fourier series of sine wave harmonics acting around a constant term. The frequency of the fundamental harmonic is often, for convenience, set at 24 hours with the remaining harmonics having diminishing periods such as 12, 6, 3, 1.5 hours and so on. As a function of harmonic frequency, the thermal factors can be determined and applied, as operators, to the individual terms of the corresponding climatic harmonic.

The method can handle any of the energy transfers within buildings, although some may be crudely approximated. For example, longwave radiation exchanges are normally handled by the environmental temperature method (Danter 1973); with a low number of harmonics, casual gains are subject to Gibbs' Phenomenon (Gower and Baker 1974) since their usually square profile will be 'clipped' when represented by a family of sine waves, and internal solar models are typically crude. Also, by accepting the principle of superimposition it is difficult (if not impossible) to preserve spatial integrity and, in addition, the simultaneous treatment of building and plant components is an intractable problem.

Much of the early work on harmonic prediction techniques (Alford *et al* 1939, Mackey and Wright 1944, 1946) concentrated on the estimation of fabric heat flow under assumptions of constant internal temperatures. External air temperatures were combined with the prevailing shortwave solar radiation to give the single sol–air temperature term, and internal radiant loads (shortwave solar and radiant casual for example) were considered as instantaneous fluxes applied directly as an air point load—an assumption no longer deemed valid, since no account is taken of the interaction with structure and content capacity.

Later workers removed the limitation imposed by the assumption of constant internal air temperature (Nottage and Parmelee 1955, Muncey 1953) and increased the accuracy of internal surface radiative and convective modelling (Pipes 1957, Gupta 1964). In recent years the 'means and swings' technique has been developed in the UK (Danter 1960) which, in its simplest form, has been adopted by the Chartered Institute of Building Services (CIBS) and is commonly referred to as the *admittance method* (Louden 1968). This method allows the estimation of flux transfers under steady cyclic (periodic) conditions, where external flux or temperature variations are repeated over a period of time, by reliance only on thermal response factors relating to some fundamental harmonic (usually taken as the 24 hour frequency harmonic) *but applied to the actual temperature or flux excitation.*

2.3.1 Multi-layered constructions

We recall the matrix relationship of equation (2.10) which expresses the Laplace transforms of the temperature and flux at one extreme boundary of a multi-layered construction in terms of the corresponding transforms at the other extreme boundary:

$$\begin{bmatrix} \theta(L,p) \\ q(L,p) \end{bmatrix} = \begin{bmatrix} A(p) & B(p) \\ C(p) & D(p) \end{bmatrix} \times \begin{bmatrix} \theta(0,p) \\ q(0,p) \end{bmatrix}. \tag{2.39}$$

Expanding the overall transmission matrix to include boundary layers of combined convective/radiative resistance R_L and R_0 (m^2 °C W^{-1}), then

$$\begin{bmatrix} E(p) & F(p) \\ G(p) & H(p) \end{bmatrix} = \begin{bmatrix} 1 & R_L \\ 0 & 1 \end{bmatrix} \times \begin{bmatrix} A(p) & B(p) \\ C(p) & D(p) \end{bmatrix} \times \begin{bmatrix} 1 & R_0 \\ 0 & 1 \end{bmatrix} \quad (2.40)$$

since

$$\begin{bmatrix} m_{11} & m_{12} \\ m_{21} & m_{22} \end{bmatrix} = \begin{bmatrix} 1 & R \\ 0 & 1 \end{bmatrix} \qquad \text{if } \varrho C \to 0.$$

Equation (2.39) now becomes:

$$\begin{bmatrix} \theta(L,p) \\ q(L,p) \end{bmatrix} = \begin{bmatrix} A(p) + R_L C(p) & R_0 A(p) + R_L R_0 C(p) + B(p) + R_L D(p) \\ C(p) & R_0 C(p) + D(p) \end{bmatrix} \times \begin{bmatrix} \theta(0,p) \\ q(0,p) \end{bmatrix}$$

$$= \begin{bmatrix} E(p) & F(p) \\ G(p) & H(p) \end{bmatrix} \times \begin{bmatrix} \theta(0,p) \\ q(0,p) \end{bmatrix} \quad (2.41)$$

and by similar reasoning the transforms of the temperature and heat flux at $x = 0$ are given by:

$$\begin{bmatrix} \theta(0,p) \\ q(0,p) \end{bmatrix} = \begin{bmatrix} A(p) + R_0 C(p) & R_L A(p) + R_0 R_L C(p) + B(p) + R_0 D(p) \\ C(p) & R_L C(p) + D(p) \end{bmatrix} \times \begin{bmatrix} \theta(L,p) \\ q(L,p) \end{bmatrix}$$

$$= \begin{bmatrix} I(p) & J(p) \\ K(p) & L(p) \end{bmatrix} \times \begin{bmatrix} \theta(L,p) \\ q(L,p) \end{bmatrix}. \quad (2.42)$$

Both matrix equations involve four unknown values and in order to permit solution some assumptions must be made concerning the problem boundary conditions. Setting one boundary layer surface at 0 °C while the other surface is subjected to a periodic temperature state dictated by sin (ωt), where $\omega = 2\pi/L$ is the angular frequency of the temperature variation, gives the frequency response function of the multi-layered construction. For $\theta(0,p) = 0$, equation (2.41) yields

$$\theta(L,p) = F(p)q(0,p) \quad (2.43)$$
$$q(L,p) = H(p)q(0,p)$$

and for $\theta(L,p) = 0$, equation (2.42) yields

$$\theta(0,p) = J(p)q(L,p) \quad (2.44)$$
$$q(0,p) = L(p)q(L,p).$$

Now if $p = j\omega$, where j is the imaginary operator defined by $j^2 = -1$, then the Laplace transform becomes the Fourier transform, for which the inverse transforms of equations (2.43) and (2.44) are

$$\theta_h(L,t) = F(j\omega)q_h(0,t) \quad (2.45)$$

$$q_h(L,t) = H(j\omega)q_h(0,t) \quad (2.46)$$

$$\theta_h(0,t) = J(j\omega)q_h(L,t) \quad (2.47)$$

$$q_h(0,t) = L(j\omega)q_h(L,t) \quad (2.48)$$

where θ_h and q_h are the temperature and flux harmonics. The matrix elements F, H, J and L are determined from the basic homogeneous element relationships with $p = j\omega$; that is, for a homogeneous element under steady periodic conditions equation (2.9) can be rewritten as

$$\begin{bmatrix} \theta(l,t) \\ q(l,t) \end{bmatrix} = \begin{bmatrix} m_{11}(j\omega) & m_{12}(j\omega) \\ m_{21}(j\omega) & m_{22}(j\omega) \end{bmatrix} \times \begin{bmatrix} \theta(0,t) \\ q(0,t) \end{bmatrix}$$

where

$$m_{11}(j\omega) = m_{22}(j\omega) = \cosh{[(j\omega/\alpha)^{\frac{1}{2}}l]}$$
$$m_{12}(j\omega) = - \sinh{[(j\omega/\alpha)^{\frac{1}{2}}l]}/k(j\omega/\alpha)^{\frac{1}{2}}$$
$$m_{21}(j\omega) = - k(j\omega/\alpha)^{\frac{1}{2}} \sinh{[(j\omega/\alpha)^{\frac{1}{2}}l]}$$

and the overall transmission matrix can be established by the usual matrix multiplication process.

Recalling the earlier assumption that $\theta(x,p) = 0$, it can now be seen that this corresponds to setting the surface (or adjacent air) temperature harmonic to zero. This implies that the temperature is held constant but not necessarily at zero.

Thus, from equation (2.45) the flux at surface $x = 0$ due to a steady periodic condition $\theta_h(L, t) = \sin \omega t$ is given by

$$q_h(0,t) = \sin \omega t/F(j\omega)$$

and at surface $x = L$ due to $\theta_h(0,t) = \sin \omega t$, from equation (2.47) by

$$q_h(L,t) = \sin \omega t/J(j\omega)$$

where F and J are complex numbers and $q_h(0,t)$ and $q_h(L,t)$ are the heat fluxes at outermost and innermost boundaries, corresponding to unit sinusoidal opposite boundary temperature variations of angular frequency ω.

In the 'means and swings' technique three separate response factors have been identified: decrement response, surface response, and admittance response. Each of these factors will possess a corresponding phase angle or time lag which dictates the response time of the process the factor addresses, that is, the time delay between cause and effect (if the phase angle is positive).

The *decrement response factor* is defined as the ratio of the cyclic flux transmission to the steady state flux transmission and is applied to fluctuations (about the mean) in external temperature or flux harmonics impinging on exposed opaque surfaces undergoing transient conduction processes. This gives the related fluctuation within a building at some later point in time depending on the decrement factor time lag. Figure 2.9 shows an arbitrary building element exposed to a sinusoidal external air temperature or solar radiation time-series. The corresponding cyclic heat flux at the inside surface is also shown after time-series modification by the decrement factor of the intermediate element (or series of elements) and application of the appropriate time shift.

The *surface response factor* defines the portion of the heat flux at any internal surface which is readmitted to the internal environmental point when

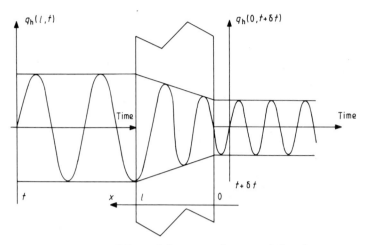

Figure 2.9 Effect of decrement factor and time lag.

temperatures are held constant. The factor is applied to cyclic energy inputs at a surface to give the corresponding cyclic energy variations at the environmental point; the concept is analogous to time shifted reflections and is illustrated in figure 2.10. Typical applications include the modification of the transmitted component of solar radiation through windows and the radiant component of casual gains.

The *admittance response factor* is defined as the amount of energy entering a surface for each degree of temperature swing at the environmental point. It is used to represent exclosure response to give the equivalent swing in temperature about some mean value due to a cyclic load on an enclosure.

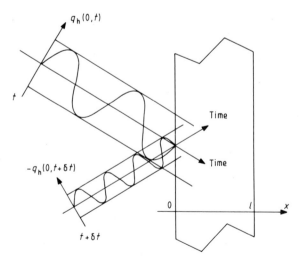

Figure 2.10 Effect of surface factor and time lag.

Any final temperature or flux prediction is obtained by summing the cyclic contribution from each harmonic and expressing the result with respect to the contribution made by the steady-state term. To simplify the procedure, Danter has proposed that the factors for the fundamental 24 hour cycle be applied to the actual climatic profile. Milbank and Harrington–Lynn (1974) have stated that this is not an unreasonable approximation and Gupta *et al* (1974) have advised that the fundamental frequency should be greater than twice the thermal time constant and not merely 24 hours if a steady periodic regime is required.

Each of the factors is now derived in turn.

Decrement response factor

Recalling the definition, the decrement factor d is the ratio of the cyclic flux transmission (usually from outside to inside) to the steady state flux transmission, namely

$$d_{0,L} = q_h(L,t)/U\theta_h(0,t)$$

where U is the steady-state overall thermal transmittance (W m^{-2} °C^{-1}).

Thus, from equation (2.47),

$$d_{0,L} = 1/J(j\omega)U$$
$$= 1/[R_L A(j\omega) + R_L R_0 C(j\omega) + B(j\omega) + R_0 D(j\omega)] U.$$

This expression will have both real and imaginary parts, r_1 and v_1 respectively, and so the magnitude of the decrement factor is

$$d_{0,L} = |r_1 + jv_1| = (r_1^2 + v_1^2)^{\frac{1}{2}}$$

with a corresponding time lag (radians) of

$$\phi_d = \tan^{-1}(v_1/r_1).$$

Surface response factor

This is defined as the ratio of the heat flux readmitted from a surface (usually internal) to the total flux absorbed. This ratio is equivalent to the the ratio of the overall impedance less the impedance of the surface layer at the surface in question to the overall impedance. For an internal surface and with reference to figure 2.11:

$$s_{L,0} = [Z(L,t) - R_L]/Z(L,t)$$
$$= 1 - R_L Z(L,t)$$

where $Z(L,t)$ is the overall impedance from $x = 0$ to $x = L$, and equals $q_h(L,t)/\theta_h(L,t)$. This implies that

$$s_{L,0} = 1 - \frac{R_L q_h(L,t)}{\theta_h(L,t)}$$

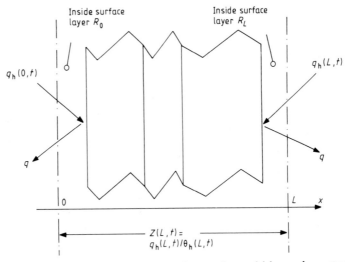

Figure 2.11   The overall thermal impedance of a multi-layered construction.

and, from equations (2.45) and (2.46),

$$= 1 - \frac{R_L H(j\omega)}{F(j\omega)}$$

$$= \frac{R_0 A(j\omega) + B(j\omega)}{R_0 A(j\omega) + R_0 R_L C(j\omega) + B(j\omega) + R_L D(j\omega)}.$$

As before this expression will have real and imaginary parts r_2 and v_2 and so

$$s_{L,0} = |r_2 + jv_2| = (r_2^2 + v_2^2)^{1/2}$$

with a time lag of

$$\phi_s = \tan^{-1}(v_2/r_2).$$

Admittance response factor
This is defined as the ratio of the energy entering an internal surface to the corresponding temperature swing:

$$a_{L,0} = q_h(L,t)/\theta_h(L,t)$$

$$= \frac{R_0 C(j\omega) + D(j\omega)}{R_0 A(j\omega) + R_0 R_L C(j\omega) + B(j\omega) + R_L D(j\omega)} = r_3 + jv_3.$$

Thus

$$a_{L,0} = (r_3^2 + v_3^2)^{1/2}$$

with the temperature swing *leading* the absorption by

$$\phi_a = \tan^{-1}(v_3/r_3).$$

Similar reasoning gives rise to the response factors for the reverse ordered sense, that is:

$$d_{L,0} = 1/[R_0A(\mathrm{j}\omega) + R_0R_LC(\mathrm{j}\omega) + B(\mathrm{j}\omega) + R_LD(\mathrm{j}\omega)]\,U$$

$$s_{0,L} = [R_LA(\mathrm{j}\omega) + B(\mathrm{j}\omega)]/[R_LA(\mathrm{j}\omega) + R_0R_LC(\mathrm{j}\omega) + B(\mathrm{j}\omega) + R_0D(\mathrm{j}\omega)]$$

$$a_{0,L} = [R_LC(\mathrm{j}\omega) + D(\mathrm{j}\omega)]/[R_LA(\mathrm{j}\omega) + R_0R_LC(\mathrm{j}\omega) + B(\mathrm{j}\omega) + R_0D(\mathrm{j}\omega)].$$

2.3.2 Zone energy balance

In determining the overall transmission matrix of equations (2.41) and (2.42) surface boundary layers were represented by combined surface resistances determined from

$$R = 1/(h_c + 1.2Eh_r)$$

where h_c is the surface convection coefficient (W m^{-2} $^\circ$C^{-1}), h_r is the radiative heat transfer coefficient (W m^{-2} $^\circ$C^{-1}) and E is the emissivity factor.

This treatment assumes that the inter-surface longwave radiation exchange can be approximated by some linearised exchange between each surface and an index temperature, determined as a weighting between zone air and mean radiant temperatures. Environmental and dry resultant temperatures are often used for this purpose. Since all heat flows through this fictitious point no inter-surface energy balance is required.

2.3.3 Response function application

The factors derived in §2.3.1 can now be applied to the component harmonics of some real climatic excitation to determine the response of some enclosure. This section demonstrates the use of the factors to determine internal temperatures, heating/cooling requirements and the effects of variable ventilation and intermittent plant operation.

Internal temperature prediction
The magnitudes of the various load fluctuations at each point in time, for each harmonic considered, and about the mean condition, are determined and modified by the decrement and surface factors appropriate to their type and frequency. This gives the energy fluctuation imposed on the enclosure at each frequency and experienced at some time after the initial excitation, a delay which depends on the time lag related to the factor used. The total fluctuating load at each point in time and at each frequency is then obtained by summing the individual load fluctuations realised at each time-row over the interval in question. These fluctuations are then modified by the appropriate admittance factor to give the contribution by each harmonic to the fluctuation about the mean of

internal environmental temperature at some later time depending on the lag associated with the admittance factor.

The mean internal environmental temperature is determined from

$$\bar{\theta}_{ei} = \bar{\theta}_{ao} + \bar{Q}_T / (\textstyle\sum AU + \bar{C}_v) \qquad (2.49)$$

where $\bar{\theta}_{ei}$ is the daily mean internal environmental temperature ($^\circ$C), $\bar{\theta}_{ao}$ is the mean external temperature ($^\circ$C), $\bar{Q}_T$ is the mean total heat flux from all sources (W), $\bar{C}_v$ is the mean ventilation conductance (W $^\circ$C^{-1}) and $\sum AU$ represents the sum of product of areas and overall thermal transmittance values (W $^\circ$C^{-1}).

The mean total heat flux is given by

$$\bar{Q}_T = \bar{Q}_{fs} + \bar{Q}_s + \bar{Q}_c$$

where $\bar{Q}_{fs}$ is the mean solar gain through opaque surfaces, $\bar{Q}_s$ the mean solar gain through transparent surfaces, and $\bar{Q}_c$ the mean gain from casual sources (all measured in W).

These terms are now considered in turn.

$$\bar{Q}_{fs} = \sum_{i=1}^{N} [A_i U_i R_0 (\alpha_i \bar{I}_{so} - \epsilon_i I_0)]$$

where N is the total number of exposed opaque surfaces, α_i the surface shortwave absorptivity, $\bar{I}_{so}$ the mean shortwave flux incident on opaque surface, and is just $\sum_{t=1}^{m} I_{so}(t)/m$, where $m = 24$ for daily mean; $I_{so}(t)$ is the instantaneous shortwave solar flux impinging on opaque surface at any time t (W m^{-2}) and $\epsilon_i I_0$ is the longwave radiation loss from surface to surroundings (W m^{-2}).

$$\bar{Q}_s = \sum_{t=1}^{m} Q_s(t)/m$$

where $Q_s(t)$ is the instantaneous solar gain at time t (W).

Section 5.3 describes a detailed solar modelling procedure which will give $I_{so}(t)$ and $Q_s(t)$ as a function of location, time of year and day, building geometry, site obstructions and thermophysical properties. As a simplification, the use of 'solar gain factors' has been proposed (Louden 1968). These factors are predetermined values which depend on solar incidence angles and the thermophysical properties of window/blind arrangements. They define the portion of the incident solar flux which penetrates the boundary as

$$Q_s(t) = \sum_{i=1}^{L} A_i S_i(t) I_{si}(t)$$

where L is the total number of transparent surfaces, $S_i(t)$ the solar gain factor at time t and $I_{si}(t)$ the instantaneous intensity at time t (see §5.3).

$$Q_c(t) = \sum_{j=1}^{K} Q_{cj}$$

where K is the number of casual sources and Q_{cj} the magnitude of any source (W).

The contribution by each harmonic to the swing in internal environmental temperature about the mean value is then computed from

$$\tilde{\theta}_{ei}(t) = \frac{\tilde{Q}_T(t - \phi_a)}{\sum Aa + C_v} \tag{2.50}$$

where $\tilde{\theta}_{ei}(t)$ is the fluctuation of internal environmental temperature about the mean at some time t ($^\circ$C), $\tilde{Q}_T(t - \phi_a)$ is the total fluctuating load at the environmental point at time $(t - \phi_a)$ and due to any harmonic (W). ϕ_a is the time lag associated with the admittance factor (h), and $\sum Aa$ is the sum of product of areas and admittance for all internal surfaces (W $^\circ$C^{-1}).

The total fluctuating energy gain at the environmental point and due to any particular excitation frequency is given by

$$\tilde{Q}_T(t) = \tilde{Q}_{fs}(t) + \tilde{Q}_s(t) + \tilde{Q}_c(t) + \tilde{Q}_{fc}(t) + \tilde{Q}_{gc}(t) + \tilde{Q}_v(t)$$

where $\tilde{Q}_{fs}(t)$ is the opaque surface solar gain fluctuation at time t, $\tilde{Q}_s(t)$ the transparent surface solar gain fluctuation, $\tilde{Q}_c(t)$ the casual gain fluctuation, $\tilde{Q}_{fc}(t)$ the opaque surface conduction gain fluctuation, $\tilde{Q}_{gc}(t)$ the transparent surface conduction gain fluctuation, and $\tilde{Q}_v(t)$ the ventilation or infiltration fluctuation (all measured in W).

Each load fluctuation is now considered in turn.

$$\tilde{Q}_{fs}(t) = \sum_{i=1}^{N} [A_i U_i R_0 d_i \alpha \tilde{I}_{so}(t - \phi_d)]$$

where d_i is the decrement factor $(d_{L,0})$ for layers behind surface i, ϕ_d is the associated time lag, $\tilde{I}_{so}(t - \phi_d)$ the fluctuation about the mean of solar intensity incident on opaque surfaces at some time $(t - \phi_d)$ (measured in W m^{-2} and equal to $I_{so}(t - \phi_d) - \bar{I}_{so}$).

$$\tilde{Q}_s(t) = \tilde{Q}_{s1}(t - \phi_s) + \tilde{Q}_{s2}(t)$$

where $\tilde{Q}_{s1}(t - \phi_s)$ is the fluctuation due to the directly transmitted component (shortwave) of the incident solar radiation through transparent surfaces (W) and $\tilde{Q}_{s2}(t)$ the fluctuation due to the absorbed component of the incident solar radiation which is subsequently retransmitted inwards to the environmental point (longwave + convection) (W). Section 5.3 gives a rigorous derivation of these terms.

The simplified admittance procedure proposes the use of 'alternating solar gain' factors which allow the fluctuation in energy at the environmental point due to the fluctuation in solar gain through transparent surfaces to be determined directly from

$$\tilde{Q}_s(t) = \sum_{i=1}^{L} A_i \tilde{S}_i(t - \phi_s) \tilde{I}_{si}(t - \phi_s)$$

where $\tilde{S}_i(t - \phi_s)$ is the alternating solar gain factor which includes the effect of the surface response factor.

$$\tilde{Q}_c(t) = Q_c(t) - \bar{Q}_c$$

where $\tilde{Q}_c(t)$ is the total instantaneous casual load (W).

With the use of environmental temperature, internally generated casual gains are only immediately realised as cooling loads if the heat input is in the proportion of two-thirds radiation to one-third convection. In many cases casual gains will not conform to this split and so they must be separated into convective and radiant components. The convective component can then be assumed to act immediately at the air point with the radiant component treated in a manner similar to solar gains and modified by the internal mass of the enclosure. In general terms

$$\widetilde{Q}_c(t) = S(t - \phi_s)\widetilde{Q}_R(t - \phi_s) + L_c\widetilde{Q}_c(t)$$

where $\widetilde{Q}_R(t - \phi_s)$ is the fluctuating radiant portion of the casual load (W), $\widetilde{Q}_c(t)$ is the fluctuating convective portion (W), L_c is the convective load correction factor which adjusts this component to account for injection at the environmental point rather than at the air point, given by $h\sum A/(h\sum A + C_v)$, and h is the hypothetical conductance associated with the environmental point (W m^{-2} $^\circ$C^{-1}).

$$\widetilde{Q}_{fc}(t) = \sum_{i=1}^{o} A_i U_i d_i \widetilde{\theta}_{Ao}(t - \phi_d)$$

where $\widetilde{\theta}_{Ao}$ is the fluctuation in outside temperature ($^\circ$C).

$$\widetilde{Q}_{gc}(t) = \sum_{i=1}^{T} A_i U_i \widetilde{\theta}_{Ao}(t).$$

It is usual practice to assume that window conduction processes undergo negligible time delay.

$$\widetilde{Q}_v(t) = C_v(t)\widetilde{\theta}_{Ao}(t)$$

where again it is usual to assume a zero time lag.

Heating/cooling requirements
The fluctuations in energy at the environmental point due to the various temperature and energy fluctuations are computed relative to some fixed internal temperature. This is done by reversing the procedure for internal temperature prediction to compute the plant capacity to maintain specified temperature conditions.

Variable ventilation
It is possible to combine the foregoing equations in such a way that the effects of known variations in the infiltration/ventilation rate can be assessed. This is of particular use in the analysis of the intermittent operation of ventilation plant where the ventilation rate, although variable, is known. To a lesser extent, perhaps, the formulation allows an examination of the effects of increasing or reducing the ventilation rate relative to some design value in order to gain insight into the sensitivity of a system under real conditions.

The variable ventilation harmonic method (Harrington-Lynn 1974a) is obtained by combining the heat balance equations involving the mean and fluctuating energy gains at the environmental point. The method allows the internal temperature to be established for any known or hypothetical ventilation scheme.

Combining equations (2.49) and (2.50) gives

$$\overline{Q}_T + \widetilde{Q}_T(t) = (\sum AU + C_v)(\overline{\theta}_{ei} - \overline{\theta}_{ao}) + (\sum Aa + C_v)\widetilde{\theta}_{ei}(t)$$

and assuming that the ventilation conductance is time-dependent gives

$$Q_x(t) = \sum AU(\overline{\theta}_{ei} - \overline{\theta}_{Ao}) + \sum Aa\widetilde{\theta}_{ei}(t) + C_v(\widetilde{\theta}_{ei}(t) + \overline{\theta}_{ei} - \overline{\theta}_{Ao}) \qquad (2.51)$$

where

$$Q_x(t) = \overline{Q}_T(t) + \widetilde{Q}_T(t).$$

Now for $\Delta\theta_{ei} = \overline{\theta}_{ei} - \overline{\theta}_{Ao}$

$$Q_x(t) = \sum AU\,\Delta\theta_{ei} + \sum Aa\widetilde{\theta}_{ei} + C_v(t)(\widetilde{\theta}_{ei}(t) + \Delta\theta_{ei}).$$

For a daily analysis, taking hourly values, there are 24 equations of this type. These equations contain 25 unknown quantities ($\widetilde{\theta}_{ei}(t)$ at each hour and the daily value of $\Delta\theta_{ei}$) and consequently another equation is required. Summation of the 24 equations gives

$$\sum_{t=1}^{24} Q_x(t) = 24 \sum AU\Delta\theta_{ei} + \sum_{t=1}^{24}\sum Aa\widetilde{\theta}_{ei}(t) + \sum_{t=1}^{24} C_v(t)[\widetilde{\theta}_{ei}(t) + \Delta\theta_{ei}]$$

and since

$$\frac{1}{24}\sum_{t=1}^{24}\sum Aa\widetilde{\theta}_{ei}(t) = 0$$

this reduces to

$$\overline{Q}_x = \Delta\theta_{ei}(\sum AU + \overline{C}_v) + \frac{1}{24}\sum_{t=1}^{24} C_v(t)\widetilde{\theta}_{ei}(t)$$

where

$$\overline{Q}_x = \frac{1}{24}\sum_{t=1}^{24} Q_x(t)$$

$$\overline{C}_v = \frac{1}{24}\sum_{t=1}^{24} C_v(t). \qquad (2.52)$$

Intermittent plant operation
Harmonic methods can be used to analyse predefined plant operational schemes (Harrington-Lynn 1974b). Consider equation (2.51) which relates to the variable ventilation case. With the addition of a time-dependent plant input or extract term this equation becomes

$$Q_x(t) + Q_p(t) = \sum AU\Delta\theta_{ei} + \sum Aa\widetilde{\theta}_{ei}(t) + C_v(t)[\widetilde{\theta}_{ei}(t) + \Delta\theta_{ei}] \qquad (2.53)$$

where $Q_p(t)$ is the plant exchange with environmental point (W).

For any given ON period the environmental temperature control state is given by

$$\theta_c = \tilde{\theta}_{ei} + \bar{\theta}_{ei} = \tilde{\theta}_{ei} + \Delta\theta_{ei} + \bar{\theta}_{Ao}.$$

Substitution in equation (2.53) gives, for an ON period

$$Q_x(t) + Q_p(t) = \sum AU\,\Delta\theta_{ei} + \sum Aa(\theta_c - \Delta\theta_{ei} - \bar{\theta}_{Ao}) + C_v(t)(\theta_c - \bar{\theta}_{Ao}) \quad (2.54)$$

and, for an OFF period, equation (2.53) reduces to:

$$Q_x(t) = \sum AU\,\Delta\theta_{ei} + \sum Aa\tilde{\theta}_{ei}(t) + C_v(t)[\tilde{\theta}_{ei}(t) + \Delta\theta_{ei}]. \quad (2.55)$$

Again considering a 24 hour period, assume the plant is ON for X hours giving X equations of type (2.54) and $(24 - X)$ equations of type (2.55). As before, the 24 equations incorporate 25 unknowns ($\tilde{\theta}_{ei}(t)$ at each OFF hour, $Q_p(t)$ at each ON hour, and the daily value of $\Delta\theta_{ei}$) and another equation is required. Summation of the 24 equations gives

$$\sum_{ON} Q_p(t) + \sum_{t=1}^{24} Q_x(t) = 24\sum AU\,\Delta\theta_{ei} + \sum_{OFF}\sum Aa\tilde{\theta}_{ei}(t) + \sum_{OFF} C_v(t)[\tilde{\theta}_{ei}(t) + \Delta\theta_{ei}]$$

$$+ \sum_{ON}\sum Aa(\theta_c - \Delta\theta_{ei} - \bar{\theta}_{Ao}) + \sum_{ON} C_v(t)(\theta_c - \bar{\theta}_{Ao}). \quad (2.56)$$

Now from equation (2.52)

$$\sum_{t=1}^{24} Q_x(t) = 24\bar{Q}_x$$

and by definition

$$\sum_{OFF}\sum Aa\tilde{\theta}_{ei}(t) + \sum_{ON}\sum Aa(\theta_c - \Delta\theta_{ei} - \bar{\theta}_{Ao}) = 0$$

and so equation (2.56) becomes

$$\sum_{ON} Q_p(t) + 24\bar{Q}_x = 24\sum AU\,\Delta\theta_{ei} + \sum_{OFF} C_v(t)(\theta_{ei}(t) + \Delta\theta_{ei}) + \sum_{ON} C_v(t)(\theta_c - \bar{\theta}_{Ao}) \tag{2.57}$$

and equations (2.54), (2.55) and (2.57) can be solved to obtain the plant capacity to maintain any ON environmental temperature θ_c.

2.4   Numerical methods

Numerical methods are playing an increasingly important role in the analysis of heat transfer problems; not only for problems which are insoluble by analytical techniques, but also for problems to which analytical methods, under certain simplifying assumptions, can be applied. Although this class of method must be regarded as representing approximate solutions, their accuracy can, by careful design, be made to satisfy even the most demanding criteria. In the field of building energy analysis two related branches are of relevance: finite difference and finite element methods. Both methods offer powerful techniques for the solution of the partial differential equation which governs the heat transfer problem. They can be used to handle problems of almost any degree of complexity such as transient heat conduction within multi-layered constructions under non-linear boundary conditions or where thermal properties are considered to be temperature- and therefore time-dependent. Whereas analytical solutions usually result in equations which permit the calculation of temperature and heat flux at any point within the system considered, numerical methods are valid only at pre-selected and discrete points having properties that may be considered as characteristic of some finite region—usually a volume of homogeneous or mixed materials possessing heat capacity and heat flow attributes.

Finite difference methods are concerned with approximating the derivatives of the heat equation either directly by a truncated Taylor series expansion, for example, or indirectly by application of the principle of conservation of energy to small control volumes.

Finite element methods are based on the calculus of variations by which the solution of the heat equation is equivalent to minimising some related integral quantity. Its inherent power lies in the ability of the method to represent complex geometries often found in engineering problems involving complex stress or fluid flow phenomena. The method is based largely on the weighted residuals technique (Finlayson 1972) and is comprehensively covered in a number of texts (for example, Zienkiewicz and Cheung 1965, Wilson and Nickell 1966).

This section (along with chapters 3 and 4) is concerned with the former method only: a method which is conceptually simple to implement and is appropriate to most of the problems encountered in building energy analysis.

2.4.1 Finite differencing by Taylor series expansion

Consider figure 2.12 which shows a continuous function $f(\gamma)$ over the range $(\gamma - \delta\gamma) \leqslant \gamma \leqslant (\gamma + \delta\gamma)$. The replacement of the derivatives of $f(\gamma)$ by finite differences involves expressing these derivatives in terms of a truncated Taylor series expansion. Taylor's theorem applied to this function gives

$$f(\gamma + \delta\gamma) = f(\gamma) + \delta\gamma f^1(\gamma) + \frac{(\delta\gamma)^2}{2} f^2(\gamma) + \frac{(\delta\gamma)^3}{6} f^3(\gamma) + \dots \quad (2.58)$$

and

$$f(\gamma - \delta\gamma) = f(\gamma) - \delta\gamma f^1(\gamma) + \frac{(\delta\gamma)^2}{2} f^2(\gamma) - \frac{(\delta\gamma)^3}{6} f^3(\gamma) + \dots \quad (2.59)$$

where $f^n(\gamma)$ is $d^n f(\gamma)/d\gamma^n$. Adding these equations through those terms involving $(\delta\gamma)^3$ gives

$$f^2(\gamma) = \frac{f(\gamma + \delta\gamma) - 2f(\gamma) + f(\gamma - \delta\gamma)}{(\delta\gamma)^2} + \epsilon[(\delta\gamma)^2] \quad (2.60)$$

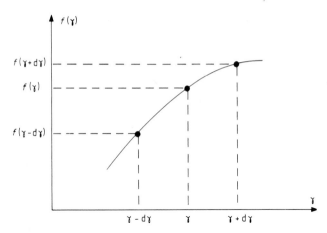

Figure 2.12 A continuous function of γ.

where $\epsilon[(\delta\gamma)^2]$ is the truncation error resulting from the approximate representation of the second order derivative.

Subtracting equation (2.59) from (2.58) through those terms involving $(\delta\gamma)^2$ gives

$$f^1(\gamma) = \frac{f(\gamma + \delta\gamma) - f(\gamma - \delta\gamma)}{2\delta\gamma} + \epsilon[(\delta\gamma)^2]. \qquad (2.61)$$

Equations (2.60) and (2.61) are termed central difference approximations to the second and first order derivatives respectively. Truncation of equation (2.58) after the term involving $\delta\gamma$ gives the first forward difference representation

$$f^1(\gamma) = \frac{f(\gamma + \delta\gamma) - f(\gamma)}{\delta\gamma} + \epsilon[(\delta\gamma)] \qquad (2.62)$$

and application of a similar truncation to equation (2.59) gives the first backward difference representation

$$f^1(\gamma) = \frac{f(\gamma) - f(\gamma - \delta\gamma)}{\delta\gamma} + \epsilon[(\delta\gamma)]. \qquad (2.63)$$

The truncation error in equations (2.62) and (2.63) is of the order of $\delta\gamma$, that is, halving the discretisation step will only approximately halve the error. On the other hand, the central difference approximation has a truncation error of order $(\delta\gamma)^2$ and so halving the discretisation step will approximately quarter the error.

Alternative mixes of these approximation schemes can be employed and will lead to explicit and implicit difference formulations. Consider the Fourier equation in one space variable and with heat generation

$$\frac{\partial^2\theta(x,t)}{\partial x^2} = \frac{1}{\alpha}\frac{\partial\theta(x,t)}{\partial t} + \frac{q}{\varrho C}. \qquad (2.64)$$

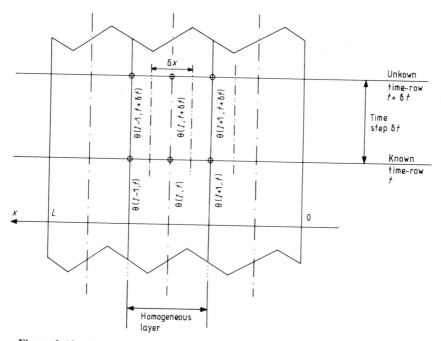

Figure 2.13 A homogeneous layer within a homogeneous region: space and time nodal scheme.

Figure 2.13 shows a homogeneous material layer arbitrarily located within a larger homogeneous region. Two finite difference schemes are possible: explicit and implicit. A scheme of explicit enumeration is obtained by representing the second-order derivative of equation (2.64) in the central difference form of equation (2.60) and the first-order derivative in first forward difference form as given by equation (2.62). Ignoring the error term and assuming (for the present) that thermophysical properties are time-independent gives

$$\frac{\theta(I+1,t) - 2\theta(I,t) + \theta(I-1,t)}{(\delta x)^2} = \frac{1}{\alpha} \frac{\theta(I,t+\delta t) - \theta(I,t)}{\delta t} + \frac{q(t)}{\varrho C} \Rightarrow \theta(I,t+\delta t)$$

$$= \frac{k\delta t}{\varrho C(\delta x)^2}\theta(I+1,t) + \left(1 - \frac{2k\,\delta t}{\varrho C(\delta x)^2}\right)\theta(I,t) + \frac{k\,\delta t}{\varrho C(\delta x)^2}\theta(I-1,t) + \frac{q(t)\delta t}{\varrho C}. \quad (2.65)$$

Note that the sum of the present time-row temperature terms is unity. This implies that, in the absence of heat generation ($q = 0$), the future time-row nodal temperature of any region is a weighted average of the present time-row temperature in the region and the temperatures in adjacent regions in thermal contact. If an equation of this form can be written for every region within a system then, given initial and boundary conditions (usually time-dependent), the discrete temperature history over any required period can be determined. Explicit schemes of this type are very easy to formulate and solve but can become unstable in certain circumstances. Consider the coefficient of the present time-row temperature term of some region I. If this coefficient should become negative, that is

$$1 - \frac{2k\delta t}{\varrho C(\delta x)^2} < 0$$

this can be interpreted as implying that the warmer the region I is now, the colder it will be after some time-step δt. This is clearly absurd and so the stability criterion becomes

$$\frac{2k\delta t}{\varrho C(\delta x)^2} \leqslant 1$$

more usually expressed as

$$\frac{\alpha\delta t}{(\delta x)^2} \leqslant \frac{1}{2}$$

where $\alpha\delta t/(\delta x)^2$ is the Fourier number, which is the ratio of the rate of heat conduction to the rate of heat storage.

Simple operations performed on equation (2.65) allow the formulation of rudimentary graphical techniques for the assessment of transient conduction in homogeneous systems. Equation (2.65) can be rewritten to give

$$\theta(I,t+\delta t) = F[\theta(I+1,t) + (1/F-2)\theta(I,t) + \theta(I-1,t)]$$

where F is the Fourier number. Setting $F = \frac{1}{2}$, which from the stability criterion is the largest value permitted, gives

$$\theta(I,t+\delta t) = \tfrac{1}{2}[\theta(I+1,t) + \theta(I-1,t)].$$

This is the basis of the Binder–Schmidt graphical method (see Simonson 1967).

A scheme of implicit enumeration is one in which the unknown temperature $\theta(I,t+\delta t)$ is expressed in terms of both future and present time-row temperatures prevailing in all regions in thermal contact. Any given system will therefore be represented by a connected series of algebraic equations which must be solved simultaneously for each finite time-step. The second-order derivative of equation (2.64) is replaced by the central difference formulation of equation (2.60) but when this is expressed in terms of the unknown temperature values at the future time-row rather than the known values at the present time-row as in the explicit formulation. The first-order derivative is expressed in the first backward formulation of equation (2.63). Again ignoring the error term and assuming constant thermophysical properties gives

$$\frac{\theta(I+1,t+\delta t) - 2\theta(I,t+\delta t) + \theta(I-1,t+\delta t)}{(\delta x)^2} = \frac{1}{\alpha} \frac{\theta(I,t+\delta t) - \theta(I,t)}{\delta t} + \frac{q(t+\delta t)}{\varrho C}$$

$$\Rightarrow \left(1 + \frac{2k\,\delta t}{\varrho C(\delta x)^2}\right)\theta(I,t+\delta t) = \theta(I,t) + \frac{k\,\delta t}{\varrho C(\delta x)^2}\left[\theta(I+1,t+\delta t) + \theta(I-1,t+\delta t)\right]$$

$$+ \frac{q(t+\delta t)\delta t}{\varrho C}. \quad (2.66)$$

Implicit formulations are unconditionally stable for all space and time discretisation schemes although large space or time steps will result in excessive discretisation errors. Discretisation is discussed in §3.1.

A weighted average of equations (2.65) and (2.66) can now be taken to construct a generalised formulation. Multiplying equation (2.65) by W and adding the result to equation (2.66) multiplied by $(1 - W)$ gives

$$(1 + 2WF)\theta(I,t+\delta t) = WF[\theta(I+1,t+\delta t) + \theta(I-1,t+\delta t)]$$

$$+ [W - (1-W)(1-2F)]\theta(I,t)$$

$$+ (1-W)F[\theta(I+1,t) + \theta(I-1,t)]$$

$$+ \frac{\delta t}{\varrho C}[Wq(t+\delta t) + (1-W)q(t)]. \quad (2.67)$$

Setting $W < 0.5$ gives an explicit scheme for which the stability criterion is

$$F \leqslant 1/2(1 - 2W)$$

and setting $W \geqslant 0.5$ gives an implicit scheme with $W = 0.5$ giving the commonly used Crank–Nicolson formulation much favoured because of its stability combined with ease of formulation.

By similar reasoning it is possible to devise finite difference formulations for more than one space dimension (see chapter 3), and for explicit methods corresponding stability criteria emerge (Croft and Lilley 1977).

2.4.2 The control volume heat balance method

Differencing by Taylor series expansion is the formal method of establishing a finite difference scheme from some known partial differential equation. Unfortunately the technique can prove unusually cumbersome and difficult to apply to all but simple problems. In a building design context commonly encountered complications include:

The simultaneous presence of multiple heat transfer processes (conduction, convection, radiation, advection and heat generation).

The time and positional dependency of heat generation due to solar radiation, installed plant, etc.

The use of a systematised nodal approximation procedure in which nodes might well represent regions of unequal physical dimensions.

Heat flow is often occurring in more than one dimension.

Nodal regions may require more than one set of thermophysical properties to define completely their heat flow storage characteristics—that is, the regions may not be homogeneous.

An alternative mechanism is to apply directly heat balance techniques to selected finite control volumes set up to represent some physical system. This ensures that the resulting solution satisfies the principle of conservation of energy even if the number of control volumes is small (although in such a case discretisation errors might well dominate).

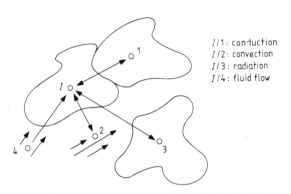

Figure 2.14   Heat exchanges in a physical system.

Consider figure 2.14 which shows a control volume I in thermal communication with four surrounding regions by the processes of conduction, convection, radiation and fluid flow. Internal heat generation is also considered to take place within the region I. Assuming that the inter-region heat exchange can be represented as a linear function of the temperature difference allows any flowpath to be represented by

$$q_{J,I} = K_{J,I}(\theta_J - \theta_I) \qquad J = 1,2,3,4$$

where $K_{J,I}$ is the linearised heat flow conductance between region J and I (W m^{-2} $^\circ$C^{-1}).

The conductance term is not considered further here since it is treated, in detail, in the equation derivations of chapter 3. Also the assumption of linearity is made here for convenience only and chapter 4 demonstrates the treatment of non-linear processes.

The heat generated within the region I is, in the present formulation, considered to be substantially independent of region temperature—perhaps originating outside the region as with shortwave solar flux—and is denoted simply by q_I.

The heat stored within the region over some finite time interval δt is given by

$$q_s = \frac{\varrho_I(\xi)C_I(\xi)\delta V_I(\xi)}{\delta t} \; [\theta(I,t+\delta t) - \theta(I,t)]$$

where $\varrho_I(\xi)$ is the characteristic density of region I at time ξ (kg m^{-3}), $C_I(\xi)$ is the characteristic specific heat of region I at time ξ (J kg^{-1} °C^{-1}), and $\delta V_I(\xi)$ is the volume of region I at time ξ (m^3).

Now, in the limit, the rate at which heat is being stored within the region I is equal to the net rate of heat flow to the region, and so for the system of figure 2.14 heat balance considerations yield

$$\frac{\varrho_I(\xi)C_I(\xi)\delta V_I(\xi)}{\delta t} \; [\theta(I,t+\delta t) - \theta(I,t)] = \sum_{j=1}^{N} K_{j,I}(\theta_j - \theta_I)\big|_{t=\xi} + q_I\big|_{t=\xi}. \tag{2.68}$$

Evaluation of the heat flux and generation terms at the present time-row $\xi = t$ gives the explicit formulation, evaluation at the future time-row $\xi = t + \delta t$ gives the fully implicit formulation and, as before, any explicit/implicit mix can be obtained as a weighted summation of both schemes.

From a mathematical classification viewpoint there are three types of differential equation required to describe energy exchanges in building and plant systems:

1 First-order ordinary differential equations used to represent physical regions possessing 'lumped' thermophysical properties.
2 Second-order parabolic partial differential equations used to describe insulation/capacity regions requiring detailed modelling.
3 First- and higher-order hyperbolic partial differential equations used to describe fluid flow and convective coupling.

Obviously equation (2.68) becomes identical to the first type as the time increment approaches the limit. It can also be proved (Lambert 1973) that equation (2.68) will become identical to types 2 and 3 if the so-called 'semi-discretisation' is applied to the space variables of these partial differential equations.

Chapter 3 demonstrates the application of this equation to multi-zone building systems when subjected to real climatic excitations causing heat flow transients and time-dependent system properties.

2.4.3 Numerical solution techniques

Explicit schemes result in a set of independent equations—one for each node—for which solution is relatively straightforward. At time $t = 0$ the nodal temperature field is prescribed and the solution consists of establishing the future time-row values at each consecutive time-step from the independent nodal equations.

With implicit schemes each nodal equation will contain present and future

Advanced modelling techniques

time-row temperature terms relating to the node in question and all surrounding nodes in thermal contact. For this reason the system of equations must be solved simultaneously at each time-step. Two main solution techniques exist to achieve this: direct and iterative. A direct method yields a solution in a finite number of computational steps which can be determined in advance. Iterative techniques will generally commence with some guessed solution which, when inserted into the equation set, will give a residue. This residue is then used to modify the initial guess in such a way as to cause convergence on the true solution. The number of computational steps is therefore dependent on the convergence criterion used and on the level of accuracy required.

Iterative methods are often applied to equation structures which can be termed sparse or which are known to converge rapidly since solution by direct means would demand, in the former case, high storage space and, in the latter case, unnecessary computation. The overall equation set to emerge from the application of numerical techniques is invariably sparse (see chapter 4) and so iterative methods would seem appropriate. However this sparseness can be removed by matrix partitioning techniques allowing each partitioned sub-matrix to be processed—by a customised direct method—at any frequency depending on the building component to which it relates and the degree to which its contents (nodal coefficients) change with time. Thus the greater part of the matrix processing can be achieved by rapid direct techniques with, in some cases, iterative potential superimposed. This is the strategy underlying the matrix processing techniques of chapter 4 which demonstrates the partitioning and mixed solution scheme concepts. For this reason, and since detailed descriptions are given elsewhere (for example Kreyszig 1979), only a brief summary is outlined here.

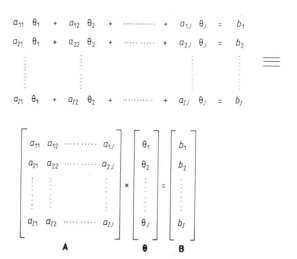

Figure 2.15 Matrix representation of I simultaneous linear equations in J unknowns.

Direct Methods

Consider the system of I simultaneous linear equations in J unknown temperature terms (θ) as shown in figure 2.15. Also shown is the equivalent matrix representation $\mathbf{A}\theta = \mathbf{B}$. Simultaneous solution is basically a five-stage process:

Step 1

Denoting the first equation as the primary and the second equation as the secondary, each of the temperature terms of the primary are multiplied by the ratio of the first existing coefficient in the secondary (which shares a temperature term with the primary) to the corresponding coefficient in the primary (the pivotal coefficient), that is a_{21}/a_{11}. Note that if a pivotal coefficient is small, the subtracted number can become large, causing numerical problems or amplifying any uncertainty in the coefficients. It is therefore important to ensure that pivotal coefficients do not become too small in absolute value. This can be accomplished by reordering the equations and/or variables as necessary; a process known as pivoting.

Step 2

The primary is now subtracted from the secondary to eliminate the temperature term (from the secondary) for which the ratio was established.

Step 3

The process of steps 1 and 2 is now repeated for all primary/secondary pairs. In fact there are

$$\sum_{k=1}^{I-1} (I - k)$$

possible pairings: the first equation with $2,3,4,\ldots,I$; the second equation with $3,4,5,\ldots,I$; up to equation $I - 1$ with I.

Step 4

The last secondary to be processed will contain only one temperature term (but only if $I \geqslant J$) and so this temperature value is known.

Step 5

The complete solution is given by backward substitution of this known temperature in the equation in two temperature variables and so on.

Figure 2.16 demonstrates this process in terms of an example equation set. Important points to note are: the primary/secondary combinations can be handled in any order; in any secondary equation any temperature term can be selected for elimination as long as the corresponding term exists in the primary; all non-eliminated terms are carried through to other equations; and the final equation in the forward reduction process ($18z = 18$ in this case) embodies the characteristics of the entire system. The matrix processing techniques of chapter 4 employ these points to achieve computationally efficient solutions of large and sparse equation systems by direct means.

The foregoing technique, known as Gaussian elimination, therefore consists of

$$4x - y + 2z = 15 \quad (a)$$
$$- x + 2y + 3z = 5 \quad (b)$$
$$5x - 7y + 9z = 8 \quad (c)$$

Step 1 and 2: $(a) \div (b)$ ratio $1/4$
$$x + 0.25y - 0.5z = -3.75$$
$$1.75y + 3.5z = 8.75$$
$$5x - 7.00y + 9.0z = 8.00$$

Step 3 (1): $(a) \div (c)$ ratio -5
$$5x - 1.25y + 2.5z = 18.75$$
$$1.75y + 3.5z = 8.75$$
$$- 5.75y + 6.5z = -10.75$$

Step 3 (2): $(b) \div (c)$ ratio $-5.75/1.75$
$$5x - 1.25y + 2.5z = 18.75 \quad (d)$$
$$- 5.75y - 11.5z = -28.75 \quad (e)$$
$$18.0z = 18.00 \quad (f)$$

Step 4: from (f) $z = 1$

Step 5: from (e) $y = 3$
from (d) $x = 4$

Figure 2.16 Gaussian elimination method.

organised multiplication, division and subtraction operations. For a system with characteristic matrix A square and order N, the number of such operations is of order $N^3/3$. A number of variants exist. These include the Gauss–Jordan and Cholesky methods which are described in many elementary texts.

Iterative methods

A number of iterative methods exist each with the same underlying technique: a guess is made of the nodal temperature field, the equations are evaluated to obtain updated temperature values and these updates are used to repeat the process until subsequent updated values differ only slightly from the previous iteration. In general, different methods can be differentiated by the stage at which an updated value is incorporated in subsequent equation evaluation.

The Gauss–Seidel technique makes use of newly computed temperature values as soon as they become available and for this reason is known as a method of successive corrections. The Jacobi technique is termed a method of simultaneous correction because no newly computed temperature is used until each of the equations has been processed in a particular iteration step. These iterative methods are subject to convergence criteria relating to the eigenvalues (or latent roots) of a corresponding 'iteration matrix' (Kreyszig 1979).

In practice, iteration methods often employ the method of over-relaxation to improve convergence. When the guessed temperature field is applied to any particular nodal equation a residual will result since the temperature values will not, in general, represent the actual solution. The objective of over-relaxation is to so adjust the newly computed temperature values that any new residual is not zeroised but changed in sign in anticipation that subsequent operations on neighbouring nodal equations will have a favourable effect on the over-relaxed residual. Other relaxation techniques such as block and group relaxation can also be used to improve convergence. These techniques are described in detail elsewhere (Croft and Lilley 1977).

With non-linear equation sets the Newton–Raphson method and the method of false position are often used. The former method is described in some detail in §6.6 where it is applied to the non-linear equation structure representing a building's distributed air flow network.

2.5 Which method?

It is extremely difficult to establish the relative merits of the three methods previously outlined. If carefully implemented, each should provide accurate dynamic modelling capabilities although the assumptions inherent in each (linearity, invariability, difference approximation etc) may, under certain circumstances, invalidate one or more method.

The two-pole criterion of accuracy and flexibility in use is difficult to apply in practice. In the former case the complexity of the interrelating heat transfer processes makes model testing difficult; indeed in the many validation exercises conducted to date (see chapter 8) no technique has emerged as a clear candidate for declaration as a standard. And, in the latter case, it has not yet been possible to assess rationally the performance of the different methods when applied in the real-world, real-time context of design practice (see chapter 9).

The response function method (in both the time and frequency domain) is a specific analytical technique, mathematically elegant and the outcome of many years of accumulated research and development. The finite difference method, no less respectable, is very general in concept, offers a better physical insight but, perhaps, produces models whose quality will depend, to a greater extent, on the care taken in the particular implementation. It may be the case then that models based on the highly prescribed analytical methods will often exhibit closer agreement when inter-compared and that models based on the finite difference technique may, more often, disagree. However no definitive statement has yet been made on method performance vis-à-vis the reality. Indeed it is likely that each method, if carefully harnessed, will perform equitably.

The finite difference technique has, in the author's opinion, one distinct advantage. Its generality allows direct application to combined building and plant systems incorporating time-dependent heat transfer properties (the subject matter of chapters 3–6). This capability, combined with the current emergence of powerful, low cost computing power will undoubtedly lead to widespread use of the technique in many future simulation systems.

It is interesting to note the recurring conclusions of a number of validation exercises:

Several programs (based on the different methods) may or may not give the same results, when applied to the same problem, depending on the complexity of the simulation task.

Several users will probably obtain different results when using the *same* program on the *same* problem.

It is evident from these conclusions that the lack of consensus between models is due in part to the issues surrounding problem interpretation and user interface design. For this reason it is important to pursue building description and program interface standards in tandem with attempts to resolve discrepancies at the theoretical level.

In conclusion, and at the present time, each method can be regarded as a state-of-the-art predictor, certainly well adapted for a comprehensive design appraisal role. Any particular modelling system must however be judged on the basis of validation information and details concerning trial or actual use.

References

ASHRAE 1971 *Procedures for Determining Heating and Cooling Loads for Computerised Energy Calculations* Task Group on Energy Requirements for Heating and Cooling, USA

Alford J S, Ryan J E and Urban F O 1939 Effects of Heat Storage and Variation in Outdoor Temperature and Solar Intensity on Heat Transfer Through Walls *ASHVE Trans.* (1123) 369–96

Brisken W R and Reque S G 1956 Heat Load Calculations by Thermal Response *ASHRAE Trans.* **62** 391

Carslaw H S and Jaeger J C 1959 *Conduction of Heat in Solids* (2nd edn) (Oxford: Oxford University Press)

Churchill R V 1958 *Operational Mathematics* (New York: McGraw-Hill)

Croft D R and Lilley D G 1977 *Heat Transfer Calculations Using Finite Difference Equations* (London: Applied Science Ltd)

Danter E 1960 Periodic Heat Flow Characteristics of Simple Walls and Roofs *J. IHVE (CIBS)* **28** 136–46

—— 1973 Heat Exchanges in a Room and the Definition of Room Temperature *Proc. IHVE (CIBS) Symp.* (June 1973)

Davies B 1978 *Integral Transforms and Their Applications* (New York: Springer-Verlag)

Finlayson B A 1972 *The Method of Weighted Residuals and Variational Principles* (New York: Academic)

Gough M 1982 Modelling Heat Flow in Buildings: An Eigenfunction Approach *PhD Thesis* University of Cambridge

—— 1984 *Private Communication*

Gower N W and Baker J E 1974 *Fourier Series* (London: Chatto and Windus and Collins)

Gupta C L 1964 A Matrix Method for Predicting the Thermal Response of Unconditioned Buildings *J. IHVE* **32** 159

Gupta C L, Spencer J W and Muncey R W R 1974 A Conceptual Survey of Computer-Oriented Thermal Calculation Methods *Proc. 2nd Symp. Use of Computers for Environ. Eng. Related to Build.*

Harrington-Lynn J 1974a The Admittance Procedure: Variable Ventilation *J. IHVE (CIBS)* **42** 199–200

—— 1974b The Admittance Procedure: Intermittent Plant Operation *J. IHVE (CIBS)* **42** 219–21

Healey M 1967 *Tables of Laplace, Heaviside, Fourier and Z transforms* (Edinburgh: Chambers)

Hittle D C 1979 Building Loads Analysis and System Thermodynamics (BLAST) Users Manual Version 2 *Technical Report E-153* (Champaign, Ill: US Army Construction Eng. Research Lab. (CERL))

—— 1981 An Improved Root-Finding Procedure for Use in Calculating Transient Heat Flow Through Multilayered Slabs *Preprint* (Champaign, Ill: US Army Construction Eng. Research Lab. (CERL))

Jury E I 1964 *Theory and Application of the Z-Transform Method* (New York: Wiley)

Kimura K 1977 *Scientific Basis of Air Conditioning* (London: Applied Science Ltd)

Kreyszig E 1979 *Advanced Engineering Mathematics* (New York: Wiley)

Kusuda T 1969 Thermal Response Factors for Multi-Layer Structures of Various Heat Conduction Systems *ASHRAE Trans.* **75** 246

Lambert J D 1973 *Computational Methods in Ordinary Differential Equations* (New York: Wiley)

Louden A G 1968 Summertime Temperatures in Buildings Without Air Conditioning *BRS CP 46* (Garston: Building Research Establishment)

Mackey C O and Wright L T 1944 Periodic Heat Flow—Homogeneous Walls or Roofs *ASHVE Trans.* **50** 293

—— 1946 Periodic Heat Flow—Composite Walls or Roofs *Heating, Piping and Air Conditioning* **18**(6) 107–10

Milbank N O and Harrington-Lynn J 1974 Thermal Response and the Admittance Procedure *BRS CP 61* (Garston: Building Research Establishment)

Mitalas G P 1965 An Assessment of Common Assumptions in Estimating Cooling Loads and Space Temperatures *ASHRAE Trans.* **71**(2) 72

Mitalas G P and Arseneault J G 1970 *Z* Transfer Functions for the Calculation of Transient Heat Transfer through Walls and Roofs *Proc. 1st Symp. Use of Computers for Environ. Eng. Related to Build.*

Muncey R W R 1953 The Calculation of Temperatures Inside Buildings Having Variable External Conditions *J. Appl. Sci.* **4** 189

—— 1979 Heat Transfer Calculations for Buildings *Appl. Sci.*

Nottage H B and Parmelee G V 1955 Circuit Analysis Applied to Load Estimating (pt 2) *ASHRAE Trans.* **61** 125

Pipes L A 1957 Matrix Analysis of Heat Transfer Problems *J. Franklin Inst.* **263** 195

Pratt A W and Ball E F 1963 Transient Cooling of a Heated Enclosure *J. Heat Mass Transfer* **6** 703–18

Saito H and Kimura K 1974 Computerised Calculation Procedures of Dynamic Air Conditioning Load Developed by SHASE of Japan *Proc. 2nd Symp. Use of Computers for Environ. Eng. Related to Build.*

Simonson J R 1967 *An Introduction to Engineering Heat Transfer* (London: McGraw-Hill)

Stephenson D G and Mitalas G P 1967a Room Thermal Response Factors *ASHVE Trans.* **73** no 2019

—— 1967b Cooling Load Calculations by Thermal Response Factor Method *ASHVE Trans.* **73** no 2018

—— 1971 Calculation of Heat Conduction Transfer Functions for Multilayer Slabs *ASHRAE Trans.* **2** 117–26

Wilson E L and Nickell R E 1966 Application of Finite Element Method to Heat Conduction Analysis *Nucl. Eng. Des.* **4** 1–11

York D A and Tucker E F (eds) 1980 *DOE 2 Reference Manual Version 2.1* (*Rep. LA 7689 M*)(Los Alamos: Los Alamos Scientific Laboratory)

Zienkiewicz O C and Cheung Y K 1965 Finite Elements in the Solution of Field Problems *The Engineer* 507–10 (Sept)

3

Numerical simulation by finite differences

The previous chapter derived the various methods by which the governing partial differential heat equation can be solved exactly or by approximation. This chapter demonstrates the application of one selected method—a finite difference formulation of implicit enumeration based on control volume heat balance—to the construction of a rigorous first principle energy model which is capable of simulating any building/plant system whilst respecting the spatial and temporal integrity of the multi-component system.

In essence the formulation of such a model is a three stage process:

First, the continuous building system is made discrete by the placement of 'nodes' at pre-selected points of interest. These nodes represent homogeneous or mixed physical volumes, such as discrete portions of fluid volumes, opaque and transparent boundary surfaces, constructional elements, plant components, distribution systems, and so on.

Second, for each one of these nodes in turn, and in terms of all surrounding nodes representing regions deemed to be in thermal contact, a heat flow simulation equation is derived which directly replaces the governing partial differential heat equation to link all inter-nodal energy flows over spatial and temporal dimensions.

Last, the entire equation set (one for each node in the system) must be solved simultaneously and repeatedly to obtain future time-row nodal temperatures and energy flows as the model steps through time.

However, before such a procedure can be implemented, a number of fundamental difficulties must be considered.

It is difficult (if not impossible) to prescribe the appropriate spatial discretisation scheme for each building component. Undoubtedly different physical components demand different treatments; some requiring fine subdivision (many nodes) whereas others may be more crudely modelled (few nodes). One way to resolve this dilemma is to utilise the developed model as the basis of a parametric study to examine alternative schemes in an attempt to establish an optimum.

Also, the generated heat flow equations will have a variable number of coefficients depending on the node type to which they relate. This in turn will dictate the need for a carefully designed coefficient indexing scheme to facilitate efficient matrix processing.

Last, because different system components will have different time constants, the matrix processing must be structured to allow fundamental temporal mismatches to be reconciled whilst not enforcing small time-step processing for all components.

The concepts underlying the formulations which follow will accommodate these problems by allowing any spatial discretisation, coefficient indexing or matrix processing frequency as required by the problem in hand (although specific recommendations are given where appropriate). The objective of this chapter is therefore threefold: to discuss system discretisation; to derive the simulation equations for the various regions (or nodes) found in buildings; and to demonstrate the formulation of the integrated systems (in matrix form) for multiple component systems. Chapter 4 describes, in detail, matrix formulation and the mechanisms for variable frequency, fast, simultaneous solution of the entire equation set as the model is driven by boundary climatic influences conducting a simulation.

Chapter 6 considers the formulation of nodal equation sets for various plant systems in more detail and demonstrates the integration of such systems with their building counterpart.

3.1   System discretisation

There are two main types of error associated with finite differencing schemes: rounding errors and discretisation errors.

The former occur in cases where computations include an insufficient number of significant figures. Any tendency towards an accumulation of such errors can rapidly become critical, especially in large numerical schemes involving many computational operations. Fortunately, errors of this type can be reduced to insignificance by the careful design of the numerical scheme and by operating, where appropriate, in double precision.

Discretisation errors result from the replacement of derivatives by finite differences. Although unavoidable, such errors can usually be minimised by reducing the space and time increments (see §2.6.1 for a general discussion of stability and convergence). Whilst accuracy considerations dictate that such increments be small, alternative considerations of computational speed and cost require that they be made relatively large. Although it is impossible to predetermine analytically, for all possible building/plant components, the exact space and time increments for any given accuracy level, optimum values can be ascertained from simple parametric studies which utilise the developed model. This of course implies that a model must first be developed against the assumption that any increment is possible. This greatly promotes the use of implicit formulations which are unconditionally stable and, if well designed, convergent and consistent with the original partial differential equation.

Such a parametric study (Clarke 1978) has been conducted with a simulation model based on the theory to be derived in this chapter. For the case of uni-directional transient conduction within a multi-layered building construction (such as a wall, floor or roof), figure 3.1 shows the temperature and heat flux variations at the internal and external boundary surfaces as the number of thermally uniform regions (nodes) representing each homogeneous layer is varied; that is, the space-step in the differencing scheme is varied whilst the time-step is held constant. These results are given for a moderately heavyweight construction. Figure 3.2 shows the corresponding variations as the time-step is varied with the space-step held constant.

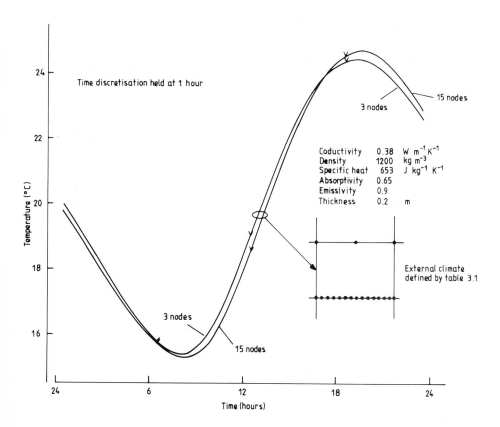

Figure 3.1 Effect of space discretisation.

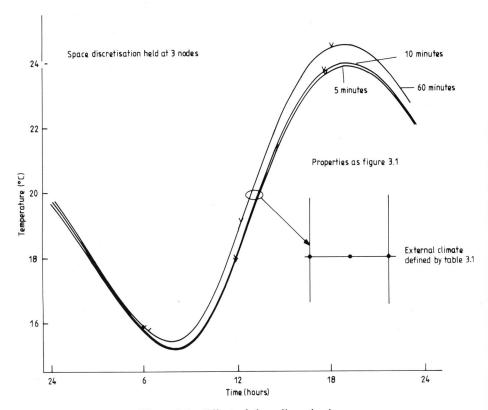

Figure 3.2   Effect of time discretisation.

The results suggest that, for the purpose of transient conduction modelling within multi-layered composites, a spatial discretisation scheme equal to or exceeding 3 nodes per homogeneous element is consistent with acceptable accuracy. Any nodes situated at the boundary between different homogeneous elements will represent mixed thermal property regions and nodes situated at extreme surfaces—undergoing convective, conductive and radiative heat exchange—will have associated thermal capacity equal to some fraction of the capacity of the next-to-surface element. For the same accuracy reasons time-steps in excess of one hour should be avoided with no lower limit imposed.

There are, of course, alternative nodal placement strategies which attempt to subdivide multi-layered constructions as a function of thermal rather than physical criteria and so reduce nodal subdivision to minimise processing. Appendix C describes some of these alternative node location criteria for the case of transient conduction modelling within multi-layered building constructions.

Table 3.1 External climate definition for figures 3.1 and 3.2.

Hour	D.B. Temp °C	Dr. N. Rad W m^{-2}	Df. H. Rad. W m^{-2}	Wind vel. m s^{-1}	Wind direction ° from N	Rel. humidity %
1	16.30	0	0	0.00	0	81
2	16.20	0	0	0.00	0	85
3	15.20	0	0	0.30	45	86
4	15.90	0	0	0.80	85	81
5	15.20	1	10	1.00	95	81
6	15.90	7	41	1.30	110	80
7	18.20	140	77	2.20	130	78
8	20.60	405	95	4.40	155	68
9	22.30	575	105	5.40	165	64
10	23.60	622	130	5.70	170	60
11	25.00	634	158	6.40	165	55
12	26.20	605	217	7.00	160	50
13	26.70	557	241	7.20	165	48
14	27.10	568	214	7.20	170	48
15	28.00	610	224	7.20	170	45
16	28.70	585	218	7.00	160	43
17	28.20	475	172	6.70	155	45
18	27.50	390	123	6.20	155	45
19	26.70	235	81	4.90	150	49
20	25.80	49	40	3.10	150	52
21	23.80	4	8	2.10	160	62
22	23.60	0	0	2.20	190	64
23	22.50	0	0	2.10	210	69
24	21.80	0	0	2.10	225	70

However, regardless of the criteria used, most schemes will involve the problem of homogeneous and mixed property region modelling.

In transient conduction schemes involving more than one space dimension it is impossible to prescribe the nodal placement criterion since this will depend on such factors as internal and external surface insolation, the existence of localised convection, the presence of corner effects and thermal bridges, and the shape of capacity/insulation systems being modelled: all factors causing position-dependent transient effects. Nevertheless, in many applications *n*-dimensional transient conduction schemes will become necessary with mixed-dimensional schemes proving useful. Figure 3.3, for example, gives some mixed schemes and their corresponding application. In the following derivations the full 3-dimensional scheme is assumed for transient conduction nodes with the reduction to lower dimensions demonstrated where appropriate.

The spatial subdivision of fluid volumes (for example some portion of air

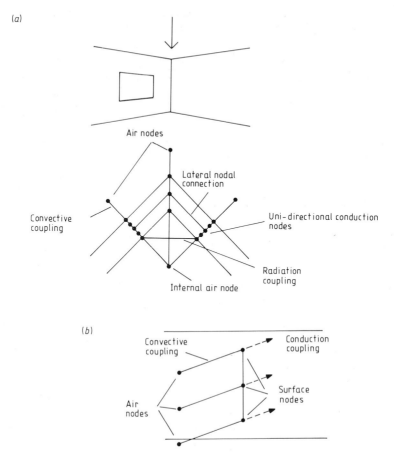

Figure 3.3 Some mixed nodal schemes and their typical application. (*a*) Corner effects: a combined one- and two-dimensional scheme. (*b*) Surface temperature gradient: a two-dimensional scheme.

within a room or the combustion chamber of a boiler) is also difficult to prescribe since, again, this will depend on the problem in hand. However a number of general points can be identified:

It is usually convenient to subdivide the volume vertically to include the buoyancy effects of density variations causing fluid stratification.

Additional finite volumes are often required adjacent to opaque or transparent surfaces to allow for localised surface phenomena such as solar patch movement.

Further global subdivision may be required to facilitate the modelling of more complex convective fields.

And it is usual to create distinct nodal regions at any point within the building system which, it is anticipated, will undergo some control action (for example at a pump, near the inlet to a radiator, at a shading device undergoing solar control, or at the compressor of a chiller unit).

In general terms, then, it is necessary to subdivide the system for modelling into a number of discrete finite regions. The subdivision criterion will depend on the expected variations of fundamental thermophysical properties and heat fluxes throughout the system, on the extent to which distinct regions will be subjected to control functions, and the ultimate output objectives of subsequent simulation processing. Particular discretisation schemes are given in §3.3 where equations derived in §3.2 are combined, in an indicative manner only, to indicate the complete equation set for multi-component systems.

3.2 Finite difference energy flow equations

This section applies the finite volume heat balance method introduced in chapter 2 to the energy flows commonly encountered in building energy analysis. For each possible type of discrete finite region (represented by a node), a characteristic simulation equation is derived which links the region with those other regions that are in thermal contact by one of the fundamental energy transport processes: conduction, convection, radiation and fluid flow.

It is important to appreciate that the equations to emerge have an underlying similarity: in each, the terms grouped on the left-hand side of the equality relate to the future (as yet unknown) time-row of some arbitrary time-step, whereas those of the right-hand side address the present (known) time-row. In the terminology the coefficient of the node, for which an equation is derived, is termed the 'self-coupling' coefficient. The remaining nodal coefficients are termed 'cross-coupling' coefficients since they link (through space) the source node with coupled regions. All terms relating to boundary condition excitations are gathered on the right-hand side since they can be assumed known for all time regardless of the time-row to which they relate. It is this similarity and symmetry of the combined equation sets which permit the matrix partitioning techniques of chapter 4 and hence efficient matrix processing.

Figure 3.4 shows the various energy flowpaths commonly occurring within buildings and so candidates for inclusion within a simulation model. Chapter 1 described each process and discussed the complexities of flowpath interaction. Recalling the primitive finite volume heat balance relationship of §2.4.2 and assuming region I properties are time-dependent:

$$\frac{\varrho_I(\xi)C_I(\xi)\,\delta V_I(\xi)}{\delta t}\,[\theta(I,t+\delta t)-\theta(I,t)] = \sum_{i=1}^{N} K_{i,I}[\theta(i,\xi)-\theta(I,\xi)] + q_I(\xi) + \epsilon \tag{3.1}$$

where $\varrho_I(\xi)$ is the representative density of region I at some time ξ (kg m^{-3}), $C_I(\xi)$ is the representative specific heat capacity of region I at time ξ (J kg^{-1} $^\circ$C^{-1}), $\delta V_I(\xi)$ is the volume of region I at time ξ (m^3), δt is the discretisation time-step (s), $\theta(I,\xi)$ is the representative temperature of region I at time ξ

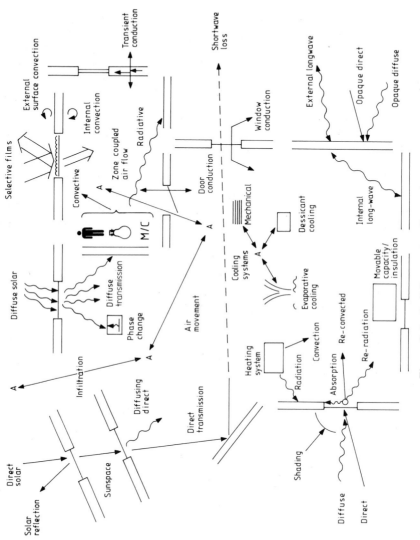

Figure 3.4 Building energy flowpaths.

(°C), and $K_{i,I}$ is the heat flow conductance between region i and region I (W °C^{-1}). The heat generation within region I (W) is denoted by q_I, ϵ is the error resulting from the evaluation over finite space and time increments, N is the number of energy exchange flowpaths between region I and surrounding regions in thermal contact, t is the present time-row, and $t + \delta t$ the future time-row.

Evaluation of equation (3.1) at the present (and known) time-row, t, gives the fully explicit scheme in which all nodal equations are independent—since they contain only present values of all coupled nodes—and so can be solved directly. Evaluation at the future (and unknown) time-row, $t + \delta t$, gives the fully implicit scheme in which all nodes are linked at the future time-row and so the entire equation set which describes a multi-node system (with one equation for each node) must be solved simultaneously.

Chapter 2 discussed the main advantages and disadvantages of both formulations and outlined the concatenation of implicit and explicit schemes to provide unconditional stability and so flexibility in application. Such a technique, based on equation (3.1), can now be applied to the main characteristic node types found in buildings, namely:

Nodes which represent the energy balance of regions located within capacity/insulation systems such as the material comprising the building fabric (multi-layered constructions) and its contents (furnishings etc).

Nodes which represent the energy balance at exposed (opaque and transparent) surface layers such as room wall and window surfaces and solar collector back-plates.

Nodes representing the energy balance within fluid volumes such as portions of room air.

The derived equations are general and can be applied equally to building and plant components. Chapter 6 applies the derived equations to a number of example plant configurations and so only cursory plant treatment is given here and in chapter 4.

3.2.1 Energy balance: capacity/insulation systems

Consider figure 3.5, which shows a number of discrete regions in conductive communication. In this scheme node I represents the discrete finite volume given by

$$(\delta_{I,I-1} + \delta_{I,I+1})(\delta_{I,J-1} + \delta_{I,J+1})(\delta_{I,K-1} + \delta_{I,K+1}).$$

The heat flux by conduction towards node I is given by

$$q_{I-1,I} = k_{I-1,I}^{i}(\delta_{I,J-1} + \delta_{I,J+1})(\delta_{I,K-1} + \delta_{I,K+1})(\theta_{I-1} - \theta_I)/\delta x_{I-1,I}$$
$$q_{I+1,I} = k_{I+1,I}^{i}(\delta_{I,J-1} + \delta_{I,J+1})(\delta_{I,K-1} + \delta_{I,K+1})(\theta_{I+1} - \theta_I)/\delta x_{I+1,I}$$
$$q_{J-1,I} = k_{J-1,I}^{j}(\delta_{I,I-1} + \delta_{I,I+1})(\delta_{I,K-1} + \delta_{I,K+1})(\theta_{J-1} - \theta_I)/\delta x_{J-1,I}$$
$$q_{J+1,I} = k_{J+1,I}^{j}(\delta_{I,I-1} + \delta_{I,I+1})(\delta_{I,K-1} + \delta_{I,K+1})(\theta_{J+1} - \theta_I)/\delta x_{J+1,I}$$
$$q_{K-1,I} = k_{K-1,I}^{k}(\delta_{I,I-1} + \delta_{I,I+1})(\delta_{I,J-1} + \delta_{I,J+1})(\theta_{K-1} - \theta_I)/\delta x_{K-1,I}$$
$$q_{K+1,I} = k_{K+1,I}^{k}(\delta_{I,I-1} + \delta_{I,I+1})(\delta_{I,J-1} + \delta_{I,J+1})(\theta_{K+1} - \theta_I)/\delta x_{K+1,I}$$

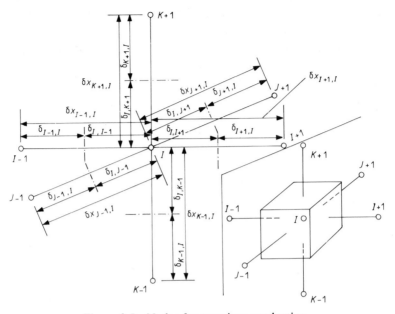

Figure 3.5 Nodes for transient conduction.

where k' is the average inter-nodal conductivity (W m^{-1} °C^{-1}), δx is the inter-nodal distance in direction of heat flow (m), and $\delta_{i\pm1,I} = \delta_{I,i\pm1} = \delta x_{i\pm1,I}/2$; $i = I, J, K$.

The average conductivity value is necessary to account for the possibility that the inter-nodal connections may not be homogeneous but be comprised of different materials. Also, since the region I may not be homogeneous, it is necessary to express representative thermophysical properties as a volumetric weighting of the different materials which comprise the region. It follows from equation (3.1) that the 3-dimensional energy balance relationship for a node undergoing transient conduction with a potential for heat generation is given by

$$
\begin{aligned}
[W_I(t + \delta t)&\theta(I, t + \delta t) - W_I(t)\theta(I,t)] \times (\delta_{I,I-1} + \delta_{I,I+1})(\delta_{I,J-1} + \delta_{I,J+1})(\delta_{I,K-1} + \delta_{I,K+1})/\delta t \\
= \quad & k_{I-1,I}(\xi)(\delta_{I,J-1} + \delta_{I,J+1})(\delta_{I,K-1} + \delta_{I,K+1})[\theta(I-1,\xi) - \theta(I,\xi)]/\delta x_{I-1,I} \\
+ \; & k_{I+1,I}(\xi)(\delta_{I,J-1} + \delta_{I,J+1})(\delta_{I,K-1} + \delta_{I,K+1})[\theta(I+1,\xi) - \theta(I,\xi)]/\delta x_{I+1,I} \\
+ \; & k_{J-1,I}(\xi)(\delta_{I,I-1} + \delta_{I,I+1})(\delta_{I,K-1} + \delta_{I,K+1})[\theta(J-1,\xi) - \theta(I,\xi)]/\delta x_{J-1,I} \\
+ \; & k_{J+1,I}(\xi)(\delta_{I,I-1} + \delta_{I,I+1})(\delta_{I,K-1} + \delta_{I,K+1})[\theta(J+1,\xi) - \theta(I,\xi)]/\delta x_{J+1,I} \\
+ \; & k_{K-1,I}(\xi)(\delta_{I,I-1} + \delta_{I,I+1})(\delta_{I,J-1} + \delta_{I,J+1})[\theta(K-1,\xi) - \theta(I,\xi)]/\delta x_{K-1,I} \\
+ \; & k_{K+1,I}(\xi)(\delta_{I,I-1} + \delta_{I,I+1})(\delta_{I,J-1} + \delta_{I,J+1})[\theta(K-1,\xi) - \theta(I,\xi)]/\delta x_{K+1,I} \\
+ \; & q_I(\xi) + \epsilon
\end{aligned}
\tag{3.2}
$$

where W_I is the volume weighted product of density and specific heat capacity of region $I(\text{J m}^{-3}\text{K}^{-1}) = \sum_{i=1}^{N} \varrho_i C_i \delta V_i / \sum_{i=1}^{N} \delta V_i$, and N is the number of different materials comprising region I.

This is the fundamental 3-dimensional relationship for transient conduction within all capacity/insulation systems. As the space- and time-steps ($\delta, \delta x$ and δt) approach zero, the resulting partial differential heat equation is termed the Fourier field equation (FFE) with heat generation:

$$\varrho_i C_I \frac{\partial \theta_I}{\partial t} = \frac{\partial}{\partial x_I}\left(k_{I-1,I+1}\frac{\partial \theta_I}{\partial x_I}\right) + \frac{\partial}{\partial x_J}\left(k_{J-1,J+1}\frac{\partial \theta_I}{\partial x_J}\right) + \frac{\partial}{\partial x_K}\left(k_{K-1,K+1}\frac{\partial \theta_I}{\partial x_K}\right) + q_I'$$

(3.3)

where q_I' is the heat generation per unit volume of region I (W m^{-3}). For an isotropic material this reduces to

$$\frac{1}{\alpha}\frac{\partial \theta_I}{\partial t} = \nabla^2 \theta_I + q_I'/k_I$$

where α is the thermal diffusivity (m^2 s^{-1}).

Equation (3.2) is now used to obtain the general form of the simulation for transient conduction nodes. Evaluation of the equation at the present time-row ($\xi = t$) yields (after rearrangement) the following *temperature explicit* formulation:

$$\theta(I, t+\delta t) = \theta(I,t)\left(\frac{W_I(t)}{W_I(t+\delta t)} - \frac{k_{I-1,I}'(t)}{S_{I-1,I}(t+\delta t)(\delta_{I,I-1}+\delta_{I,I+1})}\right.$$

$$-\frac{k_{I+1,I}'(t)}{S_{I+1,I}(t+\delta t)(\delta_{I,I-1}+\delta_{I,I+1})} - \frac{k_{J-1,I}'(t)}{S_{J-1,I}(t+\delta t)(\delta_{I,J-1}+\delta_{I,J+1})}$$

$$-\frac{k_{J+1,I}'(t)}{S_{J+1,I}(t+\delta t)(\delta_{I,J-1}+\delta_{I,J+1})} - \frac{k_{K-1,I}'(t)}{S_{K-1,I}(t+\delta t)(\delta_{I,K-1}+\delta_{I,K+1})}$$

$$\left.-\frac{k_{K+1,I}'(t)}{S_{K+1,I}(t+\delta t)(\delta_{I,K-1}+\delta_{I,K+1})}\right) + \frac{k_{I-1,I}'(t)\theta(I-1,t)}{S_{I-1,I}(t+\delta t)(\delta_{I,I-1}+\delta_{I,I+1})}$$

$$+\frac{k_{I+1,I}'(t)\theta(I+1,t)}{S_{I+1,I}(t+\delta t)(\delta_{I,I-1}+\delta_{I,I+1})} + \frac{k_{J-1,I}'(t)\theta(J-1,t)}{S_{J-1,I}(t+\delta t)(\delta_{I,J-1}+\delta_{I,J+1})}$$

$$+\frac{k'_{J+1,I}(t)\theta(J+1,t)}{S_{J+1,I}(t+\delta t)(\delta_{I,J-1}+\delta_{I,J+1})} + \frac{k_{K-1,I}'(t)\theta(K-1,t)}{S_{K-1,I}(t+\delta t)(\delta_{I,K-1}+\delta_{I,K+1})}$$

$$+\frac{k_{K+1,I}'(t)\theta(K+1,t)}{S_{K+1,I}(t+\delta t)(\delta_{I,K-1}+\delta_{I,K+1})}$$

$$+\frac{q_I(t)\,\delta t}{W_I(t+\delta t)(\delta_{I,I-1}+\delta_{I,I+1})(\delta_{I,J-1}+\delta_{I,J+1})(\delta_{I,K-1}+\delta_{I,K+1})} + \epsilon$$

where $S_{i\pm1,I}(t+\delta t) = W_I(t+\delta t)\delta x_{i\pm1,I}/\delta t$; $i = I, J, K$.

Note that, although the formulation is explicit in the temperature variable, each present time-row temperature coefficient contains region thermophysical properties which must be evaluated at the future time-row. The expression will, however, become fully explicit if the usual assumption can be made that region properties are invariant in the time dimension.

If ξ is now set to $t + \delta t$ in equation (3.2) then the fully implicit formulation is obtained in which

$$
\theta(I, t + \delta t) = \frac{W_I(t)}{W_I(t + \delta t)}\, \theta(I, t) - \left(\frac{k_{I-1,I}^i(t + \delta t)}{S_{I-1,I}(t + \delta t)(\delta_{I,I-1} + \delta_{I,I+1})} \right.
$$

$$
+ \frac{k_{I+1,I}^i(t + \delta t)}{S_{I+1,I}(t + \delta t)(\delta_{I,I-1} + \delta_{I,I+1})} + \frac{k_{J-1,I}^j(t + \delta t)}{S_{J-1,I}(t + \delta t)(\delta_{I,J-1} + \delta_{I,J+1})}
$$

$$
+ \frac{k_{J+1,I}^j(t + \delta t)}{S_{J+1,I}(t + \delta t)(\delta_{I,J-1} + \delta_{I,J+1})} + \frac{k_{K-1,I}^k(t + \delta t)}{S_{K-1,I}(t + \delta t)(\delta_{I,K-1} + \delta_{I,K+1})}
$$

$$
+ \left. \frac{k_{K+1,I}^k(t + \delta t)}{S_{K+1,I}(t + \delta t)(\delta_{I,K-1} + \delta_{I,K+1})} \right) \theta(I, t + \delta t)
$$

$$
+ \frac{k_{I-1,I}^i(t + \delta t)\theta(I - 1, t + \delta t)}{S_{I-1,I}(t + \delta t)(\delta_{I,I-1} + \delta_{I,I+1})} + \frac{k_{I+1,I}^i(t + \delta t)\theta(I + 1, t + \delta t)}{S_{I+1,I}(t + \delta t)(\delta_{I,I-1} + \delta_{I,I+1})}
$$

$$
+ \frac{k_{J-1,I}^j(t + \delta t)\theta(J - 1, t + \delta t)}{S_{J-1,I}(t + \delta t)(\delta_{I,J-1} + \delta_{I,J+1})} + \frac{k_{J+1,I}^j(t + \delta t)\theta(J + 1, t + \delta t)}{S_{J+1,I}(t + \delta t)(\delta_{I,J-1} + \delta_{I,J+1})}
$$

$$
+ \frac{k_{K-1,I}^k(t + \delta t)\theta(K - 1, t + \delta t)}{S_{K-1,I}(t + \delta t)(\delta_{I,K-1} + \delta_{I,K+1})} + \frac{k_{K+1,I}^k(t + \delta t)\theta(K + 1, t + \delta t)}{S_{K+1,I}(t + \delta t)(\delta_{I,K-1} + \delta_{I,K+1})}
$$

$$
+ \frac{\delta t q_I(t + \delta t)}{W_I(t + \delta t)(\delta_{I,I-1} + \delta_{I,I+1})(\delta_{I,J-1} + \delta_{I,J+1})(\delta_{I,K-1} + \delta_{I,K+1})} + \epsilon.
$$

Adding the explicit and implicit formulations and grouping future time-row temperature terms on the left-hand side gives

$$
\left(2 + \frac{k_{I-1,I}^i(t + \delta t)}{S_{I-1,I}(t + \delta t)(\delta_{I,I-1} + \delta_{I,I+1})} + \frac{k_{I+1,I}^i(t + \delta t)}{S_{I+1,I}(t + \delta t)(\delta_{I,I-1} + \delta_{I,I+1})} \right.
$$

$$
+ \frac{k_{J-1,I}^j(t + \delta t)}{S_{J-1,I}(t + \delta t)(\delta_{I,J-1} + \delta_{I,J+1})} + \frac{k_{J+1,I}^j(t + \delta t)}{S_{J+1,I}(t + \delta t)(\delta_{I,J-1} + \delta_{I,J+1})}
$$

$$
+ \left. \frac{k_{K-1,I}^k(t + \delta t)}{S_{K-1,I}(t + \delta t)(\delta_{I,K-1} + \delta_{I,K+1})} + \frac{k_{K+1,I}^k(t + \delta t)}{S_{K+1,I}(t + \delta t)(\delta_{I,K-1} + \delta_{I,K+1})} \right) \theta(I, t + \delta t)
$$

$$
- \frac{k_{I-1,I}^i(t + \delta t)\theta(I - 1, t + \delta t)}{S_{I-1,I}(t + \delta t)(\delta_{I,I-1} + \delta_{I,I+1})} - \frac{k_{I+1,I}^i(t + \delta t)\theta(I + 1, t + \delta t)}{S_{I+1,I}(t + \delta t)(\delta_{I,I-1} + \delta_{I,I+1})}
$$

$$
- \frac{k_{J-1,I}^j(t + \delta t)\theta(J - 1, t + \delta t)}{S_{J-1,I}(t + \delta t)(\delta_{I,J-1} + \delta_{I,J+1})} - \frac{k_{J+1,I}^j(t + \delta t)\theta(J + 1, t + \delta t)}{S_{J+1,I}(t + \delta t)(\delta_{I,J-1} + \delta_{I,J+1})}
$$

$$- \frac{k_{K-1,I}^k(t+\delta t)\theta(K-1,t+\delta t)}{S_{K-1,I}(t+\delta t)(\delta_{I,K-1}+\delta_{I,K+1})} - \frac{k_{K+1,I}^k(t+\delta t)\theta(K+1,t+\delta t)}{S_{K+1,I}(t+\delta t)(\delta_{I,K-1}+\delta_{I,K+1})}$$

$$- \frac{\delta t q_I(t+\delta t)}{W_I(t+\delta t)(\delta_{I,I-1}+\delta_{I,I+1})(\delta_{I,J-1}+\delta_{I,J+1})(\delta_{I,K-1}+\delta_{I,K+1})}$$

$$= \theta(I,t)\left(\frac{2W_I(t)}{W_I(t+\delta t)} - \frac{k_{I-1,I}^i(t)}{S_{I-1,I}(t+\delta t)(\delta_{I,I-1}+\delta_{I,I+1})} - \frac{k_{I+1,I}^i(t)}{S_{I+1,I}(t+\delta t)(\delta_{I,I-1}+\delta_{I,I+1})} \right.$$

$$- \frac{k_{J-1,I}^j(t)}{S_{J-1,I}(t+\delta t)(\delta_{I,I-1}+\delta_{I,I+1})} - \frac{k_{J+1,I}^j(t)}{S_{J+1,I}(t+\delta t)(\delta_{I,J-1}+\delta_{I,J+1})}$$

$$\left. - \frac{k_{K-1,I}^k(t)}{S_{K-1,I}(t+\delta t)(\delta_{I,K-1}+\delta_{I,K+1})} - \frac{k_{K+1,I}^k(t)}{S_{K+1,I}(t+\delta t)(\delta_{I,K-1}+\delta_{I,K+1})} \right)$$

$$+ \frac{k_{I-1,I}^i(t)\theta(I-1,t)}{S_{I-1,I}(t+\delta t)(\delta_{I,I-1}+\delta_{I,I+1})} + \frac{k_{I+1,I}^i(t)\theta(I+1,t)}{S_{I+1,I}(t+\delta t)(\delta_{I,I-1}+\delta_{I,I+1})}$$

$$+ \frac{k_{J-1,I}^j(t)\theta(J-1,t)}{S_{J-1,I}(t+\delta t)(\delta_{I,J-1}+\delta_{I,J+1})} + \frac{k_{J+1,I}^j(t)\theta(J+1,t)}{S_{J+1,I}(t+\delta t)(\delta_{I,J-1}+\delta_{I,J+1})}$$

$$+ \frac{k_{K-1,I}^k(t)\theta(K-1,t)}{S_{K-1,I}(t+\delta t)(\delta_{I,K-1}+\delta_{I,K+1})} + \frac{k_{K+1,I}^k(t)\theta(K+1,t)}{S_{K+1,I}(t+\delta t)(\delta_{I,K-1}+\delta_{I,K+1})}$$

$$+ \frac{q_I(t)\delta t}{W_I(t+\delta t)(\delta_{I,I-1}+\delta_{I,I+1})(\delta_{I,J-1}+\delta_{I,J+1})(\delta_{I,K-1}+\delta_{I,K+1})} + \epsilon.$$

It should be noted that the coefficients of the temperature variables of this equation are dimensionless Fourier numbers. This number defines the ratio of the heat conduction rate to the heat storage rate. High values represent good conductors with relatively poor storage potential, low values represent the converse: poor conductors with relatively good storage potential.

If a model based on this and later equations is to be used for the investigation of zero capacity systems, then the capacity term of the denominator will cause problems due to division by zero. This difficulty can be overcome by multiplying throughout by the common W_I term present in S to give

$$\left(2W_I(t+\delta t) + \frac{k_{I-1,I}^i(t+\delta t)\delta t}{\delta x_{I-1,I}(\delta_{I,I-1}+\delta_{I,I+1})} + \frac{k_{I+1,I}^i(t+\delta t)\delta t}{\delta x_{I+1,I}(\delta_{I,I-1}+\delta_{I,I+1})} \right.$$

$$+ \frac{k_{J-1,I}^j(t+\delta t)\delta t}{\delta x_{J-1,I}(\delta_{I,J-1}+\delta_{I,J+1})} + \frac{k_{J+1,I}^j(t+\delta t)\delta t}{\delta x_{J+1,I}(\delta_{I,J-1}+\delta_{I,J+1})} + \frac{k_{K-1,I}^k(t+\delta t)\delta t}{\delta x_{K-1,I}(\delta_{I,K-1}+\delta_{I,K+1})}$$

$$\left. + \frac{k_{K+1,I}^k(t+\delta t)\delta t}{\delta x_{K+1,I}(\delta_{I,K-1}+\delta_{I,K+1})} \right) \theta(I,t+\delta t) - \frac{k_{I-1,I}^i(t+\delta t)\delta t\theta(I-1,t+\delta t)}{\delta x_{I-1,I}(\delta_{I,I-1}+\delta_{I,I+1})}$$

$$- \frac{k_{I+1,I}^i(t+\delta t)\delta t\theta(I+1,t+\delta t)}{\delta x_{I+1,I}(\delta_{I,I-1}+\delta_{I,I+1})} - \frac{k_{J-1,I}^j(t+\delta t)\delta t\theta(J-1,t+\delta t)}{\delta x_{J-1,I}(\delta_{I,J-1}+\delta_{I,J+1})}$$

$$- \frac{k_{J+1,I}^{J}(t+\delta t)\,\delta t\theta(J+1,t+\delta t)}{\delta x_{J+1,I}(\delta_{I,J-1}+\delta_{I,J+1})} - \frac{k_{K-1,I}^{K}(t+\delta t)\,\delta t\theta(K-1,t+\delta t)}{\delta x_{K-1,I}(\delta_{I,K-1}+\delta_{I,K+1})}$$

$$- \frac{k_{K+1,I}^{K}(t+\delta t)\,\delta t\theta(K+1,t+\delta t)}{\delta x_{K+1,I}(\delta_{I,K-1}+\delta_{I,K+1})}$$

$$- \frac{q_I(t+\delta t)\,\delta t}{W_I(t+\delta t)(\delta_{I,I-1}+\delta_{I,I+1})(\delta_{I,J-1}+\delta_{I,J+1})(\delta_{I,K-1}+\delta_{I,K+1})}$$

$$= \left(2W_I(t) - \frac{k_{I-1,I}^{I}(t)\,\delta t}{\delta x_{I-1,I}(\delta_{I,I-1}+\delta_{I,I+1})} - \frac{k_{I+1,I}^{I}(t)\,\delta t}{\delta x_{I+1,I}(\delta_{I,I-1}+\delta_{I,I+1})}\right.$$

$$- \frac{k_{J-1,I}^{J}(t)\,\delta t}{\delta x_{J-1,I}(\delta_{I,J-1}+\delta_{I,J+1})} - \frac{k_{J+1,I}^{J}(t)\,\delta t}{\delta x_{J+1,I}(\delta_{I,J-1}+\delta_{I,J+1})} - \frac{k_{K-1,I}^{K}(t)\,\delta t}{\delta x_{K-1,I}(\delta_{I,K-1}+\delta_{I,K+1})}$$

$$\left. - \frac{k_{K+1,I}^{K}(t+\delta t)\,\delta t}{\delta x_{K+1,I}(\delta_{I,K-1}+\delta_{I,K+1})}\right)\theta(I,t) + \frac{k_{I-1,I}^{I}(t)\theta(I-1,t)}{\delta x_{I-1,I}(\delta_{I,I-1}+\delta_{I,I+1})}$$

$$+ \frac{k_{I+1,I}^{I}(t)\theta(I+1,t)}{\delta x_{I+1,I}(\delta_{I,I-1}+\delta_{I,I+1})} + \frac{k_{J-1,I}^{J}(t)\,\delta t\theta(J-1,t)}{\delta x_{J-1,I}(\delta_{I,J-1}+\delta_{I,J+1})} + \frac{k_{J+1,I}^{J}(t)\,\delta t\theta(J+1,t)}{\delta x_{J+1,I}(\delta_{I,J-1}+\delta_{I,J+1})}$$

$$+ \frac{k_{K-1,I}^{K}(t)\,\delta t\theta(K-1,t)}{\delta x_{K-1,I}(\delta_{I,K-1}+\delta_{I,K+1})} + \frac{k_{K+1,I}^{K}(t)\,\delta t\theta(K+1,t)}{\delta x_{K+1,I}(\delta_{I,K-1}+\delta_{I,K+1})}$$

$$+ \frac{q_I(t)\,\delta t}{W_I(t+\delta t)(\delta_{I,I-1}+\delta_{I,I+1})(\delta_{I,J-1}+\delta_{I,J+1})(\delta_{I,K-1}+\delta_{I,K+1})} + \epsilon. \tag{3.4}$$

This equation—the general transient conduction node formulation—is equivalent to a Crank–Nicolson difference formulation: a weighted average of the fully explicit scheme—in which the second-order space derivative of the FFE (equation (3.3)) is expressed in central difference form with the first-order time derivative expressed as a forward difference—and the implicit scheme—in which the first-order time derivative is expressed as a backward difference with the second-order space derivative in central difference form. This method can be shown to be consistent, convergent and A-stable, providing the possibility of variable time stepping and well adapted for the solution of the so-called 'stiff' problem (Mitchell 1969).

In condensed form equation (3.4) gives for any internal material mix represented by node I (homogeneous or non-homogeneous):

$$C_s(t+\delta t)\theta(I,t+\delta t) - \sum_{i=1}^{N} C_{ci}(t+\delta t)\theta(i,t+\delta t) - \frac{\delta t q_I(t+\delta t)}{W_I(t+\delta t)\delta V_I}$$

$$= C_s(t)\theta(I,t) + \sum_{i=1}^{N} C_{ci}(t)\theta(i,t) + \frac{\delta t q_I(t)}{W_I(t+\delta t)\delta V_I}$$

where $C_s(\xi)$ is the self-coupling coefficient at time ξ, $C_c(\xi)$ is the cross-coupling coefficient at time ξ, and N is the number of inter-nodal contacts.

In any integrated building/plant simulation model, economy of nodal discretisation is essential and so it is important to distinguish between plant and building components with respect to transient conduction modelling.

With plant components it is usually the flow processes and surface heat transfers that are of prime concern and the formulations of §§3.2.2 and 3.2.3 are extant. It is, therefore, only necessary to represent the material of such components by a small number of nodes to represent adequately thermal storage and time lagged conduction. Section 3.3 and chapter 6 demonstrate the application of equation (3.4) to represent, in a non-complex manner, the material of typical plant components.

With the range of building-side components (walls, windows, furniture and the range of emerging passive solar features such as mass walls, phase change materials, remote storage facilities etc), the detailed modelling of transient conduction is crucial to an accurate simulation of system time shifts and hence energy flow. It is important, therefore, to devise a mathematical model of transient conduction which conserves the integrity of the components in terms of the relative positions of the constituent elements. For this reason, equation (3.4) is now applied to each of the characteristic node types, representing regions found within such components, to generate node-specific simulation equations for later use in §3.3 and chapter 4.

Consider figure 3.6, which shows some multi-layered construction with a three-dimensional nodal mesh imposed.

Opaque homogeneous element nodes
Assume node I is situatated at the centre plane of an opaque homogeneous element located within some multi-layered construction (wall, ceiling, floor, furniture, window system etc). The distance $I-1 \rightarrow I+1$ defines the thickness of the element in a direction normal to the layered construction. For this case

$$\delta x_{I-1,I} = \delta x_{I+1,I} = \delta x_I$$
$$\delta x_{J-1,I} = \delta x_{J+1,I} = \delta x_J$$
$$\delta x_{K-1,I} = \delta x_{K+1,I} = \delta x_K.$$

Noting that

$$k_{I-1,I}^i = k_{I+1,I}^i = k_I$$
$$k_{J-1,I}^j = k_{J+1,I}^j = k_J$$
$$k_{K-1,I}^k = k_{K+1,I}^k = k_K$$
$$W_I = \varrho_I C_I$$
$$\delta x_{i \pm 1}(\delta_{I,i-1} + \delta_{I,i+1}) = \delta x_i^2 \qquad i = I, J, K$$

then from equation (3.4) it follows that

$$\left(2\varrho_I(t+\delta t)C_I(t+\delta t) + \frac{2k_I(t+\delta t)\delta t}{\delta x_I^2} + \frac{2k_J(t+\delta t)\delta t}{\delta x_J^2} + \frac{2k_K(t+\delta t)\delta t}{\delta x_K^2} \right)\theta(I,t+\delta t)$$

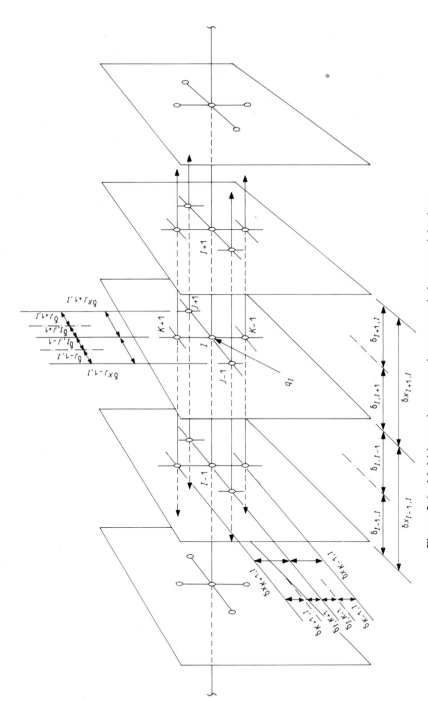

Figure 3.6 Multi-layered construction energy balance nodal scheme.

$$-\frac{k_I(t+\delta t)\,\delta t}{\delta x_I^2}\theta(I-1,t+\delta t)-\frac{k_I(t+\delta t)\,\delta t}{\delta x_I^2}\theta(I+1,t+\delta t)$$

$$-\frac{k_J(t+\delta t)\,\delta t}{\delta x_J^2}\theta(J-1,t+\delta t)-\frac{k_J(t+\delta t)\,\delta t}{\delta x_J^2}\theta(J+1,t+\delta t)$$

$$-\frac{k_K(t+\delta t)\,\delta t}{\delta x_K^2}\theta(K-1,t+\delta t)-\frac{k_K(t+\delta t)\,\delta t}{\delta x_K^2}\theta(K+1,t+\delta t)-\frac{q_I(t+\delta t)\,\delta t}{\delta x_I\,\delta x_J\,\delta x_K}$$

$$=\left(2\varrho_I(t)C_I(t)-\frac{2k_I(t)\,\delta t}{\delta x_I^2}-\frac{2k_J(t)\,\delta t}{\delta x_J^2}-\frac{2k_K(t)\,\delta t}{\delta x_K^2}\right)\theta(I,t)+\frac{k_I(t)\,\delta t}{\delta x_I^2}\theta(I-1,t)$$

$$+\frac{k_I(t)\,\delta t}{\delta x_I^2}\theta(I+1,t)+\frac{k_J(t)\,\delta t}{\delta x_J^2}\theta(J-1,t)+\frac{k_J(t)\,\delta t}{\delta x_J^2}\theta(J+1,t)$$

$$+\frac{k_K(t)\,\delta t}{\delta x_K^2}\theta(K-1,t)+\frac{k_K(t)\,\delta t}{\delta x_K^2}\theta(K+1,t)+\frac{q_I(t)\,\delta t}{\delta x_I\,\delta x_J\,\delta x_K}. \tag{3.5}$$

If the usual assumption of isotropic behaviour is made, then

$$k_I=k_J=k_K=k.$$

This equation can be utilised to represent any homogeneous medium by simply dividing the medium into a number of finite volumes. In the case of wall constructions accuracy levels can be improved by simple element subdivision as demonstrated in figure 3.7. The heat generation term permits the direct nodal injection or extraction of energy when modelling, for example, underfloor heating systems, electrical storage units or isolated low temperature thermal stores.

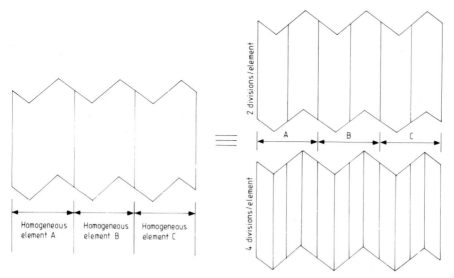

Figure 3.7 Homogeneous element subdivision to improve accuracy.

The uni-directional counterpart of equation (3.5) is given, for the x direction, by

$$\left(2\varrho_I(t+\delta t)C_I(t+\delta t)+\frac{2k(t+\delta t)\,\delta t}{\delta x_I^2}\right)\theta(I,t+\delta t)-\frac{k(t+\delta t)\,\delta t}{\delta x_I^2}\,\theta(I-1,t+\delta t)$$

$$-\frac{k(t+\delta t)\,\delta t}{\delta x_I^2}\,\theta(I+1,t+\delta t)-\frac{q_I(t+\delta t)\,\delta t}{\delta x_I\,\delta x_J\,\delta x_K}$$

$$=\left(2\varrho_I(t)C_I(t)-\frac{2k(t)\,\delta t}{\delta x_I^2}\right)\theta(I,t)+\frac{k(t)\,\delta t}{\delta x_I^2}\,\theta(I-1,t)+\frac{k(t)\,\delta t}{\delta x_I^2}\,\theta(I+1,t)$$

$$+\frac{q_I(t)\,\delta t}{\delta x_I\,\delta x_J\,\delta x_K}. \tag{3.6}$$

Transparent homogeneous element nodes
Equation (3.5) also holds for this node type, but here the heat generation term will also include the absorption of shortwave energy as it travels through the transparent medium. The derivation of algorithmic techniques for shortwave radiation prediction is given in §5.3.

Phase change material nodes
When modelling materials which undergo some change of phase, to absorb or release the latent heat of vaporisation or fusion at constant temperature, it is possible to utilise equation (3.5). When the temperature of phase change is reached the heat generation term can be used to maintain a constant node temperature until that quantity of energy has been absorbed or released at which temperature change will recommence. A simple counter mechanism can be established to keep record of the latent energy in store at any time.

Boundary nodes separating two homogeneous elements
Again referring to figure 3.6, assume node I is situated at the boundary between two homogeneous elements (opaque and/or transparent) of (perhaps) different thermophysical properties. The distance $I-1 \rightarrow I$ defines the half thickness of one element denoted by subscript A (node $I-1$ is located at the centre plane of this element) and $I \rightarrow I+1$ defines the half thickness of the second element denoted by subscript B. Note that if the elements A and B have undergone subdivision as shown in figure 3.7, then nodes $I-1$ and $I+1$ will be relocated closer to the interface between the elements.

For boundary nodes it is necessary to implement a volumetric weighting to establish representative thermophysical properties for each inter-nodal flowpath and the region represented by node I. A contact resistance, R_c, is also introduced to impose an additional resistance to heat flow at the interface.

Noting that

$$k_{I-1,I}^t = k_A$$
$$k_{I+1,I}^t = k_B$$

$$kj_{-1,I} = kj_{+1,I} = kk_{-1,I} = kk_{+1,I} = (\delta_{I-1,I}k_A + \delta_{I+1,I}k_B)/(\delta_{I-1,I} + \delta_{I+1,I})$$

$$= k_{AB} \text{ (assuming isotropic behaviour)}$$

$$W_I = (\varrho_A C_A \delta V_A + \varrho_B C_B \delta V_B)/(\delta V_A + \delta V_B)$$

$$(\delta_{i-1,I} + \delta_{i+1,I}) = \delta_{i-1,i+1} \qquad i = I, J, K$$

then equation (3.4) gives

$$\left(2W_I(t+\delta t) + \frac{[k_A(t+\delta t)R_c(t+\delta t) + 2\delta x_{I-1,I}]\,\delta t}{\delta x_{I-1,I}R_c(t+\delta t)\delta_{I-1,I+1}}\right.$$

$$+ \frac{[k_B(t+\delta t)R_c(t+\delta t) + 2\delta x_{I+1,I}]\,\delta t}{\delta x_{I+1,I}R_c(t+\delta t)\delta_{I-1,I+1}} + \frac{k_{AB}(t+\delta t)\delta t}{\delta x_{J-1,I}\delta_{J-1,J+1}} + \frac{k_{AB}(t+\delta t)\delta t}{\delta x_{J+1,I}\delta_{J-1,J+1}}$$

$$\left. + \frac{k_{AB}(t+\delta t)\delta t}{\delta x_{K-1,I}\delta_{K-1,K+1}} + \frac{k_{AB}(t+\delta t)\delta t}{\delta x_{K+1,I}\delta_{K-1,K+1}}\right)\theta(I,t+\delta t)$$

$$- \frac{[k_A(t+\delta t)R_c(t+\delta t) + 2\delta x_{I-1,I}]\,\delta t}{\delta x_{I-1,I}R_c(t+\delta t)\delta_{I-1,I+1}}\theta(I-1,t+\delta t)$$

$$- \frac{[k_B(t+\delta t)R_c(t+\delta t) + 2\delta x_{I+1,I}]\,\delta t}{\delta x_{I+1,I}R_c(t+\delta t)\delta_{I-1,I+1}}\theta(I+1,t+\delta t) - \frac{k_{AB}(t+\delta t)\delta t}{\delta x_{J-1,I}\delta_{J-1,J+1}}\theta(J-1,t+\delta t)$$

$$- \frac{k_{AB}(t+\delta t)\delta t}{\delta x_{J+1,I}\delta_{J-1,J+1}}\theta(J+1,t+\delta t) - \frac{k_{AB}(t+\delta t)\delta t}{\delta x_{K-1,I}\delta_{K-1,K+1}}\theta(K-1,t+\delta t)$$

$$- \frac{k_{AB}(t+\delta t)\delta t}{\delta x_{K+1,I}\delta_{K-1,K+1}}\theta(K+1,t+\delta t) - \frac{q_I(t+\delta t)\delta t}{\delta_{I-1,I+1}\delta_{J-1,J+1}\delta_{K-1,K+1}}$$

$$= \left(2W_I(t) - \frac{[k_A(t)R_c(t) + 2\delta x_{I-1,I}]\,\delta t}{\delta x_{I-1,I}R_c(t)\delta_{I-1,I+1}} - \frac{[k_B(t)R_c(t) + 2\delta x_{I+1,I}]\,\delta t}{\delta x_{I+1,I}R_c(t)\delta_{I-1,I+1}}\right.$$

$$\left. - \frac{k_{AB}(t)\delta t}{\delta x_{J-1,I}\delta_{J-1,J+1}} - \frac{k_{AB}(t)\delta t}{\delta x_{J+1,I}\delta_{J-1,J+1}} - \frac{k_{AB}(t)\delta t}{\delta x_{K-1,I}\delta_{K-1,K+1}} - \frac{k_{AB}(t)\delta t}{\delta x_{K+1,I}\delta_{K-1,K+1}}\right)\theta(I,t)$$

$$+ \frac{[k_A(t)R_c(t) + 2\delta x_{I-1,I}]\,\delta t}{\delta x_{I-1,I}R_c(t)\delta_{I-1,I+1}}\theta(I-1,t) + \frac{[k_B(t)R_c(t) + 2\delta x_{I+1,I}]\,\delta t}{\delta x_{I+1,I}R_c(t)\delta_{I-1,I+1}}\theta(I+1,t)$$

$$+ \frac{k_{AB}(t)\delta t}{\delta x_{J-1,I}\delta_{J-1,J+1}}\theta(J-1,t) + \frac{k_{AB}(t)\delta t}{\delta x_{J+1,I}\delta_{J-1,J+1}}\theta(J+1,t)$$

$$+ \frac{k_{AB}(t)\delta t}{\delta x_{K-1,I}\delta_{K-1,K+1}}\theta(K-1,t) + \frac{k_{AB}(t)\delta t}{\delta x_{K+1,I}\delta_{K-1,K+1}}\theta(K+1,t)$$

$$+ \frac{q_I(t)\delta t}{\delta_{I-1,I+1}\delta_{J-1,J+1}\delta_{K-1,K+1}}. \tag{3.7}$$

The heat generation term, q_I, will permit, in addition to plant interaction potential, the absorption of shortwave radiant energy if either element is transparent and exposed to some shortwave source (as for example in a window/blind system).

The uni-directional form of equation (3.7) is given by:

$$
\left(2W_I(t+\delta t) + \frac{[k_A(t+\delta t)R_c(t+\delta t) + 2\,\delta x_{I-1,I}]\,\delta t}{\delta x_{I-1,I}R_c(t+\delta t)\,\delta_{I-1,I+1}} \right.
$$

$$
\left. + \frac{[k_B(t+\delta t)R_c(t+\delta t) + 2\,\delta x_{I+1,I}]\,\delta t}{\delta x_{I+1,I}R_c(t+\delta t)\,\delta_{I-1,I+1}} \right)\theta(I,t+\delta t)
$$

$$
- \frac{[k_A(t+\delta t)R_c(t+\delta t) + 2\,\delta x_{I-1,I}]\,\delta t}{\delta x_{I-1,I}R_c(t+\delta t)\,\delta_{I-1,I+1}}\theta(I-1,t+\delta t)
$$

$$
- \frac{[k_B(t+\delta t)R_c(t+\delta t) + 2\,\delta x_{I+1,I}]\,\delta t}{\delta x_{I+1,I}R_c(t+\delta t)\,\delta_{I-1,I+1}}\theta(I+1,t+\delta t) - \frac{q_I(t+\delta t)\,\delta t}{\delta_{I-1,I+1}\,\delta_{J-1,J+1}\,\delta_{K-1,K+1}}
$$

$$
= \left(2W_I(t) - \frac{[k_A(t)R_c(t) + 2\,\delta x_{I-1,I}]\,\delta t}{\delta x_{I-1,I}R_c(t)\,\delta_{I-1,I+1}} - \frac{[k_B(t)R_c(t) + 2\,\delta x_{I+1,I}]\,\delta t}{\delta x_{I+1,I}R_c(t)\,\delta_{I-1,I+1}} \right)\theta(I,t)
$$

$$
+ \frac{[k_A(t)R_c(t) + 2\,\delta x_{I-1,I}]\,\delta t}{\delta x_{I-1,I}R_c(t)\,\delta_{I-1,I+1}}\theta(I-1,t) + \frac{[k_B(t)R_c(t) + 2\,\delta x_{I+1,I}]\,\delta t}{\delta x_{I+1,I}R_c(t)\,\delta_{I-1,I+1}}\theta(I+1,t)
$$

$$
+ \frac{q_I(t)\,\delta t}{\delta_{I-1,I+1}\,\delta_{J-1,J+1}\,\delta_{K-1,K+1}}. \tag{3.8}
$$

Lumped Material Nodes

When applying equation (3.4) to capacity regions contained by some environment (as opposed to separating two environments)—such as room contents, structural beams or columns, or an underground thermal store—it is convenient to use a modified form of equation (3.5) given by

$$
2W_I(t+\delta t) + \sum_i \frac{k'_{i,I}(t+\delta t)\,\delta t\,\delta A_{i,I}}{\delta V_I\delta x_{i,I}}\theta(I,t+\delta t) - \sum_i \frac{k'_{i,I}(t+\delta t)\,\delta t\,\delta A_{i,I}}{\delta V_I\delta x_{i,I}}\theta(i,t+\delta t)
$$

$$
- \frac{q_I(t+\delta t)\,\delta t}{\delta V_I} = 2W_I(t) - \sum_i \frac{k'_{i,I}(t)\,\delta t\,\delta A_{i,I}}{\delta V_I\delta x_{i,I}}\theta(I,t)
$$

$$
+ \sum_i \frac{k'_{i,I}(t)\,\delta t\,\delta A_{i,I}}{\delta V_I\delta x_{i,I}}\theta(i,t) + \frac{q_I(t)\,\delta t}{\delta V_I} \tag{3.9}
$$

where $\delta A_{i,I}$ is the area normal to the direction of heat flow between nodes i and I (m^2), δV_I the volume of the region represented by node I (m^3), and i the number of inter-region (node) thermal connections.

This allows a simple nodal network to be established to represent capacity and so introduce inertia. Figure 3.8 shows two simple nodal schemes to represent furnishings and thermal storage respectively. With such systems it is usual to maximise the quantity $\delta A/\delta V$ to represent the high exposed area to contained volume ratio typical of such components. More detailed formulations are possible and usually desirable in the case of, for example, thermal stores with horizontal and vertical stratification.

Taken together equations (3.4)–(3.9) allow the construction of a discrete nodal network representing transient energy flow within any capacity/insulation system containing opaque and transparent regions. Section 3.3 demonstrates the technique of structuring these equations (and those that follow) to represent real multi-component building systems.

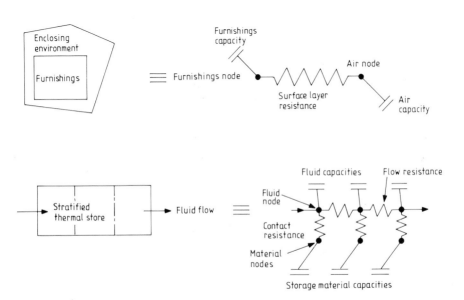

Figure 3.8   Nodal scheme for furnishings and remote thermal store.

3.2.2   Energy balance: exposed surface layers

Consider figure 3.9 which shows some node I located at the exposed surface of some capacity/insulation system such as a multi-layered wall construction, the inside or outside surface of a pipe, or the surface of a radiator. Node $I-1$ is the next-to-surface node buried within the material of the next-to-surface layer and node $I+1$ represents the fluid volume immediately beyond the surface boundary layer. Assuming that the boundary layer thickness is negligible, compared with the other dimensions of the region represented by node I, then the finite volume represented by I is given by

$$\delta_{I,I-1}(\delta_{I,J-1} + \delta_{I,J+1})(\delta_{I,K-1} + \delta_{I,K+1}).$$

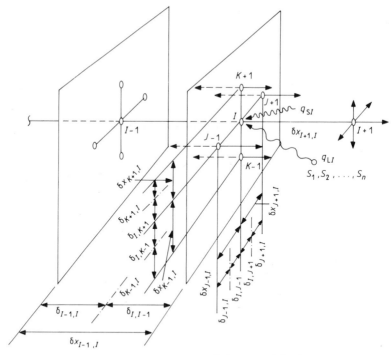

Figure 3.9   Surface energy balance nodal scheme.

For each inter-nodal flowpath the heat flux is given by

$$q_{I-1,I} = k_{I-1,I}^i(\delta_{I,J-1} + \delta_{I,J+1})(\delta_{I,K-1} + \delta_{I,K+1})(\theta_{I-1} - \theta_I)/\delta x_{I-1,I}$$

$$q_{I+1,I} = h_{cI+1,I}(\delta_{I,J-1} + \delta_{I,J+1})(\delta_{I,K-1} + \delta_{I,K+1})(\theta_{I+1} - \theta_I)$$

$$q_{J-1,I} = k_{J-1,I}^j\delta_{I,I-1}(\delta_{I,K-1} + \delta_{I,K+1})(\theta_{J-1} - \theta_I)/\delta x_{J-1,I}$$

$$q_{J+1,I} = k_{J+1,I}^j\delta_{I,I-1}(\delta_{J,K-1} + \delta_{I,K+1})(\theta_{J+1} - \theta_I)/\delta x_{J+1,I}$$

$$q_{K-1,I} = k_{K-1,I}^k\delta_{I,I-1}(\delta_{I,J-1} + \delta_{I,J+1})(\theta_{K-1} - \theta_I)/\delta x_{K-1,I}$$

$$q_{K+1,I} = k_{K+1,I}^k\delta_{I,I-1}(\delta_{I,J-1} + \delta_{I,J+1})(\theta_{K+1} - \theta_I)/\delta x_{K+1,I}$$

and noting that for this node type

$$q_I = q_{SI} + q_{LI} + q_{RI} + q_{PI}$$

$$q_{LI} = \sum_{s=1}^{N} h_{rs,I}(\delta_{I,J-1} + \delta_{I,J+1})(\delta_{I,K-1} + \delta_{I,K+1})(\theta_s - \theta_I)$$

where q_{SI} is the shortwave energy absorption, q_{LI} the longwave energy exchange with the surroundings, q_{RI} the radiant energy from casual sources and q_{PI} the radiant component of plant input (all measured in W); h_r is the 'grey' body radiation coefficient (W m^{-2} °C^{-1}), N the number of surfaces in longwave contact, and h_c the surface convection coefficient (W m^{-2} °C^{-1}).

Note that the longwave radiation term has been linearised to preserve, for the present, linear equation structures essential for matrix processing. Chapter 4 discusses the impact of linearity assumptions and describes techniques to include non-linear terms.

Substituting these heat fluxes in equation (3.1) gives the exposed surface energy balance relationship:

$$[W_I(t + \delta t)\theta(I, t + \delta t) - W_I(t)\theta(I, t)]\,\delta_{I, I-1}(\delta_{I, J-1} + \delta_{I, J+1})(\delta_{I, K-1} + \delta_{I, K+1})/\delta t$$

$$= k^i_{I-1, I}(\xi)(\delta_{I, J-1} + \delta_{I, J+1})(\delta_{I, K-1} + \delta_{I, K+1})[\theta(I-1, \xi) - \theta(I, \xi)]/\delta x_{I-1, I}$$

$$+ h_{cI+1, I}(\xi)(\delta_{I, J-1} + \delta_{I, J+1})(\delta_{I, K-1} + \delta_{I, K+1})[\theta(I+1, \xi) - \theta(I, \xi)]$$

$$+ k^j_{J-1, I}(\xi)\,\delta_{I, I-1}(\delta_{I, K-1} + \delta_{I, K+1})[\theta(J-1, \xi) - \theta(I, \xi)]/\delta x_{J-1, I}$$

$$+ k^j_{J+1, I}(\xi)\delta_{I, I-1}(\delta_{I, K-1} + \delta_{I, K+1})[\theta(J+1, \xi) - \theta(I, \xi)]/\delta x_{J+1, I}$$

$$+ k^k_{K-1, I}(\xi)\,\delta_{I, I-1}(\delta_{I, J-1} + \delta_{I, J+1})[\theta(K-1, \xi) - \theta(I, \xi)]/\delta x_{K-1, I}$$

$$+ k^k_{K+1, I}(\xi)\,\delta_{I, I-1}(\delta_{I, J-1} + \delta_{I, J+1})[\theta(K+1, \xi) - \theta(I, \xi)]/\delta x_{K+1, I}$$

$$+ \sum_{s=1}^{N} h_{rs, I}(\xi)(\delta_{I, J-1} + \delta_{I, J+1})(\delta_{I, K-1} + \delta_{I, K+1})[\theta(s, \xi) - \theta(I, \xi)] + q_{SI}(\xi)$$

$$+ q_{RI}(\xi) + q_{PI}(\xi).$$

As with the transient conduction formulation this equation can be evaluated at some present time-row, t, some future time-row, $t + \delta t$, and the two formulations concatenated to give the characteristic space- and time-dependent simulation equation for a surface node exposed to an adjacent fluid region.

The temperature explicit formulation is

$$\theta(I, t + \delta t) = \theta(I, t)\left(\frac{W_I(t)}{W_I(t + \delta t)} - \frac{k^i_{I-1, I}(t)}{S_{I-1, I}(t + \delta t)\,\delta_{I, I-1}} - \frac{h_{cI+1, I}(t)\,\delta t}{W_I(t + \delta t)\,\delta_{I, I-1}}\right.$$

$$- \frac{k^j_{J-1, I}(t)}{S_{J-1, I}(t + \delta t)\,\delta_{J-1, J+1}} - \frac{k^j_{J+1, I}(t)}{S_{J+1, I}(t + \delta t)\,\delta_{J-1, J+1}}$$

$$- \frac{k^k_{K-1, I}(t)}{S_{K-1, I}(t + \delta t)\,\delta_{K-1, K+1}} - \frac{k^k_{K+1, I}(t)}{S_{K+1, I}(t + \delta t)\,\delta_{K-1, K+1}}$$

$$\left. - \frac{\delta t \sum\limits_{s=1}^{N} h_{rs, I}(t)}{W_I(t + \delta t)\,\delta_{I, I-1}}\right) + \frac{k^i_{I-1, I}(t)}{S_{I-1, I}(t + \delta t)\,\delta_{I, I-1}}\theta(I-1, t)$$

$$+ \frac{h_{cI+1, I}(t)\,\delta t}{W_I(t + \delta t)\,\delta_{I, I-1}}\theta(I+1, t) + \frac{k^j_{J-1, I}(t)}{S_{J-1, I}(t + \delta t)\,\delta_{J-1, J+1}}\theta(J-1, t)$$

$$+ \frac{k^j_{J+1, I}(t)}{S_{J+1, I}(t + \delta t)\,\delta_{J-1, J+1}}\theta(J+1, t) + \frac{k^k_{K-1, I}(t)}{S_{K-1, I}(t + \delta t)\,\delta_{K-1, K+1}}\theta(K-1, t)$$

$$+ \frac{k^k_{K+1, I}(t)}{S_{K+1, I}(t + \delta t)\,\delta_{K-1, K+1}}\theta(K+1, t) + \frac{\delta t \sum\limits_{s=1}^{N} h_{rs, I}(t)\theta(s, t)}{W_I(t + \delta t)\,\delta_{I, I+1}}$$

$$+ \frac{[q_{SI}(t) + q_{RI}(t) + q_{PI}(t)]\,\delta t}{W_I(t + \delta t)\,\delta_{I, I-1}\,\delta_{J-1, J+1}\,\delta_{K-1, K+1}} + \epsilon$$

where S is as defined in §3.2.1.

And the implicit formulation is

$$
\theta(I, t+\delta t) = \frac{W_I(t)}{W_I(t+\delta t)}\theta(I,t) - \theta(I, t+\delta t)\left(\frac{k_{I-1,I}^i(t+\delta t)}{S_{I-1,I}(t+\delta t)\delta_{I,I-1}} + \frac{h_{cI+1,I}(t+\delta t)\delta t}{W_I(t+\delta t)\delta_{I,I-1}}\right.
$$

$$
+ \frac{k_{J-1,I}^j(t+\delta t)}{S_{J-1,I}(t+\delta t)\delta_{J-1,J+1}} + \frac{k_{J+1,I}^j(t+\delta t)}{S_{J+1,I}(t+\delta t)\delta_{J-1,J+1}}
$$

$$
+ \frac{k_{K-1,I}^k(t+\delta t)}{S_{K-1,I}(t+\delta t)\delta_{K-1,K+1}} + \frac{k_{K+1,I}^k(t+\delta t)}{S_{K+1,I}(t+\delta t)\delta_{K-1,K+1}}
$$

$$
\left. + \frac{\delta t \sum_{s=1}^{N} h_{rs,I}(t+\delta t)}{W_I(t+\delta t)\delta_{I,I-1}}\right) + \frac{k_{I-1,I}^i(t+\delta t)}{S_{I-1,I}(t+\delta t)\delta_{I,I-1}}\theta(I-1, t+\delta t)
$$

$$
+ \frac{h_{cI+1,I}(t+\delta t)}{W_I(t+\delta t)\delta_{I,I-1}}\theta(I+1, t+\delta t) + \frac{k_{J-1,I}^j(t+\delta t)}{S_{J-1,I}(t+\delta t)\delta_{J-1,J+1}}\theta(J-1, t+\delta t)
$$

$$
+ \frac{k_{J+1,I}^j(t+\delta t)}{S_{J+1,I}(t+\delta t)\delta_{J-1,J+1}}\theta(J+1, t+\delta t)
$$

$$
+ \frac{k_{K-1,I}^k(t+\delta t)}{S_{K-1,I}(t+\delta t)\delta_{K-1,K+1}}\theta(K-1, t+\delta t)
$$

$$
+ \frac{k_{K+1,I}^k(t+\delta t)}{S_{K+1,I}(t+\delta t)\delta_{K-1,K+1}}\theta(K+1, t+\delta t)
$$

$$
+ \frac{\delta t \sum_{s=1}^{N} h_{rs,I}(t+\delta t)\theta(s, t+\delta t)}{W_I(t+\delta t)\delta_{I,I-1}}
$$

$$
+ \frac{[q_{SI}(t+\delta t) + q_{RI}(t+\delta t) + q_{PI}(t+\delta t)]\,\delta t}{W_I(t+\delta t)\delta_{I,I-1}\delta_{J-1,J+1}\delta_{K-1,K+1}} + \epsilon.
$$

An equal weighting is now performed and all unknown future time-row quantities grouped on the left-hand side. Future time-row quantities which are known—such as boundary condition solar injections and casual radiant energy—are moved to the right-hand side and therefore assumed known (or independently calculable) for all time.

$$
\left(2W_I(t+\delta t) + \frac{k_{I-1,I}^i(t+\delta t)\delta t}{\delta x_{I-1,I}\delta_{I,I-1}} + \frac{h_{cI+1,I}(t+\delta t)\delta t}{\delta_{I,I-1}} + \frac{k_{J-1,I}^j(t+\delta t)\delta t}{\delta x_{J-1,I}\,\delta_{J-1,J+1}}\right.
$$

$$
+ \frac{k_{J+1,I}^j(t+\delta t)\delta t}{\delta x_{J+1,I}\delta_{J-1,J+1}} + \frac{k_{K-1,I}^k(t+\delta t)\delta t}{\delta x_{K-1,I}\delta_{K-1,K+1}} + \frac{k_{K+1,I}^k(t+\delta t)\delta t}{\delta x_{K+1,I}\delta_{K-1,K+1}}
$$

$$
\left. + \frac{\delta t \sum_{s=1}^{N} h_{rs,I}(t+\delta t)}{\delta_{I,I-1}}\right)\theta(I, t+\delta t) - \frac{k_{I-1,I}^i(t+\delta t)\delta t}{\delta x_{I-1,I}\delta_{I,I-1}}\theta(I-1, t+\delta t)
$$

$$-\frac{h_{c\,I+1,I}(t+\delta t)\,\delta t}{\delta_{I,I-1}}\theta(I+1,t+\delta t)-\frac{k_{J-1,I}'(t+\delta t)\,\delta t}{\delta x_{J-1,I}\,\delta_{J-1,J+1}}\theta(J-1,t+\delta t)$$

$$-\frac{k_{J+1,I}'(t+\delta t)\,\delta t}{\delta x_{J+1,I}\,\delta_{J-1,J+1}}\theta(J+1,t+\delta t)-\frac{k_{K-1,I}'(t+\delta t)\,\delta t}{\delta x_{K-1,I}\,\delta_{K-1,K+1}}\theta(K-1,t+\delta t)$$

$$-\frac{k_{K+1,I}'(t+\delta t)\,\delta t}{\delta x_{K+1,I}\,\delta_{K-1,K+1}}\theta(K+1,t+\delta t)-\frac{\delta t\sum\limits_{s=1}^{N}h_{rs,I}(t+\delta t)\theta(s,t+\delta t)}{\delta_{I,I-1}}$$

$$-\frac{q_{PI}(t+\delta t)\,\delta t}{\delta_{I,I-1}\,\delta_{J-1,J+1}\,\delta_{K-1,K+1}}$$

$$=\left(2W_I(t)-\frac{k_{I-1,I}'(t)\,\delta t}{\delta x_{I-1,I}\,\delta_{I,I-1}}-\frac{h_{c\,I+1,I}(t)\,\delta t}{\delta_{I,I-1}}-\frac{k_{J-1,I}'(t)\,\delta t}{\delta x_{J-1,I}\,\delta_{J-1,J+1}}-\frac{k_{J+1,I}'(t)\,\delta t}{\delta x_{J+1,I}\,\delta_{J-1,J+1}}\right.$$

$$\left.-\frac{k_{K-1,I}'(t)\,\delta t}{\delta x_{K-1,I}\,\delta_{K-1,K+1}}-\frac{k_{K+1,I}'(t)\,\delta t}{\delta x_{K+1,I}\,\delta_{K-1,K+1}}-\frac{\delta t\sum\limits_{s=1}^{N}h_{rs,I}(t)}{\delta_{I,I-1}}\right)\theta(I,t)$$

$$+\frac{k_{I-1,I}'(t)\,\delta t}{\delta x_{I-1,I}\,\delta_{I,I-1}}\theta(I-1,t)+\frac{h_{c\,I+1,I}(t)\,\delta t}{\delta_{I,I-1}}\theta(I+1,t)+\frac{k_{J-1,I}'(t)\,\delta t}{\delta x_{J-1,I}\,\delta_{J-1,J+1}}\theta(J-1,t)$$

$$+\frac{k_{J+1,I}'(t)\,\delta t}{\delta x_{J+1,I}\,\delta_{J-1,J+1}}\theta(J+1,t)+\frac{k_{K-1,I}'(t)\,\delta t}{\delta x_{K-1,I}\,\delta_{K-1,K+1}}\theta(K-1,t)$$

$$+\frac{k_{K+1,I}'(t)\,\delta t}{\delta x_{K+1,I}\,\delta_{K-1,K+1}}\theta(K+1,t)+\frac{\delta t\sum\limits_{s=1}^{N}h_{rs,I}(t)\theta(s,t)}{\delta_{I,I-1}}$$

$$+\frac{[q_{PI}(t)+q_{SI}(t)+q_{RI}(t)]\,\delta t}{\delta_{I,I-1}\,\delta_{J-1,J+1}\,\delta_{K-1,K+1}}+\frac{[q_{SI}(t+\delta t)+q_{RI}(t+\delta t)]\,\delta t}{\delta_{I,I-1}\,\delta_{J-1,J+1}\,\delta_{K-1,K+1}}. \tag{3.10}$$

This simulation equation relates to nodes located at some exposed opaque or transparent surface experiencing heat exchange by conduction within the next-to-surface material, by convection with the adjacent fluid layer, by longwave radiation connections with surrounding surfaces and by absorption of shortwave flux. Chapter 5 discusses the theoretical considerations and algorithmic approach to the computation of surface heat generation terms and convection and longwave radiation coefficients at each time-row. The uni-directional formulation of equation (3.10) is given by:

$$\left(2W_I(t+\delta t)+\frac{k_{I-1,I}'(t+\delta t)\,\delta t}{\delta x_{I-1,I}\,\delta_{I,I-1}}+\frac{h_{c\,I+1,I}(t+\delta t)\,\delta t}{\delta_{I,I-1}}\right.$$

$$\left.+\frac{\delta t\sum\limits_{s=1}^{N}h_{rs,I}(t+\delta t)}{\delta_{I,I-1}}\right)\theta(I,t+\delta t)-\frac{k_{I-1,I}'(t+\delta t)\,\delta t}{\delta x_{I-1,I}\,\delta_{I,I-1}}\theta(I-1,t+\delta t)$$

$$-\frac{h_{cI+1,I}(t+\delta t)\,\delta t}{\delta_{I,I-1}}\theta(I+1,t+\delta t)-\frac{\delta t\sum_{s=1}^{N}h_{rs,I}(t+\delta t)\theta(s,t+\delta t)}{\delta_{I,I-1}}$$

$$=\left(2W_I(t)-\frac{k_{I-1,I}'(t)\,\delta t}{\delta x_{I-1,I}\delta_{I,I-1}}+\frac{h_{cI+1,I}(t)\,\delta t}{\delta_{I,I-1}}-\frac{\delta t\sum_{s=1}^{N}h_{rs,I}(t)}{\delta_{I,I-1}}\right)\theta(I,t)$$

$$+\frac{k_{I-1,I}'(t)\,\delta t}{\delta x_{I-1,I}\delta_{I,I-1}}\theta(I-1,t)+\frac{h_{cI+1,I}(t)\,\delta t}{\delta_{I,I-1}}\theta(I+1,t)+\frac{\delta t\sum_{s=1}^{N}h_{rs,I}(t)\theta(s,t)}{\delta_{I,I-1}}$$

$$+\frac{[q_{PI}(t)+q_{SI}(t)+q_{RI}(t)+q_{SI}(t+\delta t)+q_{RI}(t+\delta t)]\,\delta t}{\delta_{I,I-1}\delta_{J-1,J+1}\delta_{K-1,K+1}}. \tag{3.11}$$

Some special applications of equation (3.10) are now described.

Multi-layered construction surface nodes

For such an application, node I represents the exposed surface of the construction and node $I + 1$ represents the adjacent zone air, internally (same zone or adjacent zone) or externally located. If externally located, the $I + 1$ nodal term can be relocated on the equation right-hand side since it is an ambient boundary condition known for all time.

Ground contact nodes

For the special case of node I being located at the outermost surface of a floor slab in contact with the ground, node $I + 1$ represents some finite ground volume. This simulation equation can be handled by one of two mechanisms: the ground temperature can be treated as known with (say) only seasonal variation and the $I + 1$ nodal term relocated on the right-hand side as before or, alternatively, node $I + 1$ can be treated as an unknown and connected to a subterranean nodal network (using the formulations of §3.2.1) which links node $I + 1$ through to ambient conditions and to a depth where temperatures become relatively stable and substantially independent of ambient fluctuations (see figure 3.3).

Air gap boundary nodes

If node I is situated at the interface of an air gap (within a multi-layered construction) and some adjacent homogeneous (opaque or transparent) element then equation (3.10) can be used to represent the usual processes of conduction, convection and inter-surface longwave radiation exchange. One simplification can, however, be introduced without substantial loss of accuracy. By assuming a combined convective/radiative boundary layer given by

$$R = 1/(h_c + \epsilon h_r)$$

where ϵ is the surface emissivity, then the convection coefficient h_c of equation (3.10) can be replaced with a combined coefficient h_T given by

$$h_T = h_c + \epsilon h_r$$

and the longwave radiation terms removed. Such a trade-off between accuracy and flexibility has the useful effect of ensuring that the number of cross-coupling coefficients is the same as for the transient conduction nodes of §3.2.1: a useful feature which will reduce matrix processing requirements as explained in chapter 4.

3.2.3 Energy balance: fluid volumes

Figure 3.10 shows some node I located at the centroid of some finite fluid volume bounded by a collection of fictitious and/or real facets comprising the interfaces with adjacent fluid volumes, exposed surface layers (opaque or transparent), or other remote regions (ambient conditions, other plant components etc). The heat flow conductances of equation (3.1) will depend on the type of heat flow connection between region I and its thermal neighbours. Two possibilities exist: fluid-to-fluid exchange due to pressure, temperature and/or mechanical effects; and fluid-to-surface exchange due to convective effects.

For the case of fluid-to-fluid exchange the conductance is given by

$$K_{i,I}(\xi) = v_{i,I}(\xi)\bar{\varrho}_{i,I}(\xi)\bar{C}_{i,I}(\xi)$$

where $v_{i,I}$ is the inter-region volume flowrate relative to volume I (m^3 s^{-1}), and $\bar{\varrho}_{i,I}$, $\bar{C}_{i,I}$ are the density and specific heat of the fluid evaluated at the mean temperature of the two regions (kg m^{-3} and J kg^{-1} $^{\circ}$C^{-1}).

Note that the flowrate term is a vector quantity and therefore fluid leaving region I will not impose a load, only fluid entering the region being significant as it is raised or lowered to the region temperature.

For fluid-to-surface exchange the conductance is

$$K_{i,I}(\xi) = h_{ci,I}(\xi)\delta A_{i,I}$$

where $\delta A_{i,I}$ is the area normal to flowpath direction (m^2).

Evaluation of equation (3.1) at the present and future time-rows and combination of the formulations as before gives

$$\left(2W_I(t+\delta t) + \frac{\delta t \sum_{i=1}^{N} h_{ci,I}(t+\delta t)\delta A_{i,I}}{\delta V_I} + \frac{\delta t \sum_{j=1}^{M} v_{j,I}(t+\delta t)\bar{\varrho}_{j,I}(t+\delta t)\bar{C}_{j,I}(t+\delta t)}{\delta V_I}\right)$$

$$\times \theta(I,t+\delta t) - \frac{\delta t \sum_{i=1}^{N} h_{ci,I}(t+\delta t)\delta A_{i,I}\theta(i,t+\delta t)}{\delta V_I}$$

$$-\frac{\delta t \sum_{j=1}^{M} v_{j,I}(t+\delta t)\bar{\varrho}_{j,I}(t+\delta t)\bar{C}_{j,I}(t+\delta t)\theta(j,t+\delta t)}{\delta V_I} - \frac{q_I(t+\delta t)\delta t}{\delta V_I}$$

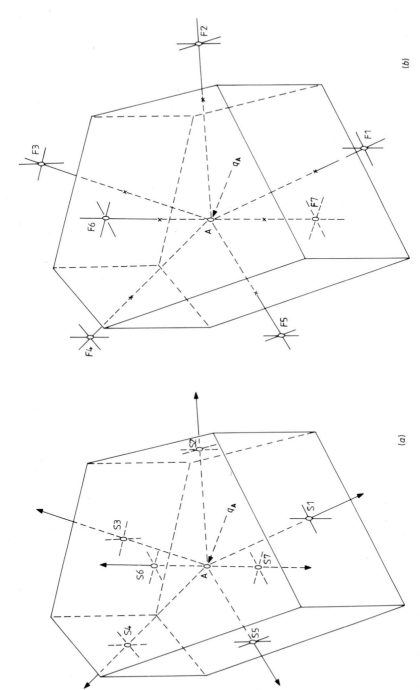

Figure 3.10 Fluid volume energy balance nodal scheme. (*a*) Fluid volume contained by real surfaces: convective heat transfer. (*b*) Fluid colume contained by fictitious surfaces: advective heat transfer.

$$= \left(2W_I(t) - \frac{\delta t \sum_{i=1}^{N} h_{ci,I}(t)\,\delta A_{i,I}}{\delta V_I} - \frac{\delta t \sum_{j=1}^{M} v_{j,I}(t)\bar{\varrho}_{j,I}(t)\bar{C}_{j,I}(t)}{\delta V_I} \right)\theta(I,t)$$

$$+ \frac{\delta t \sum_{i=1}^{N} h_{ci,I}(t)\,\delta A_{i,I}\theta(i,t)}{\delta V_I} + \frac{\delta t \sum_{j=1}^{M} v_{j,I}\bar{\varrho}_{j,I}(t)\bar{C}_{j,I}(t)\theta(j,t)}{\delta V_I}$$

$$+ \frac{q_I(t)\,\delta t}{\delta V_I} \tag{3.12}$$

where N is the number of flowpaths linking the fluid node I with boundary surface nodes, and M is the number of fluid flowstreams converging on node I.

Chapter 5 introduces the theoretical considerations underlying the convective and advective terms. Special cases of this equation are now considered.

Air gap nodes
Referring to figure 3.6 and for the case of node I located at the centre plane of an air gap with nodes $I-1$ and $I+1$ located at the boundaries, the heat flows are

$$q_{I-1,I} = h_{cI-1,I}(\delta_{I,J-1} + \delta_{I,J+1})(\delta_{I,K-1} + \delta_{I,K+1})(\theta_{I-1} - \theta_I)$$

$$q_{I+1,I} = h_{cI+1,I}(\delta_{I,J-1} + \delta_{I,J+1})(\delta_{I,K-1} + \delta_{I,K+1})(\theta_{I+1} - \theta_I)$$

$$q_{J-1,I} = v_{J-1,I}\bar{\varrho}_{J-1,I}\bar{C}_{J-1,I}(\theta_{J-1} - \theta_I)$$

$$q_{J+1,I} = v_{J+1,I}\bar{\varrho}_{J+1,I}\bar{C}_{J+1,I}(\theta_{J+1} - \theta_I)$$

$$q_{K-1,I} = v_{K-1,I}\bar{\varrho}_{K-1,I}\bar{C}_{K-1,I}(\theta_{K-1} - \theta_I)$$

$$q_{K+I,I} = v_{K+1,I}\bar{\varrho}_{K+1,I}\bar{C}_{K+1,I}(\theta_{K+1} - \theta_I)$$

and therefore equation (3.12) becomes

$$\left(2W_I(t+\delta t) + \frac{h_{cI-1,I}(t+\delta t)\,\delta t}{\delta_{I-1,I+1}} + \frac{h_{cI+1,I}(t+\delta t)\,\delta t}{\delta_{I-1,I+1}} \right.$$

$$\left. + \frac{\sum_i v_{i\pm1,I}(t+\delta t)\bar{\varrho}_{i\pm1,I}(t+\delta t)\bar{C}_{i\pm1,I}(t+\delta t)\,\delta t}{\delta_{I-1,I+1}\,\delta_{J-1,J+1}\,\delta_{K-1,K+1}} \right)\theta(I,t+\delta t)$$

$$- \frac{h_{cI-1,I}(t+\delta t)\,\delta t}{\delta_{I-1,I+1}}\theta(I-1,t+\delta t) - \frac{h_{cI+1,I}(t+\delta t)\,\delta t}{\delta_{I-1,I+1}}\theta(I+1,t+\delta t)$$

$$- \frac{\sum_i v_{i\pm1,I}(t+\delta t)\bar{\varrho}_{i\pm1,I}(t+\delta t)\bar{C}_{i\pm1,I}(t+\delta t)\,\delta t\theta(i\pm1,t+\delta t)}{\delta_{I-1,I+1}\,\delta_{J-1,J+1}\,\delta_{K-1,K+1}}$$

$$- \frac{q_I(t+\delta t)\,\delta t}{\delta_{I-1,I+1}\,\delta_{J-1,J+1}\,\delta_{K-1,K+1}}$$

$$= \left(2W_I(t) - \frac{h_{cI-1,I}(t)\,\delta t}{\delta_{I-1,I+1}} - \frac{h_{cI+1,I}(t)\,\delta t}{\delta_{I-1,I+1}} \right.$$

$$\left. - \frac{\sum_i v_{i\pm1,I}(t)\bar{\varrho}_{i\pm1,I}(t)\bar{C}_{i\pm1,I}(t)\,\delta t}{\delta_{I-1,I+1}\,\delta_{J-1,J+1}\,\delta_{K-1,K+1}} \right)\theta(I,t) + \frac{h_{cI-1,I}(t)\,\delta t}{\delta_{I-1,I+1}}\theta(I-1,t)$$

$$+ \frac{h_{cI+1,I}(t)\,\delta t}{\delta_{I-1,I+1}}\theta(I+1,t) + \frac{\sum_i v_{i\pm1,I}(t)\bar{\varrho}_{i\pm1,I}\bar{C}_{i\pm1,I}\,\delta t}{\delta_{I-1,I+1}\,\delta_{J-1,J+1}\,\delta_{K-1,K+1}}\theta(i\pm1,t)$$

$$+ \frac{q_I(t)\,\delta t}{\delta_{I-1,I+1}\,\delta_{J-1,J+1}\,\delta_{K-1,K+1}} \qquad i = J,K. \tag{3.13}$$

Note that any coupled nodes $(J-1, J+1, K-1, K+1)$ can be located external to the air gap and in this way time-dependent cavity ventilation can be introduced.

The uni-directional formulation of equation (3.13) with cavity ventilation, q_{VI}, is

$$\left(2W_I(t+\delta t) + \frac{h_{cI-1,I}(t+\delta t)\,\delta t}{\delta_{I-1,I+1}} + \frac{h_{cI+1,I}(t+\delta t)\,\delta t}{\delta_{I-1,I+1}} \right)\theta(I,t+\delta t)$$

$$- \frac{h_{cI-1,I}(t+\delta t)\,\delta t}{\delta_{I-1,I+1}}\theta(I-1,t+\delta t) - \frac{h_{cI+1,I}(t+\delta t)\,\delta t}{\delta_{I-1,I+1}}\theta(I+1,t+\delta t)$$

$$- \frac{q_I(t+\delta t)\,\delta t}{\delta_{I-1,I+1}}$$

$$= \left(2W_I(t) - \frac{h_{cI-1,I}(t)\,\delta t}{\delta_{I-1,I+1}} - \frac{h_{cI+1,I}(t)\,\delta t}{\delta_{I-1,I+1}} \right)\theta(I,t) + \frac{h_{cI-1,I}(t)\,\delta t}{\delta_{I-1,I+1}}\theta(I-1,t)$$

$$+ \frac{h_{cI+1,I}(t)\,\delta t}{\delta_{I-1,I+1}}\theta(I+1,t) + \frac{[q_I(t) + q_{VI}(t) + q_{VI}(t+\delta t)]\,\delta t}{\delta_{I-1,I+1}}. \tag{3.14}$$

Fluid flow in ducts and pipes
Assuming uni-directional flow as indicated in figure 3.11 then the heat flows are given by

$$q_{I-1,I} = v_{I-1,I}\bar{\varrho}_{I-1,I}\bar{C}_{I-1,I}(\theta_{I-1} - \theta_I)$$
$$q_{I+1,I} = v_{I+1,I}\bar{\varrho}_{I+1,I}\bar{C}_{I+1,I}(\theta_{I+1} - \theta_I)$$
$$q_{J-1,I} = h_{fcJ-1,I}A_{J-1}(\theta_{J-1} - \theta_I)$$
$$q_{J+1,I} = 0$$
$$q_{K-1,I} = 0$$
$$q_{K+1,I} = 0$$

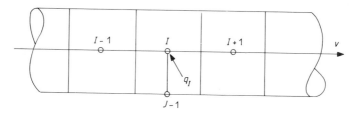

Figure 3.11 Simple nodal scheme for fluid flow in pipes and ducts.

where h_{fc} is the forced convection coefficient (W m^{-2} °C^{-1}) and A_{J-1} is the surface area of duct/pipe walls exposed to the fluid volume represented by node I (m^2).

Substitution in equation (3.12) gives

$$
\left(2W_I(t + \delta t) + \frac{h_{\text{fc}J-1,I}(t + \delta t)A_{J-1}\,\delta t}{\delta V_I} + \frac{v_{I-1,I}(t + \delta t)\bar{\varrho}_{I-1,I}(t + \delta t)\bar{C}_{I-1,I}(t + \delta t)\,\delta t}{\delta V_I} \right.
$$

$$
\left. + \frac{v_{I+1,I}(t + \delta t)\bar{\varrho}_{I+1,I}(t + \delta t)\bar{C}_{I+1,I}(t + \delta t)\,\delta t}{\delta V_I} \right) \theta(I, t + \delta t)
$$

$$
- h_{\text{fc}J-1,I}(t + \delta t)A_{J-1}\,\delta t\,\theta(J - 1, t + \delta t)
$$

$$
- \frac{v_{I-1,I}(t + \delta t)\bar{\varrho}_{I-1,I}(t + \delta t)\bar{C}_{I-1,I}(t + \delta t)\,\delta t}{\delta V_I}\theta(I - 1, t + \delta t)
$$

$$
- \frac{v_{I+1,I}(t + \delta t)\bar{\varrho}_{I+1,I}(t + \delta t)\bar{C}_{I+1,I}(t + \delta t)\,\delta t}{\delta V_I}\theta(I + 1, t + \delta t) - \frac{q_I(t + \delta t)\,\delta t}{\delta V_I}
$$

$$
= \left(2W_I(t) - \frac{h_{\text{fc}J-1,I}(t)A_{J-1}\,\delta t}{\delta V_I} - \frac{v_{I-1,I}(t)\bar{\varrho}_{I-1,I}(t)\bar{C}_{I-1,I}(t)\,\delta t}{\delta V_I} \right.
$$

$$
\left. - \frac{v_{I+1,I}(t)\bar{\varrho}_{I+1,I}(t)\bar{C}_{I+1,I}(t)\,\delta t}{\delta V_I} \right) \theta(I, t) + \frac{h_{\text{fc}J-1,I}(t)A_{J-1}\,\delta t\,\theta(J - 1, t)}{\delta V_I}
$$

$$
+ \frac{v_{I-1,I}(t)\varrho_{I-1,I}(t)C_{I-1,I}(t)\,\delta t}{\delta V_I}\theta(I - 1, t)
$$

$$
+ \frac{v_{I+1,I}(t)\varrho_{I+1,I}(t)C_{I+1,I}(t)\,\delta t}{\delta V_I}\theta(I + 1, t) + \frac{q_I(t)\,\delta t}{\delta V_I}. \tag{3.15}
$$

In use, equations of this form will have control considerations superimposed to dictate the value of v (for example) as a function of pump or fan status.

Isolated fluid volumes

In many applications it is necessary to establish volume subdivision in order to obtain information about temperature distribution. In this case many of the discrete fluid volumes will have coupling only to other volumes with no direct flowpath to opaque or transparent surfaces. For such a case equation (3.12) reduces to

$$\left(2W_I(t+\delta t) + \frac{\delta t \sum_{j=1}^{M} v_{j,I}(t+\delta t)\bar{\varrho}_{j,I}(t+\delta t)\bar{C}_{j,I}(t+\delta t)}{\delta V_I}\right)\theta(I,t+\delta t)$$

$$-\frac{\delta t \sum_{j=1}^{M} v_{j,I}(t+\delta t)\bar{\varrho}_{j,I}(t+\delta t)\bar{C}_{j,I}(t+\delta t)\theta(j,t+\delta t)}{\delta V_I} - \frac{q_I(t+\delta t)}{\delta V_I}$$

$$= \left(2W_I(t) - \frac{\delta t \sum_{j=1}^{M} v_{j,I}(t)\bar{\varrho}_{j,I}(t)\bar{C}_{j,I}(t)}{\delta V_I}\right)\theta(I,t)$$

$$+ \frac{\delta t \sum_{j=1}^{M} v_{j,I}(t)\bar{\varrho}_{j,I}(t)\bar{C}_{j,I}(t)\theta(j,t)}{\delta V_I} + \frac{q_I(t)\delta t}{\delta V_I}. \tag{3.16}$$

It is possible to establish a mesh of node points, in two or three dimensions, in such a way that, by repeated application of equation (3.16) and the more general form of equation (3.12), the energy flowpaths within the fluid volumes comprising a zone can be interconnected to define all flowpaths up to the internal surface boundary of the zone. The corresponding equations of §§3.2.1 and 3.2.2 are then employed to extend the equation structure to some defined boundary condition such as external climatic.

3.3 Structuring the equations for multi-component simulation

Utilising the formulations of §3.2 it is now possible to formulate a mathematical model of any building/plant system. The process to be followed will include: devising a suitable discretisation scheme for the multi-component system; gathering together the simulation equation set which constitutes the mathematical equivalent; and arranging for solution at each chosen time-step as the simulation proceeds.

Chapter 4 describes this process in detail when applied to multi-zone building problems with a variety of controller and plant characteristics and for variable time-stepping. The purpose of this section is to demonstrate the potential of the method to preserve the simultaneity of multiple component systems without offering, at this stage, any specific method for the solution of the system matrix equation to emerge.

The technique of the examples which follow is to show the equivalent resistance/capacitance network for each configuration cited and to give the corresponding equation structure. This process is made more explicit in chapter 5 where an equation set is established, step-by-step, and solved in some detail. Hence no attempt is made here to show the form of the individual coefficient entries in the system matrix. Instead only their relative position is indicated to demonstrate matrix sparseness and the nature of inter-component linkages. Section 4.1 demonstrates coefficient set-up and placement as matrix elements and later sections in chapter 4 develop customised solution techniques.

Single zone

Figure 3.12 shows a simple multi-component system comprising 6 multi-layered constructions, 6 surface layers and 2 fluid volumes, all combined to form a single building zone. A rudimentary nodal scheme is imposed in which only uni-directional transient conduction is permitted with the enclosed air volume being

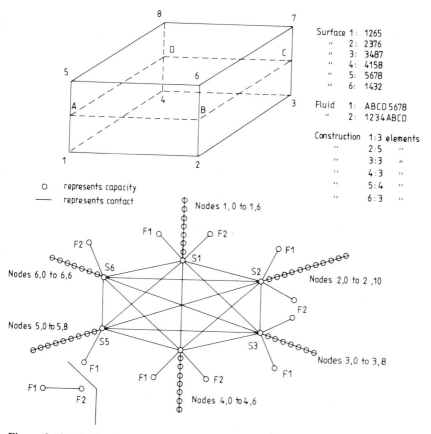

Figure 3.12 A simple multi-component (single-zone) system and equivalent nodal scheme.

divided into a lower and upper portion. In this case the entire system is represented by 50 nodes and so there will be 50 simultaneous equations each having a number of cross- and self-coupling terms evaluated at the present and future time-rows of some current computational time-step as a simulation progresses.

Figure 3.13 gives the form of the corresponding equation set which, in matrix notation, can be expressed as

$$\mathbf{A}\theta_{n+1} = \mathbf{B}\theta_n + \mathbf{C}. \qquad (3.17)$$

The coefficient matrix **A** is not square, since coefficients relating to future time-row boundary nodes are retained on the left-hand side (future time-row) of each equation even though, in any given problem, their numerical values are known. This is necessary to allow matrix interlocking when other components or building

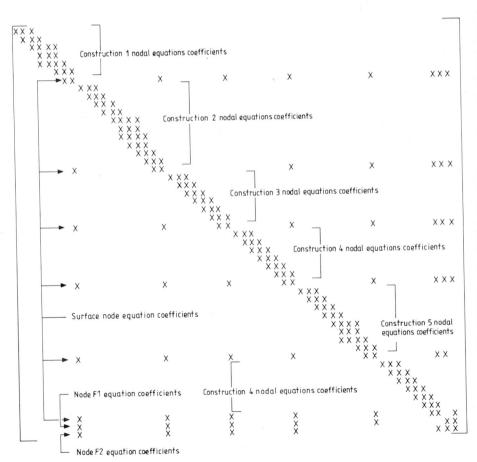

Figure 3.13 The future time-row coefficients matrix (**A**) of the single-zone matrix equation $\mathbf{A}\theta_{n+1} = \mathbf{B}\theta_n + \mathbf{C}$.

zones are added, effectively to remove the problem boundary elsewhere. Even when boundary condition terms are removed to the equation left-hand side, the **A** matrix will remain non-square for all systems in which some nodal heat injection/extraction (due to plant interaction for example) is to be predicted. This is important, since it allows nodal heat requirements to be determined as a function of nodal temperature statements, or, conversely, the single coefficient entry relating to this heat interaction term can be replaced by a whole plant matrix properly interlocked with the various building zones. The descriptive examples which follow are elaborated in chapter 6 where plant matrix interlocking is discussed at length. The solution of equation (3.17) can then be achieved by the inclusion of negative feedback (or feedforward) control as demonstrated in chapter 4.

The two-dimensional arrays **A** and **B** have the same number of rows as system equations and any existing element a_{ij} or b_{ij} ($i \neq j + 1$) is a coefficient which links two nodal regions at the future (a_{ij}) and present (b_{ij}) time-row of the time-step for which the matrix was established. Any non-existent a_{ij} or b_{ij} merely indicates that no coupling exists between the nodal regions in question. Elements a_{ii+1} and b_{ii+1} relate to self-coupling coefficients which generally represent the storage potential of the region represented by equation i.

The column matrices θ_{n+1} and θ_n contain the nodal temperature terms and heat injection/extractions at the future and present time-rows respectively. The column matrix **C** contains the known boundary condition excitations due to temperature and heat flux fluctuations which act on selected nodes to cause energy flow and so 'drive' a simulation.

Since all terms on the left-hand side of equation (3.17) relate either to the known present time-row (**B**, θ_n) or are known boundary terms (**C**) it is appropriate to generate the column matrix **Z** where

$$\mathbf{Z} = \mathbf{B}\theta_n + \mathbf{C}. \tag{3.18}$$

This allows equation (3.17) to be re-expressed as

$$\mathbf{A}\theta_{n+1} = \mathbf{Z}$$

and the solution which follows by

$$\theta_{n+1} = \mathbf{A}^{-1}\mathbf{Z}. \tag{3.19}$$

Various techniques exist to achieve this solution as discussed in §2.4.3 and chapter 4.

Solution of the matrix formulation of figure 3.13 allows accurate assessment of energy requirements, indicative surface temperatures, vertical temperature gradients and would provide detailed causal energy breakdowns listing the effects of infiltration, ventilation, shortwave and longwave radiation, casual gains and so on. It would not be suitable for the study of localised convective phenomena, corner effects, thermal bridging or the assessment of plant performance to predict energy consumption. For this purpose the nodal network must be extended.

Single zone with contents and room-side appliances

Figure 3.14 shows the single zone of figure 3.12, but with contents and heating system radiators included. In this case the uni-directional construction model is retained, but a two-dimensional radiator model is added to represent the water and metal volumes at the top, middle and bottom of the radiator. Contents are represented by a single node and, in this model, no radiative coupling is allowed between the contents and radiator surface nodes, and the zone surface nodes. The equivalent equation structure is shown in figure 3.15. As before, time-step solution is given by equation (3.19).

This formulation will allow assessment of energy requirements, surface temperatures and causal energy flows as with the previous single zone model. In addition, radiator wall and water temperature distribution can be studied and related to air temperature evolution close to the radiator. For this equation structure, boundary conditions consist of time-series temperatures and heat fluxes at the outermost surface of each facet of the zone and radiator inlet conditions. It is therefore possible to predict zone energy requirements—for any desired environmental condition—as a function of radiator operational characteristics although it is still not possible, with this model, to convert to consumption values.

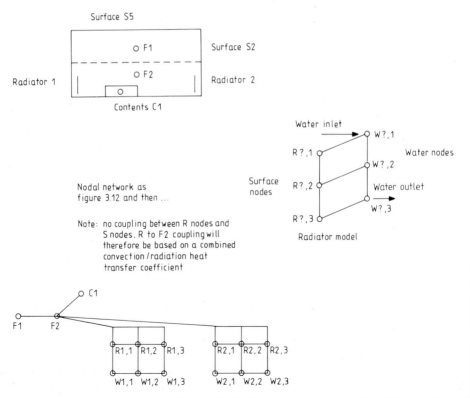

Figure 3.14   The single zone of figure 3.12 with room contents and radiators added.

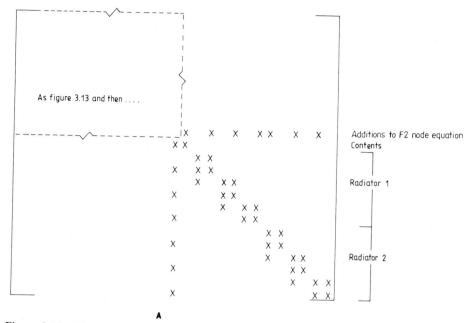

Figure 3.15   The coefficients matrix of figure 3.13 with zone contents and radiators added.

Note that radiator control can be applied as a function of any single or multiple nodal status test, depending on the anticipated controller characteristics. For example, in the case of a thermostatic valve inlet, flowrate can be throttled as a function of the nodal temperature of the air volume within which the sensor is located, perhaps augmented by some radiant component evaluated on the basis of a solid angle weighting for each visible zone surface as subtended at the valve.

Multi-zone with central plant

Figure 3.16 shows a number of single zone primitives combined to form a small multi-zone system with a central boiler and distributed radiator system superimposed. Again transient conduction is predominantly uni-directional (for clarity) but in one zone a multi-directional scheme has been incorporated to facilitate the study (say) of corner effects and thermal bridges. A simple nodal scheme is used to represent distribution losses with a more detailed scheme applied to the various boiler sections (this is discussed more fully in chapter 6).

Figure 3.17 gives the equivalent equation system. This model will allow, in addition to the analysis potential of the previous examples, a detailed investigation of the many energy and comfort related aspects of building performance: such as building and plant zoning strategies, global versus zone plant control, whole building energy requirements and consumption, boiler efficiency studies, distribution losses, priority and optimum start systems, and so on, to name but a few.

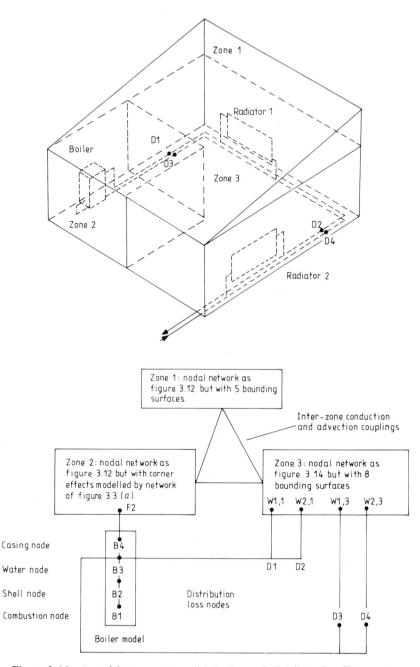

Figure 3.16   A multi-zone system with boiler and distributed radiator system.

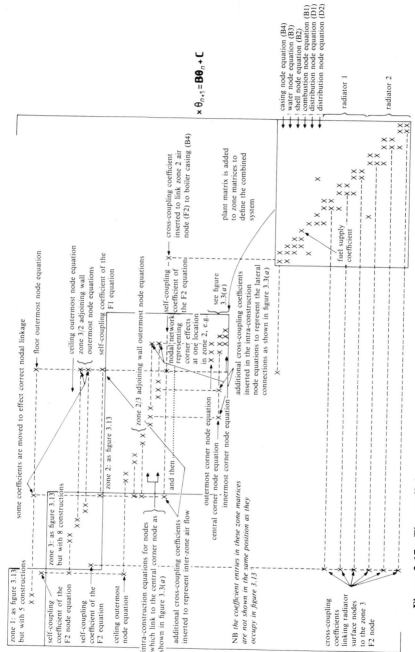

Figure 3.17 The coefficients matrix of the energy balance matrix equation for the system of figure 3.16.

Active/passive solar system with remote storage

Figure 3.18 shows a simple 7-zone active/passive solar house in which zones 1, 2, 3 and 4 are the bedroom and living zones, zone 5 is an attic space, zone 6 is an active solar collector (or an attached sunspace with a Trombe wall feature) and zone 7 is a remote rockbed thermal store. A detailed nodal scheme has been introduced to the collector (or sunspace) to facilitate accurate modelling of the asymmetric heating likely to be experienced due to solar pick-up at the collector back plate (or Trombe wall), and to permit realistic inlet and outlet air couplings between the collector (or sunspace) and the occupied zones or rockbed. This will allow the transfer of actively or passively collected energy when available. Such a nodal scheme will also accommodate the variation of surface temperature over the collector backplate (or Trombe wall) in the direction of the operating fluid flow.

Figure 3.19 gives the corresponding matrix structure, which can be subjected to any number of simple or complex control statements governing active or passive energy collection, utilisation or storage, and solved simultaneously at each time-step with the computed future time-row nodal values becoming the basis for matrix reformulation at the next time-step.

This model allows the optimisation of the elements of active and passive solar collection and utilisation as well as the study of comfort conditions in terms of temperature gradients and air movement.

Many other configurations can be addressed. Indeed any energy transport system can be treated in this manner and it is possible to build computer-based models which allow a user to interactively construct any system from primitive components (walls, windows, pipes, boilers, chillers, etc) already established in multi-nodal form and held in some database. Unfortunately at the present time, with contemporary programming languages such as FORTRAN, a user-friendly system of this kind would undoubtedly require in excess of 250 000 lines of code and for this reason many program developers will choose to customise code to particular systems of immediate interest. This is one of the major advantages of a numerical simulation approach like the one mooted here: it is not difficult to establish a system matrix for any defined system, of any complexity, and then develop the software counterpart to any required level of sophistication.

In the case of components that exhibit non-linear behaviour, the technique can still be applied by linearising, in the first instance, by re-expressing the non-linear temperature term θ^n as $\theta^m\theta$. The θ^m term can then be treated as part of the coefficient of θ and determined at the previous time-step if an 'approximate' solution is acceptable or the technique of chapter 4 invoked for exact solutions.

Chapter 4 now continues discussion of matrix processing to achieve efficient solution of the generated equation sets, chapter 5 introduces the theory appropriate to the many building subsystems such as air flow, solar and natural and forced convection, and chapter 6 covers the application of the numerical method to the modelling of plant systems.

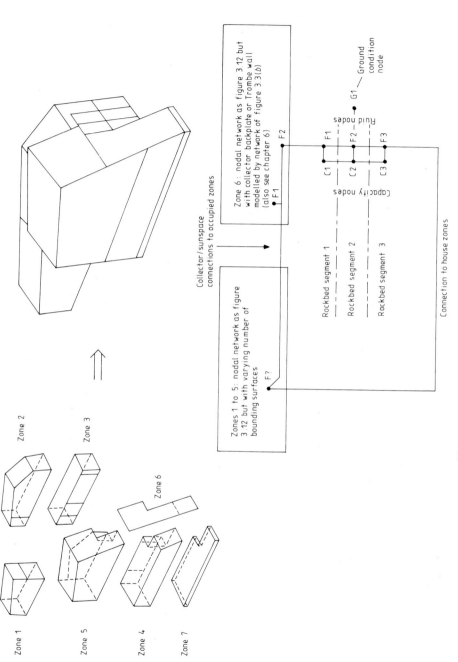

Zone 2

Zone 3

Zone 6

Zone 1

Zone 5

Zone 4

Zone 7

Collector/sunspace
connections to occupied zones

Zone 6: nodal network as figure 3.12 but
with collector backplate or Trombe wall
modelled by network of figure 3.3(b)
(also see chapter 6)

F1

F2

Zones 1 to 5: nodal network as figure
3.12 but with varying number of
bounding surfaces

F?

Rockbed segment 1

Rockbed segment 2

Rockbed segment 3

Capacity nodes

C1

C2

C3

Fluid nodes

F1

F2

F3

G1 — Ground
condition
node

Connection to house zones

Figure 3.18   An active/passive solar system.

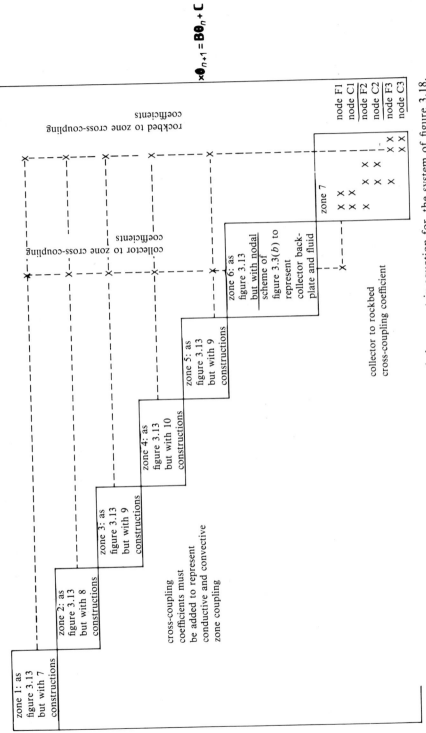

$$\mathbf{x}\boldsymbol{\theta}_{n+1} = \mathbf{B}\boldsymbol{\theta}_n + \mathbf{C}$$

rockbed to zone cross-coupling coefficients

collector to zone cross-coupling coefficients

node F1
node C1
node F2
node C2
node F3
node C3

zone 7

zone 6: as figure 3.13 but with nodal scheme of figure 3.3(b) to represent collector back-plate and fluid

zone 5: as figure 3.13 but with 9 constructions

zone 4: as figure 3.13 but with 10 constructions

zone 3: as figure 3.13 but with 9 constructions

zone 2: as figure 3.13 but with 8 constructions

zone 1: as figure 3.13 but with 7 constructions

cross-coupling coefficients must be added to represent conductive and convective zone coupling

collector to rockbed cross-coupling coefficient

Figure 3.19 The coefficients matrix of the energy balance matrix equation for the system of figure 3.18.

References

Clarke J A 1978 The Effects of Space and Time Steps in Finite Difference Computations of Building Energy Flow *ABACUS Occasional Paper* (Glasgow: University of Strathclyde)
Mitchell A R 1969 *Computational Methods in Partial Differential Equations* (New York: Wiley)

Bibliography

Dusinberre G M 1961 *Heat Transfer Calculations by Finite Differences* (International Textbook Co)
Hildebrand F B 1968 *Finite Difference Equations and Simulations* (New Jersey: Prentice-Hall)
Lambert J D 1973 *Computational Methods in Ordinary Differential Equations* (New York: Wiley)
Levy H and Lessman F 1959 *Finite Difference Equations* (London: Pitman)
John F 1982 *Partial Differential Equations* (New York: Springer-Verlag)
Mitchell A R 1969 *Computational Methods in Partial Differential Equations* (New York: Wiley)
Richtmyer R D 1957 *Difference Methods for Initial-Value Problems* (New York: Interscience)

4

Matrix processing of energy flow equation sets

Chapter 3 detailed the derivation of the energy balance simulation equations, one for each characteristic node type, which represent the flow of heat energy between all communicating regions in relation to intra-region storage. Thus the linear algebraic equation set to emerge approximates the energy flow within the corresponding distributed system over some finite period of time. In matrix notation the system of equations can be expressed as

$$\mathbf{A}\theta_{n+1} = \mathbf{B}\theta_n + \mathbf{C} = \mathbf{Z} \qquad (4.1)$$

where $\mathbf{A}$ is a non-homogeneous sparse matrix of future time-row coefficients of the nodal temperature or heat injection terms of the simulation equations, $\mathbf{B}$ is the corresponding matrix established at the present time-row, $\mathbf{C}$ is a column matrix of known boundary excitations relating to both the present and future time-rows, θ is a column matrix (or vector) of nodal temperatures and heat injections, $n + 1$ refers to the future time-row, n the present time-row, and $\mathbf{Z}$ is a column matrix $= \mathbf{B}\theta_n + \mathbf{C}$. Initial conditions are given by $\theta(0) = \theta_0$.

Because the equations contain both present and future time-row terms, they must be solved simultaneously, and, because the entire system is sparse and populated by clusters of equations (representing various components) of differing 'stiffness', any computer-based solution technique must be flexible and efficient in operation. As described in §2.4.3 there are two main solution techniques: iterative and direct. In general terms, direct elimination methods require more storage space than do iterative methods and, especially in the case of standard direct inversion packages, they can become extremely inappropriate in the case of sparse matrix systems (which tend to be the rule in energy modelling applications). Conversely, and paradoxically perhaps, direct methods hold much promise for the efficient solution of sparse simultaneous equation systems representing discrete building and plant components demanding fundamentally different processing frequencies. Obviously control decisions must be made, and thermophysical properties recomputed anew, more frequently for an item of plant such as a zone-side radiator (requiring a processing frequency of, say, 15 minutes) than for a heavyweight multi-layered construction (requiring, say, 60 minutes). This efficient solution is achieved by ensuring that only the actual physical scheme is addressed by partitioning the overall (and sparse) system matrix into a number of discrete sub-matrices so that solution can be achieved in the lowest number of computational steps. Each partitioned matrix can then be processed as far as possible by a direct method, and at any frequency, to gather inter-component information to permit the global solution stream to continue. Thus, if circumstances allow, any sub-matrix need not be reprocessed until its contents change by an appreciable amount. Taken to an extreme, where no sub-matrix is reprocessed after the first time-step, the model will assume the characteristic operation of a response factor method as described in chapter 2.

This chapter is concerned with the derivation of such a direct solution technique suitable for the efficient processing of sparse linear simultaneous equation sets. Initial sections establish, in detail, a multi-zone matrix and discuss the problems inherent in establishing time-dependent matrix coefficients at the future time-row. Later sections introduce matrix partitioning and direct reduction methods suitable for approximate (but often very accurate) solutions, and mixed direct/iterative methods for use where exact solutions are sought, where non-linear processes are active or where the solution stream is to be determined on the basis of mixed node control statement tests to accommodate controller location and response aspects.

Section 2.4.3 gives the salient features of the commonly used solution methods; this, and the referenced material, is essential reading for those uninitiated in matrix algebra.

4.1 Establishing the equation set for solution

The first step is to establish the interrelated equation set, which represents the discrete nodal network, by gathering together the simulation equations, one for each node, in some ordered manner subject to a linking protocol. Equation sets can be established for any distributed system consisting of single or multiple zones with or without contents and plant interaction. However, as discussed in §4.1.4, the treatment of future time-row temperature-dependent quantities will require special care.

4.1.1 Single zone formulation

Consider again figure 3.12 and assume the constructional details given in table 4.1. Nodes 1,0 through 6,0 are boundary conditions and as such will not have an associated equation: for each multi-layered construction, node ?,1 (where ? is the construction number) is the starting point. Node 1,1 is an outside surface node as represented by equation (3.11). This equation is now located within the system

Table 4.1 Constructional details for association with the single zone example of figure 3.12.

Construction	Element †	Thermophysical properties subscript
1: External window	Glass	11
	Air gap	12
	Glass	13
2: External wall	Rendering	21
	Brick	22
	Cavity (ventilated)	23
	Brick	24
	Light plaster	25
3: Internal wall	Light plaster	31
	Brick	32
	Light plaster	33
4: Internal wall	Light plaster	41
	Brick	42
	Light plaster	43
5: External roof	Concrete tile	51
	Sarking	52
	Air gap	53
	Insulation	54
6: Ground floor	Dense concrete	61
	Concrete screed	62
	Plastic tile	63
Fluid volume 1		1
Fluid volume 2		2

k = conductivity; ϱ = density; C = specific heat; δx = normal direction element thickness.
†All constructions are specified from the outermost to the innermost element relative to the contained volume.

matrix of figure 4.1 as the **A** matrix row 1; **B** matrix row 1 equation (hereafter referred to as the 'row n' equation), where

$$-a_{1,1}\theta_{1,0}(t + \delta t) + a_{1,2}\theta_{1,1}(t + \delta t) - a_{1,3}\theta_{1,2}(t + \delta t)$$
$$= b_{1,1}\theta_{1,0}(t) + b_{1,2}\theta_{1,1}(t) + b_{1,3}\theta_{1,2}(t) + c_1$$
$$= z_1$$

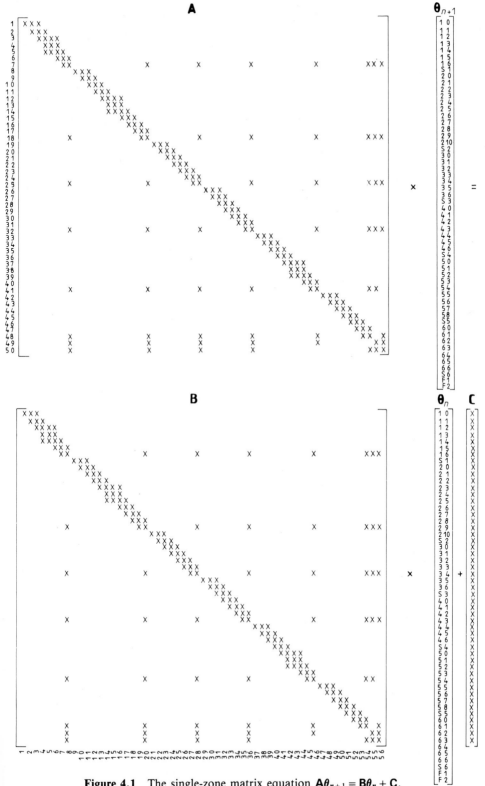

Figure 4.1   The single-zone matrix equation $\mathbf{A}\theta_{n+1} = \mathbf{B}\theta_n + \mathbf{C}$.

where

$$a_{1,1} = \frac{8k_{11}(t + \delta t)\,\delta t}{\delta x_{11}^2} \qquad\qquad b_{1,1} = a_{1,1}(t)$$

$$a_{1,2} = 2\varrho_{11}(t + \delta t)C_{11}(t + \delta t) + \frac{8k_{11}(t + \delta t)\,\delta t}{\delta x_{11}^2} + \frac{4h_c(t + \delta t)\,\delta t}{\delta x_{11}} + \frac{4h_{rs}(t + \delta t)\,\delta t}{\delta x_{11}}$$

$$+ \frac{4h_{rb}(t + \delta t)\,\delta t}{\delta x_{11}} + \frac{4h_{rg}(t + \delta t)\,\delta t}{\delta x_{11}}$$

$$b_{1,2} = -a_{1,2}(t) + 4\varrho_{11}(t)C_{11}(t)$$

$$a_{1,3} = \frac{4h_c(t + \delta t)\,\delta t}{\delta x_{11}} \qquad\qquad b_{1,3} = a_{1,3}(t)$$

$$c_1 = \{4\delta t\,[\,h_{rs}(t + \delta t)\theta_s(t + \delta t) + h_{rb}(t + \delta t)\theta_b(t + \delta t) + h_{rg}(t + \delta t)\theta_g(t + \delta t)$$

$$+ h_{rs}(t)\theta_s(t) + h_{rb}(t)\theta_b(t) + h_{rg}(t)\theta_g(t) + q'_{SI}(t + \delta t) + q'_{SI}(t)]\,\}/\delta x_{11}.$$

Here, q'_{SI} is the shortwave radiation absorption per unit area ($W\,m^{-2}$), h_{rs} is the longwave radiation coefficient for surface/sky, h_{rb} the coefficient for surface/surroundings and h_{rg} the coefficient for surface/ground (all $W\,m^{-2}\,^{\circ}C^{-1}$), θ_s is the sky temperature, θ_b the surroundings temperature and θ_g the ground temperature (all $^{\circ}C$). See §5.4.3 for a discussion of sky, surroundings and ground temperature evaluation.

It is important to note that no plant interaction is allowed for this node and so $q_p = 0$ at both time-rows. Thus row 1 of the **A** matrix will have 3 coefficient entries occupying the first 3 column positions.

Node 1,2 is a homogeneous conduction node of the transparent type as represented by equation (3.6) and can be located in row 2 of figure 4.1 in such a manner that

$$-a_{2,2}\theta_{1,1}(t + \delta t) + a_{2,3}\theta_{1,2}(t + \delta t) - a_{2,4}\theta_{1,3}(t + \delta t)$$

$$= b_{2,2}\theta_{1,1}(t) + b_{2,3}\theta_{1,2}(t) + b_{2,4}\theta_{1,3}(t) + c_2$$

$$= z_2$$

where

$$a_{2,2} = \frac{4k_{11}(t + \delta t)\,\delta t}{\delta x_{11}^2} \qquad\qquad b_{2,2} = a_{2,2}(t)$$

$$a_{2,3} = 2\varrho_{11}(t + \delta t)C_{11}(t + \delta t) + \frac{8k_{11}(t + \delta t)\delta t}{\delta x_{11}^2} \qquad b_{2,3} = -a_{2,3}(t) + 4\varrho_{11}(t)C_{11}(t)$$

$$a_{2,4} = \frac{4k_{11}(t + \delta t)\,\delta t}{\delta x_{11}^2} \qquad\qquad b_{2,4} = a_{2,4}(t)$$

$$c_2 = \frac{2\,\delta t\,[\,q'_{SI}(t + \delta t) + q'_{SI}(t + \delta t)]}{\delta x_{11}}.$$

As with node 1,1 no plant heat generation is allowed but, since the element is transparent, shortwave solar absorption will occur and so $q_r = q_{SI}$ which is a known quantity at both time-rows (see §5.3) and so the future time-row expression can be relocated on the equation right-hand side.

Row 2 of the **A** matrix also has 3 coefficients but in this case there is a column offset to preserve spatial relationships.

Node 1,3 is (say) located at the interface between the outer glass element and the inter-element air gap and is represented by equation (3.10) so that

$$-a_{3,3}\theta_{1,2}(t+\delta t) + a_{3,4}\theta_{1,3}(t+\delta t) - a_{3,5}\theta_{1,4}(t+\delta t) - a_{3,6}\theta_{1,5}(t+\delta t)$$
$$= b_{3,3}\theta_{1,2}(t) + b_{3,4}\theta_{1,3}(t) + b_{3,5}\theta_{1,4}(t) + b_{3,6}\theta_{1,5}(t) + c_3 = z_3$$

where

$$a_{3,3} = \frac{8k_{11}(t+\delta t)\delta t}{\delta x_{11}^2} \qquad b_{3,3} = a_{3,3}(t)$$

$$a_{3,4} = 2\varrho_{11}(t+\delta t)C_{11}(t+\delta t) + \frac{8k_{11}(t+\delta t)\delta t}{\delta x_{11}^2} + \frac{4h_c(t+\delta t)\delta t}{\delta x_{11}} + \frac{4h_{rA}(t+\delta t)\delta t}{\delta x_{11}}$$

$$b_{3,4} = -a_{3,4}(t) + 4\varrho_{11}(t)C_{11}(t)$$

$$a_{3,5} = \frac{4h_c(t+\delta t)\delta t}{\delta x_{11}} \qquad b_{3,5} = a_{3,5}(t)$$

$$a_{3,6} = \frac{4h_{rA}(t+\delta t)\delta t}{\delta x_{11}} \qquad b_{3,6} = a_{3,6}(t)$$

$$c_3 = \{4\delta t [q'_{SI}(t+\delta t) + q'_{SI}(t)] \}/\delta x_{11}.$$

h_{rA} is the longwave radiation coefficient for cavity inter-surface exchange (see §5.4).

Again shortwave solar absorption is possible within the region represented by $\theta_{1,3}$. Note that an additional coefficient is introduced to link nodes 1,3 and 1,5 to represent longwave radiation exchange.

Node 1,4 is located at the centre plane of the cavity and so is represented by equation (3.14) so that

$$-a_{4,4}\theta_{1,3}(t+\delta t) + a_{4,5}\theta_{1,4}(t+\delta t) - a_{4,6}\theta_{1,5}(t+\delta t)$$
$$= b_{4,4}\theta_{1,3}(t) + b_{4,5}\theta_{1,4}(t) + b_{4,6}\theta_{1,5}(t) + c_4$$
$$= z_4$$

where

$$a_{4,4} = \frac{2h_{co}(t+\delta t)\delta t}{\delta x_{12}} \qquad b_{4,4} = a_{4,4}(t)$$

$$a_{4,5} = 2\varrho_{12}(t+\delta t)C_{12}(t+\delta t) + \frac{2h_{co}(t+\delta t)\delta t}{\delta x_{12}} + \frac{2h_{ci}(t+\delta t)\delta t}{\delta x_{12}}$$

$$+ \frac{2v_o(t + \delta t)\bar{\varrho}_A(t + \delta t)\bar{C}_A(t + \delta t)\,\delta t}{\delta x_{12}A_c}$$

$$b_{4,5} = -a_{4,5}(t) + 4\varrho_{12}(t)C_{12}(t)$$

$$a_{4,6} = \frac{2h_{ci}(t + \delta t)\,\delta t}{\delta x_{12}} \qquad b_{4,6} = a_{4,6}(t)$$

$$c_4 = \frac{\{2\delta t\,[\,v_o(t)\bar{\varrho}_A(t)\bar{C}_A(t)\theta_o(t) + v_o(t + \delta t)\bar{\varrho}_A(t + \delta t)\bar{C}_A(t + \delta t)\theta_o(t + \delta t)]\,\}}{\delta x_{12}A_c}$$

Here, h_{co} is the convection coefficient for the outermost cavity boundary (W m^{-2} $^\circ$C^{-1}), h_{ci} is the convection coefficient for the innermost cavity, v_o is the cavity ventilation rate (m^3 s^{-1}) caused by the ingress of outside air.

Note that cavity air is assumed to be non-absorbing to shortwave radiation and so the heat generation term $q_I = 0$. Also, the cavity ventilation rate is assumed known at the future time-row and so is moved to the equation right-hand side. During simulation the quantity v_o is computed in tandem using the available temperature data and by the theory (say) of §5.6. It is appropriate to set $v_o = 0$, for all time, in the case of a non-ventilated cavity system.

Node 1,5 is primarily a repeat of node 1,3 but with the fourth nodal coefficient handling cavity longwave exchange by cross-linking to node 1,3. The row 5 equation therefore utilises matrix elements $a_{5,4}$, $a_{5,5}$, $a_{5,6}$, $a_{5,7}$, $b_{5,4}$, $b_{5,5}$, $b_{5,6}$, $b_{5,7}$ and c_5 where, as before, the c_5 term will contain shortwave heat generation because the multi-layered system is transparent.

Node 1,6 is a repeat of node 1,2 utilising matrix elements $a_{6,6}$, $a_{6,7}$, $a_{6,8}$, $b_{6,6}$, $b_{6,7}$, $b_{6,8}$ and c_6.

The nodal progression has now arrived at the inside surface node, S1, associated with the current multi-layered construction. The relevant simulation equation is given by equation (3.11) so that

$$- a_{7,7}\theta_{1,6}(t + \delta t) + a_{7,8}\theta_{S1}(t + \delta t) - a_{7,f}\theta_f(t + \delta t) - a_{7,20}\theta_{S2}(t + \delta t)$$

$$- a_{7,28}\theta_{S3}(t + \delta t) - a_{7,36}\theta_{S4}(t + \delta t) - a_{7,46}\theta_{S5}(t + \delta t) - a_{7,54}\theta_{S6}(t + \delta t)$$

$$= b_{7,7}\theta_{1,6}(t) + b_{7,8}\theta_{S1}(t) + b_{7,f}\theta_f(t) + b_{7,20}\theta_{S2}(t) + b_{7,28}\theta_{S3}(t) + b_{7,36}\theta_{S4}(t)$$

$$+ b_{7,46}\theta_{S5}(t) + b_{7,54}\theta_{S6}(t) + c_7 = z_7$$

where

$$a_{7,7} = \frac{8k_{13}(t + \delta t)\,\delta t}{\delta x_{13}^2} \qquad b_{7,7} = a_{7,7}(t)$$

$$a_{7,8} = 2\varrho_{13}(t + \delta t)C_{13}(t + \delta t) + \frac{8k_{13}(t + \delta t)\,\delta t}{\delta x_{13}^2} + \frac{4h_c(t + \delta t)\,\delta t}{\delta x_{13}} + \frac{4h_{r2}(t + \delta t)\,\delta t}{\delta x_{13}}$$

$$+ \frac{4h_{r3}(t + \delta t)\,\delta t}{\delta x_{13}} + \frac{4h_{r4}(t + \delta t)\,\delta t}{\delta x_{13}} + \frac{4h_{r5}(t + \delta t)\,\delta t}{\delta x_{13}} + \frac{4h_{r6}(t + \delta t)\,\delta t}{\delta x_{13}}$$

$$b_{7,8} = -a_{7,8}(t) + 4\varrho_{13}(t)C_{13}(t)$$

$$a_{7,j} = \frac{4h_c(t + \delta t)\,\delta t}{\delta x_{13}} \qquad b_{7,j} = a_{ij}(t)$$

$$a_{7,20} = \frac{4h_{r2}(t + \delta t)\,\delta t}{\delta x_{13}} \qquad b_{7,20} = a_{7,20}(t)$$

$$a_{7,28} = \frac{4h_{r3}(t + \delta t)\,\delta t}{\delta x_{13}} \qquad b_{7,28} = a_{7,28}(t)$$

$$a_{7,36} = \frac{4h_{r4}(t + \delta t)\,\delta t}{\delta x_{13}} \qquad b_{7,36} = a_{7,36}(t)$$

$$a_{7,46} = \frac{4h_{r5}(t + \delta t)\,\delta t}{\delta x_{13}} \qquad b_{7,46} = a_{7,46}(t)$$

$$a_{7,54} = \frac{4h_{r6}(t + \delta t)\,\delta t}{\delta x_{13}} \qquad b_{7,54} = a_{7,54}(t)$$

$$c_7 = \frac{4\delta t\,[q_{S1}(t) + q_{S1}(t + \delta t) + q_{R1}(t) + q_{R1}(t + \delta t)]}{\delta x_{13} A_{S1}}.$$

Here, h_{ri} is the radiation coefficient between surface i and the current surface being processed (S1), and A_{S1} is the area of surface 1 (m^2).

As before, plant interaction potential has been removed and all solar and free gain terms treated as known for all time since they can be independently determined.

The terms $a_{7,j}\theta_f(t + \delta t)$ and $b_{7,j}\theta_f(t)$ relate the surface node S1 to some adjacent air node at temperature θ_f. It is possible to subdivide surface 1 to introduce two nodal schemes allowing independent connections between one surface subdivision and air node F1 and between the other surface subdivision and air node F2. Here the choice is to connect the one surface node S1 to some contact area weighting of nodes F1 and F2 such that

$$\theta_f = \frac{A_{S1f1}\theta_{f1} + A_{S1f2}\theta_{f2}}{A_{S1f1} + A_{S1f2}}$$

where A_{S1f1} is the contact area between surface 1 and the volume represented by F1 (m^2), and A_{S1f2} is the contact area between surface 1 and the volume represented by F2 (m^2).

This allows the previous equation to be expanded to give

$$-a_{7,7}\theta_{1,6}(t + \delta t) + a_{7,8}\theta_{S1}(t + \delta t) - a_{7,55}\theta_{f1}(t + \delta t) - a_{7,56}\theta_{f2}(t + \delta t)$$
$$-a_{7,20}\theta_{S2}(t + \delta t) - a_{7,28}\theta_{S3}(t + \delta t) - a_{7,36}\theta_{S4}(t + \delta t) - a_{7,46}\theta_{S5}(t + \delta t)$$
$$-a_{7,54}\theta_{S6}(t + \delta t) = b_{7,7}\theta_{1,6}(t) + b_{7,8}\theta_{S1}(t) + b_{7,55}\theta_{f1}(t) + b_{7,56}\theta_{f2}(t)$$
$$+ b_{7,20}\theta_{S2}(t) + b_{7,28}\theta_{S3}(t) + b_{7,36}\theta_{S4}(t) + b_{7,46}\theta_{S5}(t) + b_{7,54}\theta_{S6}(t) + c_7 = z_7$$

where

$$a_{7,55} = \frac{4h_c(t + \delta t)\delta t A_{\text{S1f1}}}{\delta x_{13}(A_{\text{S1f1}} + A_{\text{S1f2}})} \qquad b_{7,55} = a_{7,55}(t)$$

$$a_{7,56} = \frac{4h_c(t + \delta t)\delta t A_{\text{S1f2}}}{\delta x_{13}(A_{\text{S1f1}} + A_{\text{S1f2}})} \qquad b_{7,56} = a_{7,56}(t)$$

and so couplings have been achieved between surface 1 and both adjacent fluid volumes. The magnitude of h_c, the convection coefficient (see §5.5) will be some averaged value for the entire surface although it will be determined on the basis of the mean fluid temperature θ_f. Should this assumption prove unacceptable then surface subdivision is necessary to create a number of distinct surface/fluid flowpaths each with an independent h_c value. For this situation the formulation of equation (3.12) is appropriate if modelling of heat flow in the plane of the surface becomes an important objective.

The indexing scheme of this surface node equation requires additional comment. As can be seen the $a_{7,7}$ and $a_{7,8}$ elements (similar reasoning holds for the b elements) are determined on the basis of simple diagonal progression. The intersurface coupling coefficients $a_{7,20}$, $a_{7,28}$, $a_{7,36}$, $a_{7,46}$ and $a_{7,56}$ are determined as a simple function of the number of nodes representing the next highest numbered multi-layered construction. In this way it can be predetermined that surface 1 will always be referenced by a coefficient in column $n + 1$ where n is the number of nodes representing the associated multi-layered construction. Thus n is related to the number of homogeneous elements; in a 3 node/element scheme $n = 2 n_e + 1$ where n_e is the number of elements. In general form the column location of surface coefficients is given by

$$\sum_{i=1}^{N} (2n_{ei} + 2)$$

where N is all surfaces up to and including the one of interest.

It is also convenient to locate cofficients which link to fluid nodes at the highest available column number in the matrix system to form a matrix pattern which is easy to interpret. Thus for an M node system with S surfaces and F fluid nodes, the column location of fluid node coefficients is given by $(N + S - F + I)$, where I is the fluid node of interest.

Matrix building is now largely repetitive as each multi-layered construction node is processed in turn until a fluid node is reached. Nevertheless, some specific problems will be encountered. Node 2,3 is located at the boundary between two elements and so represents a mixed property region for which the relevant equation is (3.7), which can be rewritten as

$$-a_{10,11}\theta_{2,2}(t + \delta t) + a_{10,12}\theta_{2,3}(t + \delta t) - a_{10,13}\theta_{2,4}(t + \delta t)$$
$$= b_{10,11}\theta_{2,2}(t) + b_{10,12}\theta_{2,3}(t) + b_{10,13}\theta_{2,4}(t) + c_{10} = z_{10}$$

where

$$a_{10,11} = \frac{8[k_{21}(t + \delta t)R_c(t + \delta t) + \delta x_{21}]\delta t}{\delta x_{21}R_c(t + \delta t)} \qquad b_{10,11} = a_{10,11}(t)$$

$$a_{10,12} = \frac{2[\delta x_{21}\varrho_{21}(t + \delta t)C_{21}(t + \delta t) + \delta x_{22}\varrho_{22}(t + \delta t)C_{22}(t + \delta t)]}{(\delta x_{21} + \delta x_{22})}$$

$$+ \frac{8[k_{21}(t + \delta t)R_c(t + \delta t) + \delta x_{21}]\delta t}{\delta x_{21}R_c(t + \delta t)} + \frac{8[k_{22}(t + \delta t)R_c(t + \delta t) + \delta x_{22}]\delta t}{\delta x_{22}R_c(t + \delta t)}$$

$$b_{10,12} = -a_{10,12}(t) + 4[\delta x_{21}\varrho_{21}(t)C_{21}(t) + \delta x_{22}\varrho_{22}(t)C_{22}(t)]/(\delta x_{21} + \delta x_{22})$$

$$a_{10,13} = \frac{8[k_{22}(t + \delta t)R_c(t + \delta t) + \delta x_{22}]\delta t}{\delta x_{22}R_c(t + \delta t)} \qquad b_{10,13} = a_{10,13}(t)$$

$$c_{10} = 0.$$

For this nodal case it has been necessary to determine region storage potential on the basis of a spatial weighting of the temperature-dependent thermal properties of the component elements.

Each of the surfaces 1–4 has equations which involve both fluid nodes. Surfaces 5 and 6 however, in keeping with the physical schema, have only one fluid node connection.

Rows 49 and 50 hold the equations for fluid nodes F1 and F2 for which case equation (3.12) can be located as follows:

$$-a_{49,8}\theta_{S1}(t + \delta t) - a_{49,20}\theta_{S2}(t + \delta t) - a_{49,28}\theta_{S3}(t + \delta t) - a_{49,36}\theta_{S4}(t + \delta t)$$
$$-a_{49,46}\theta_{S5}(t + \delta t) + a_{49,55}\theta_{f1}(t + \delta t) + a_{49,56}\theta_{f2}(t + \delta t) = b_{49,8}\theta_{S1}(t) + b_{49,20}\theta_{S2}(t)$$
$$+ b_{49,28}\theta_{S3}(t) + b_{49,36}\theta_{S4}(t) + b_{49,46}\theta_{S5}(t) + b_{49,55}\theta_{f1}(t) + b_{49,56}\theta_{f2}(t) + c_{49} = z_{49}$$

$$\text{for F1}$$

and

$$-a_{50,8}\theta_{S1}(t + \delta t) - a_{50,20}\theta_{S2}(t + \delta t) - a_{50,28}\theta_{S3}(t + \delta t) - a_{50,36}\theta_{S4}(t + \delta t)$$
$$-a_{50,54}\theta_{S6}(t + \delta t) + a_{50,55}\theta_{f1}(t + \delta t) + a_{50,56}\theta_{f2}(t + \delta t) = b_{50,8}\theta_{S1}(t) + b_{50,20}\theta_{S2}(t)$$
$$+ b_{50,28}\theta_{S3}(t) + b_{50,36}\theta_{S4}(t) + b_{50,54}\theta_{S6}(t) + b_{50,55}\theta_{f1}(t) + b_{50,56}\theta_{f2}(t) + c_{50} = z_{50}$$

$$\text{for F2}$$

where

$$a_{49,8} = \frac{h_{cf1S1}(t + \delta t)\delta A_{S1}\delta t}{\delta V_{f1}} \qquad b_{49,8} = a_{49,8}(t)$$

$$a_{49,20} = \frac{h_{cf1S2}(t + \delta t)\delta A_{S2}\delta t}{\delta V_{f1}} \qquad b_{49,20} = a_{49,20}(t)$$

$$a_{49,28} = \frac{h_{cf1S3}(t + \delta t)\delta A_{S3}\delta t}{\delta V_{f1}} \qquad b_{49,28} = a_{49,28}(t)$$

$$a_{49,36} = \frac{h_{cf1S4}(t + \delta t)\,\delta A_{S4}\,\delta t}{\delta V_{f1}} \qquad b_{49,36} = a_{49,36}(t)$$

$$a_{49,46} = \frac{h_{cf1S5}(t + \delta t)\,\delta A_{S5}\,\delta t}{\delta V_{f1}} \qquad b_{49,46} = a_{49,46}(t)$$

$$a_{49,55} = 2\varrho_{f1}(t + \delta t)C_{f1}(t + \delta t) + a_{49,8} + a_{49,20} + a_{49,28} + a_{49,36} + a_{49,46}$$

$$+ \frac{v_{f2f1}(t + \delta t)\overline{\varrho}_{f2f1}(t + \delta t)\overline{C}_{f2f1}(t + \delta t)\,\delta t}{\delta V_{f1}}$$

$$+ \frac{v_{of1}(t + \delta t)\overline{\varrho}_{of1}(t + \delta t)\overline{C}_{of1}(t + \delta t)\,\delta t}{\delta V_{f1}}$$

$$b_{49,55} = - a_{49,55}(t) + 4\varrho_{f1}(t)C_{f1}(t)$$

$$a_{49,56} = \frac{v_{f2f1}(t + \delta t)\overline{\varrho}_{f2f1}(t + \delta t)\overline{C}_{f2f1}(t + \delta t)\,\delta t}{\delta V_{f1}} \qquad b_{49,56} = a_{49,56}(t)$$

$$c_{49} = \frac{\delta t}{\delta V_{f1}}\,[q_{cf}(t) + q_{cf}(t + \delta t) + v_{of1}(t)\overline{\varrho}_{of1}(t)\overline{C}_{of1}(t)\theta_o(t)$$

$$+ v_{of1}(t + \delta t)\overline{\varrho}_{of1}(t + \delta t)\overline{C}_{of1}(t + \delta t)\theta_o(t + \delta t)]\,.$$

In these expressions, $h_{cf1S?}$ is the convection coefficient at fluid/surface interface (W m^{-2} $^\circ$C^{-1}), δV_{f1} the volume represented by F1 (m^3), v_{f2f1} the volume flow rate from F2 to F1 (m^{3}s^{-1}), $\overline{\varrho}_{f2f1}$ the mean density determined on the basis of F1 and F2 temperatures (kg m^{-3}), $\overline{\varrho}_{of1}$ the mean density determined on the basis of F1 and 'outside' source temperatures, $\delta A_{S?}$ the contact area between fluid and surface—part of area A_S(m^2), A_S the surface area (m^2), v_{of1} the flowrate from 'outside'—infiltration or zone-coupled (m^{3}s^{-1}) and $\overline{C}_{f2f1}$ the mean specific heat determined on the basis of F1 and F2 temperatures (J kg^{-1} $^\circ$C^{-1}).

Similar reasoning applies to the fluid 2 node. Note that with fluid 1 no contact path exists between node F1 and the floor node S6 since no connection exists in reality. For the same reason no contact exists between F2 and S5 in the fluid 2 equation.

In this example some air flow coupling with 'outside' is assumed. Such coupling will be caused by infiltration, natural ventilation, mechanical ventilation and/or zone-coupled air flow. If all four processes exist simultaneously then the $v_{of?}$ terms of coefficients $a_{49,55}$, $b_{49,55}$ and c_{49} should be subdivided into four distinct quantities (with similar modifications applied to the fluid 2 coefficients). Of course, with the single zone problem being processed here, the assumption is made that temperature θ_o is a known boundary condition. However, when processing multi-zone networks, the θ_o variable, if located in another zone, must be treated as a future time-row unknown quantity and determined by simultaneous solution as described later in this chapter. In this case the $\theta_o(t + \delta t)$ term in c_{49}, *where θ_o is an active node in the problem*, must be removed to the equation left-hand side to become an *a* coefficient properly located and linked: the $\theta_o(t)$ coefficient will then become a *b* coefficient. If the θ_o term relates to infiltration or natural ventilation then θ_o will represent external air temperature—a boundary condition—and so the term can remain as one of the c_{49} collection. If the θ_o term relates to some known mechanical supply air temperature then it can be treated in the same way as infiltration and natural ventilation.

The fluid volume coupling flowrate v_{f2f1} (and v_{f1f2} which can exist simultaneously) is determined from the nodal temperature data by the formulations (for example) of §5.6.

4.1.2   Zone contents and plant interaction

Having established the single zone matrix of figure 4.1, it is now possible to arrange for simultaneous equation solution to achieve zone response to any given climatic stimuli. If the system matrix is re-established at each successive time-step on the basis of the results from the previous time-step then the system response function will be dynamically dependent on the boundary condition excitations in that all system lags are preserved whilst complying with the first and second laws of thermodynamics.

Before considering the multi-zone formulations of §4.1.3 it is appropriate to examine possible variations of the single zone processed in §4.1.1.

Zone contents
Consider the zone contents of figure 3.14 and assume for simplicity that these can be combined into some equivalent volume of material such that the nodal network of figure 4.2 is obtained. Here the assumption is made that since the contents are contained by fluid F2 then the material temperature is likely to vary little with spatial position and so can be represented by a single node C1. Equation (3.9) relates to this nodal case and can be located in the zone matrix of figure 4.1 such that

$$- a_{51,56}\theta_{f2}(t + \delta t) + a_{51,57}\theta_{C1}(t + \delta t) = b_{51,56}\theta_{f2}(t) + b_{51,57}\theta_{C1}(t)$$

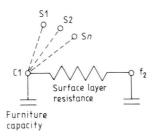

Figure 4.2 Simplified nodal scheme to represent zone contents. Note that a nodal connection between C1 and surrounding surfaces can be introduced if required (shown dotted).

where

$$a_{51,56} = \frac{k_{C1}(t + \delta t)\delta t A_{C1}}{V_{C1}\delta_{C1f2}} \qquad\qquad b_{51,56} = a_{51,56}(t)$$

$$a_{51,57} = 2\varrho_{C1}(t + \delta t)C_{C1}(t + \delta t) + \frac{k_{C1}(t + \delta t)\delta t A_{C1}}{V_{C1}\delta_{C1f2}}$$

$$b_{51,57} = -a_{51,57} + 4\varrho_{C1}(t)C_{C1}(t)$$

and where k_{C1} is the conductivity of contents material (W m^{-1}°C^{-1}), C_{C1} is the specific heat of contents material (J kg^{-1}°C^{-1}), A_{C1} is the exposed surface area of the contents volume (m^2), V_{C1} is the contents volume (m^3), δ_{C1f2} is a characteristic linear dimension (m), equal to the radius for a sphere, $1/2a$ for a cube where a is side length, or $(a + b + c)/6$ for a cuboid where a, b, c is length, width and height respectively.

This equation can now be added, as the row 51 equation, to give the matrix of figure 4.3. Since this equation has a cross-coupling coefficient ($a_{51,56}$) linking to the F2 node, a coefficient entry must also be made for the F2 equation since nodes communicating by non-flow processes must be linked in both directions. Thus coefficients $a_{50,57}$ and $b_{50,57}$ are added, where

$$a_{50,57} = \frac{h_{cf2C1}(t + \delta t)A_{C1}\delta t}{\delta V_{f2}} \qquad b_{50,57} = a_{50,57}(t).$$

Contents capacity is now an integral part of the system and zone performance will be modified accordingly.

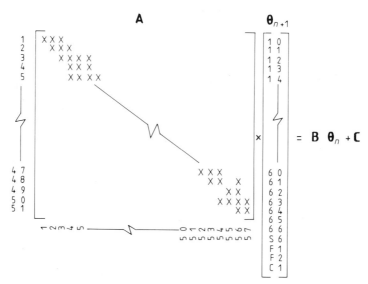

Figure 4.3 Addition of nodal coefficients (to the matrix of figure 4.1) to include zone contents.

Plant interaction

If modelling objectives are to predict zone energy requirements to provide stated environmental conditions it is necessary to allow plant interaction with one or more nodes. In this case an additional coefficient is added to the affected nodal equations and the corresponding entry in the $\boldsymbol{\theta}_{n+1}$ vector is a nodal flux term rather than nodal temperature.

Locating the plant interaction point at different system nodes has different physical interpretations:

Location at a fluid node implies that the plant input or extract is by convective means since the energy is added to or extracted from the zone air. Such a location is the appropriate mechanism for the conceptual modelling (see chapter 6 for a discussion of the mechanisms of plant modelling) of most air conditioning systems in that cooling can be added convectively at any zone position (high level, intermediate level, etc) with any time-based or mechanical constraint imposed—for example, VAV box minimum stop limit reached.

Location at a surface node implies that the plant input or extract is by convective and radiant means since subsequent energy redistribution will take place by longwave radiation to the surroundings and by convection to any adjacent fluid volume. Some of the energy will interact with the capacity associated with the surface node. This location corresponds to the conceptual modelling of radiant panels and radiators for which an independent surface description has been created in the geometry data structure.

Location at a capacity/insulation node allows the plant input or extract to be directed to some intra-material node. This device is useful for the modelling

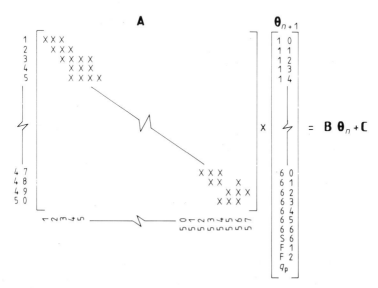

Figure 4.4 Addition of a nodal coefficient (to the matrix of figure 4.1) to include plant interaction potential.

of systems in which the input energy interacts directly with some material region as with electrical storage units or underfloor heating systems.

It is possible to introduce a number of plant interaction points to handle complex plant input schemes such as found in multiple radiator installations, distributed air conditioning systems and multi-unit off-peak electrical storage configurations. In the current example convective input to the F1 fluid volume is assumed and so coefficients $a_{49,57}$ and $b_{49,57}$ are added to the matrix of figure 4.1, where

$$a_{49,57} = -\delta t / V_{f1} \qquad b_{49,57} = -a_{49,57}$$

and the plant variable terms $q_p(t + \delta t)$ and $q_p(t)$ appended to the $\boldsymbol{\theta}_{n+1}$ and $\boldsymbol{\theta}_n$ vectors.

If plant interaction potential is to be allowed with more than one zone node then similar coefficients will be added to the affected nodal equation (as given by a row in the **A** and **B** matrices) with plant variable terms appended as above. This process is demonstrated in figure 4.4.

4.1.3 Multi-zone systems

Any collection of single-zone matrices, as exemplified by the matrix of figure 4.1, can be combined by matrix interlocking to represent some multi-zone building (or part of a building) by a single overall system matrix. Figure 4.5 illustrates the interlocking of four single-zone matrices. Important points to note are:

Each component matrix is uniquely located and will be of varying size depending on the associated zone shape (that is, the number of bounding surfaces and hence multi-layered constructions) and nodal discretisation (perhaps finer subdivision is introduced to improve modelling accuracy in some critical region).

The nodal equations relating to the outermost node(s) in each multi-layered construction nodal network will already contain a cross-coupling coefficient linking the node to some air node in an adjacent zone if the adjacent zone is included in the multi-zone system. This is done to facilitate fast simultaneous solution by matrix partitioning as described in §4.2.

Whilst intra-zone longwave radiation exchange is described in detail within each component matrix, the corresponding exchange between the outermost node and the internal surface nodes, pertaining to the equation set for some adjacent zone, must now be introduced. There are two possibilities.

The heat generation term of the outermost node can be used to introduce an exchange with the internal surfaces of the adjacent zone by operating one time-step in arrears to effectively link the outermost node at the future time-row with the adjacent zone internal surface nodes at the present time-row. This scheme is computationally attractive since it preserves the compact matrix form without introducing, in most cases, a detectable error.

If an exact solution is sought then the outermost nodal equation (communicating with some adjacent zone included in the matrix system) can be

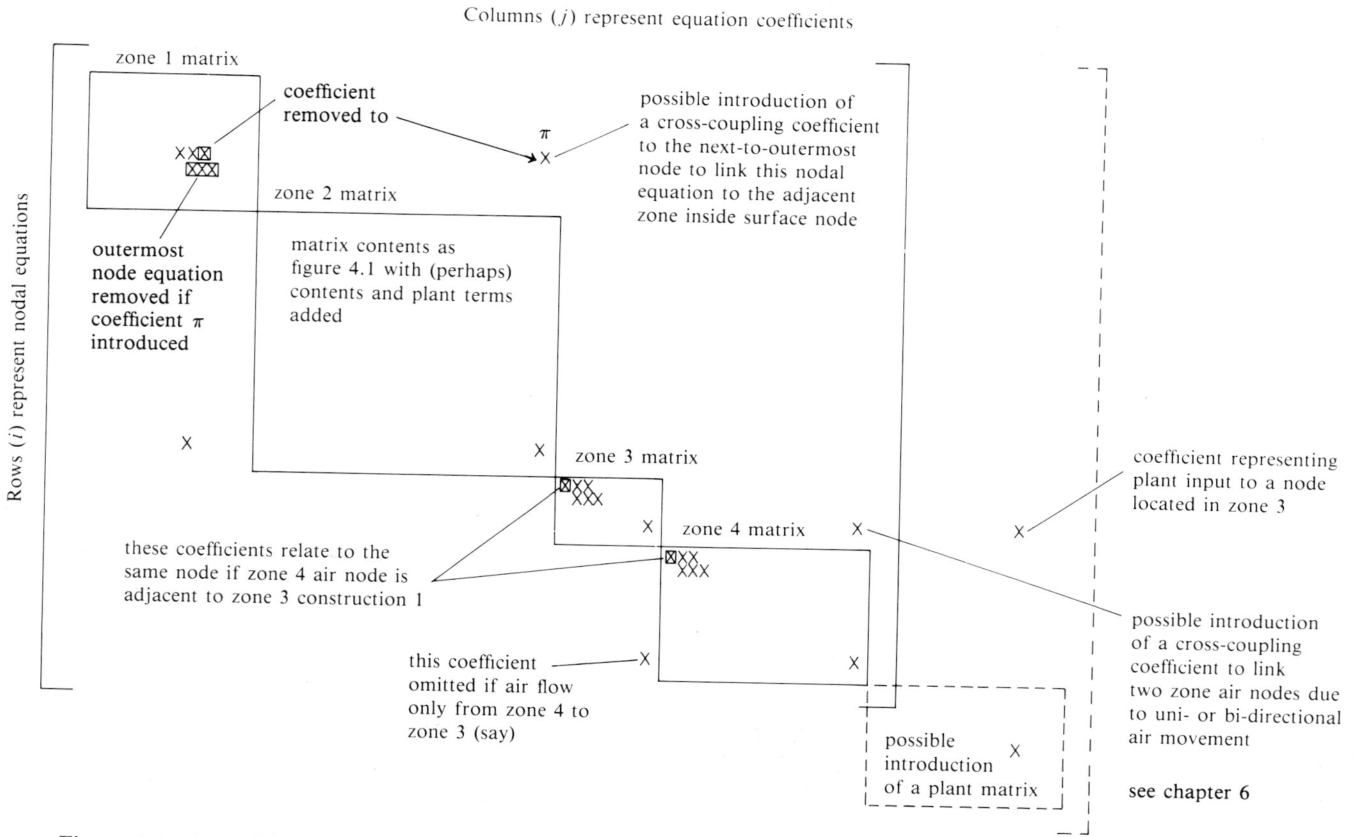

Columns (j) represent equation coefficients

Rows (i) represent nodal equations

zone 1 matrix

coefficient removed to

outermost node equation removed if coefficient π introduced

zone 2 matrix

matrix contents as figure 4.1 with (perhaps) contents and plant terms added

possible introduction of a cross-coupling coefficient to the next-to-outermost node to link this nodal equation to the adjacent zone inside surface node

π

X

zone 3 matrix

these coefficients relate to the same node if zone 4 air node is adjacent to zone 3 construction 1

this coefficient omitted if air flow only from zone 4 to zone 3 (say)

zone 4 matrix

coefficient representing plant input to a node located in zone 3

possible introduction of a cross-coupling coefficient to link two zone air nodes due to uni- or bi-directional air movement

see chapter 6

possible introduction of a plant matrix

Figure 4.5 A multi-zone matrix formed by interlocking four single-zone matrices as exemplified by figure 4.1.

removed from the component matrix equation set and the cross-coupling coefficient linking the next-to-outermost surface node to the outermost surface node can be relocated to link, instead, to the appropriate inside surface node in the adjacent zone which is already linked, at the future time-row, to the other zone surfaces.

At this point additional cross-coupling coefficients must be introduced to link zone air nodes if inter-zone air movement is to be modelled. In the case of uni-directional air flow between two zone air nodes—as in the case of flow through a small connecting orifice—only one coefficient is added to the receiving node equation. With large openings, such as doorways, where bi-directional air movement can occur, an additional coefficient must be added to both air node equations.

The annotation applied to figure 4.5 summarises this multi-zone matrix formulation procedure which can be extended to include any number of zones (as a function of the available computing power) to represent whole building energy balance whilst preserving the sensitive spatial and temporal connectivities.

It is at this stage that a further matrix equation set, representing the energy flows within some distributed plant system, can be introduced to allow the simultaneous processing of integrated building/plant systems. The formulation of plant matrices is the subject matter of chapter 6.

4.1.4 Treatment of time-dependent properties

The coefficient entries of the **A** matrix relate to the future time-row for which the nodal temperatures are unknown until simultaneous solution has been achieved. It is therefore impossible to evaluate these entries on the basis of the, as yet, uncomputed future time-row quantities. Two processing strategies are possible:

Coefficient evaluation can proceed one time-step in arrears so that the future time-row **A** matrix is established on the basis of present time-row nodal information with the present time-row **B** matrix utilising immediate past information. For most simulation applications this option will have undetectable consequences; indeed most contemporary energy models hold thermophysical properties constant and so the **A** matrix is in effect unchanging. And, of course, as the time-step is reduced, so the temperature variations between time-steps will diminish and the exact solution be approached.

Based on the 'one time-step in arrears' principle it is possible to establish and solve the system of equations to give future time-row nodal conditions which, by iteration at one time-step, can be used to achieve coefficient convergence. There are few practical applications where such rigour will be required.

4.2 Fast simultaneous solution by matrix processing

Consider, for the present, the single-zone matrix of figure 4.1 with plant interaction potential at the F1 node. The time dependence of the coefficients of the unknown nodal temperatures (**A**) dictate that the matrix be inverted ($\mathbf{A}^{-1}$) at each time-step in order to achieve the necessary repetitive simultaneous solution. The multiple inversion of such a sparse matrix is extremely inefficient from a computational viewpoint—a heating season simulation at a one hour time-step requiring approximately 5500 inversions—and for this reason a direct and rapid solution scheme is required. The symmetry of the **A** matrix (in single- or multiple-zone form) suggests a linear step-wise scheme in which the overall matrix is partitioned into a series of elemental submatrices each of which can then be processed independently, with the output information from each being used to achieve final solution at each time-step. Thus the entire zone equation set can be viewed as a number of interlocking partitioned matrices. The advantages of this approach are threefold:

Only the actual physical schema is addressed, that is, any unfilled elements within the system matrix are not processed at any point in the inversion (in fact the matrices of figures 4.1 and 4.5 are never formally defined or held in computer memory).

Any partitioned matrix, once inverted, need not be re-inverted at subsequent time-steps unless its contents have changed by a discernible amount.

And, because each sub-matrix is processed independently, it is possible to arrange for different matrices to be processed at different frequencies depending on the time constant of the physical component they represent.

The sections which follow cover single-zone and multiple-zone processing, describe techniques for matrix solution on the basis of comfort criteria, and discuss the treatment of non-linear processes and complex matrix systems.

4.2.1 Single-zone solution

For a single-zone equation set, without nodal plant interaction, partitioning is performed according to the following procedure:

A series of partitioned matrices is extracted, each one representing a multi-layered construction within the single-zone problem. These matrices contain the coefficients relating to the intra-constructional nodal equations addressing material conduction and heat storage. These sub-matrices will also contain the coefficient of the future time-row temperature of the next-to-inside surface node relating to the surface energy balance equation, to enable adjustment of the next-to-inside surface nodal equation, to facilitate later modification of the zone energy balance matrix as detailed below. Partitioned construction matrices are designated by $\mathbf{S}_n$; there are 6 in this case as shown in figure 4.6. It is important to note that only the filled elements of the **S** matrix should be held in memory to be processed during matrix reduction. The two-dimensional representations given here are adopted only to clarify the technique.

Further partitioning of the **A** matrix gives rise to a sub-matrix (designated by **R**) containing the unknown future temperature coefficients of the heat balance equations relating to each internal surface node and each fluid node within the problem. These coefficients represent the zone inter-surface radiation exchange, surface convection and fluid flow. At this stage the coefficients

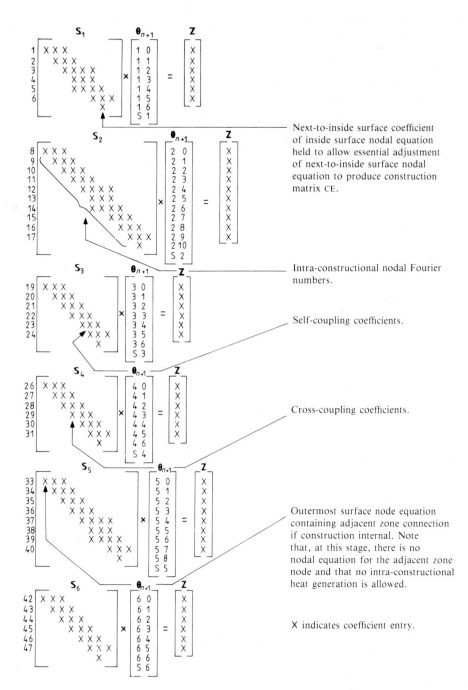

Figure 4.6 Partitioned construction matrix equations.

of the next-to-inside surface node equations are also held to define the linkage between the **R** and **S** type matrices. The contents of the **R** matrix are shown in figure 4.7.

Since any node can, in addition, have prescribed heat generation due to plant interaction—for example, convective input to air nodes, direct capacity input in the case of underfloor heating systems or electrical storage units, and mixed node injection in radiant systems—the partitioned matrices of figures 4.6 and 4.7 can be modified to incorporate nodal plant interaction terms. Figure 4.8 gives four example matrices for the case of air point injection, surface injection, intra-construction injection and a mixed scheme which incorporates all three nodal type injection possibilities.

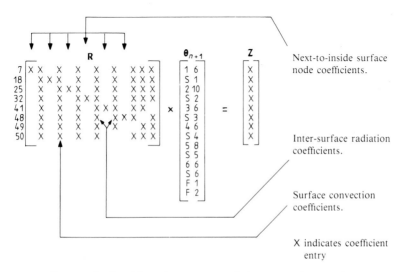

Figure 4.7 Partitioned zone energy balance matrix equation.

The solution of the entire zone equation set—now held in partitioned form—is a 6 stage process which addresses only the 'filled' matrix elements and which accommodates any spatial position of controller location and plant input or extract.

The underlying objective is to process each partitioned matrix as far as possible to allow the extraction of one or more 'characteristic equations' (hereafter referred to as a CE) which embodies the dynamics of the related component. These CEs are then gathered together for onward simultaneous reduction to produce a set of whole-system CEs which relate control nodes (whose temperature or heat flux status will control the simulation path) to the required nodal plant interaction. This CE set is then solved in terms of user-imposed control statements so that the back substitution phase of the matrix solution procedure can recommence. The primary task is to ensure that all nodal terms, representing declared control regions, are 'carried through' to that point in the

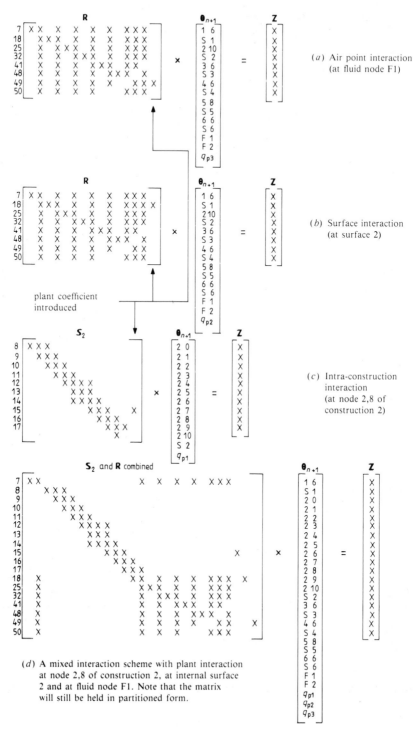

Figure 4.8 Example plant interaction schemes.

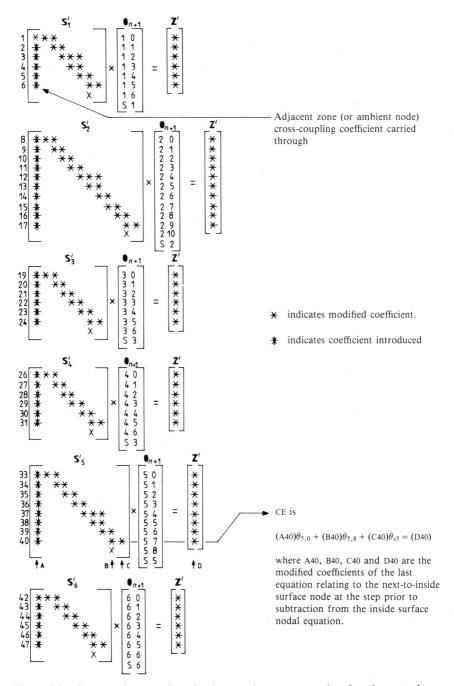

Adjacent zone (or ambient node) cross-coupling coefficient carried through

✳ indicates modified coefficient.

✸ indicates coefficient introduced

CE is

$$(A40)\theta_{5,0} + (B40)\theta_{5,8} + (C40)\theta_{s5} = (D40)$$

where A40, B40, C40 and D40 are the modified coefficients of the last equation relating to the next-to-inside surface node at the step prior to subtraction from the inside surface nodal equation.

Figure 4.9 Construction matrix reduction: no intra-constructional node control or plant interaction.

solution process where control theory is introduced. The six stages in the solution technique are:

Stage 1
Each constructional sub-matrix is processed towards the end of the forward reduction stage of a direct elimination as detailed in §2.4.3. The reduction continues until the equation relating to the next-to-inside surface node is reached. At this point sub-matrix processing is terminated and the CE, containing adjusted matrix coefficients relating to the next-to-inside surface nodal equation, extracted. Two possible reduction schemes exist, depending on whether or not intra-constructional node control is active. Figure 4.9 demonstrates this reduction process assuming that a control point is not located within the multi-layered construction and for the case of no nodal plant interaction. In this case the extracted end equation, the CE, will have terms relating to the adjacent zone air node, the next-to-inside surface node, the inside surface node and a composite term relating to the matrix equation right-hand side (a **Z** element representing the entire known time-row).

Figure 4.10 demonstrates the reduction process if a control node is located within the construction. Note that the control point coefficient is carried through and introduced to each subsequent nodal equation (matrix row) so that two CEs emerge, both of which contain control point coefficients.

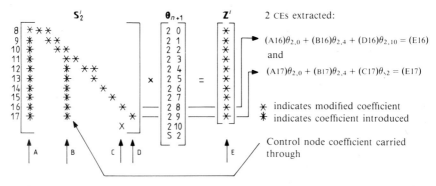

Figure 4.10 information:

S_2^i θ_{n+1} Z^i 2 CEs extracted:

$$(A16)\theta_{2,0} + (B16)\theta_{2,4} + (D16)\theta_{2,10} = (E16)$$

and

$$(A17)\theta_{2,0} + (B17)\theta_{2,4} + (C17)\theta_{s2} = (E17)$$

✳ indicates modified coefficient
✴ indicates coefficient introduced

Control node coefficient carried through

Figure 4.10 Construction matrix reduction: intra-constructional node control; no plant interaction. (Control is imposed on node 2,4.)

Figure 4.11 demonstrates matrix forward reduction and CE extraction for the case of intra-constructional node plant interaction when the control point is and is not located within a capacity/insulation system.

Stage 2
The CEs, when collected from all constructional sub-matrices, can now be used to eliminate one term from each of the internal surface node equations of the **R** matrix representing radiative, convective, advective, conductive and heat generation processes within a single zone. This is done by algebraic addition of the CE

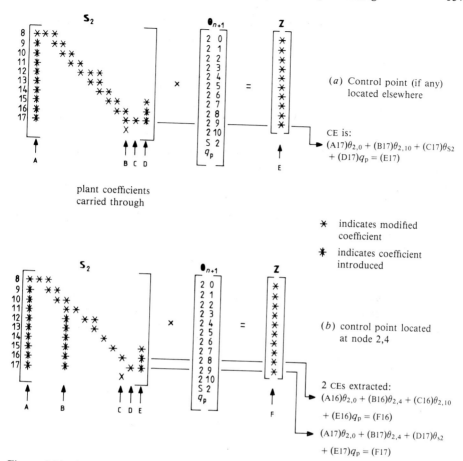

Figure 4.11 Matrix forward reduction and CE extraction for constructions with plant interaction.

and the **R** matrix row equation which relates to the same construction. Figure 4.12 demonstrates this **R** matrix modification for the case where a control point is and is not located within a multi-layered construction but when no plant interaction nodes are present.

Plant interaction nodes may already exist in the **R** matrix as shown in figure 4.8 or may be introduced to the **R**′ matrix by addition of the construction CEs containing one or more plant coefficients. Figure 4.13 demonstrates the **R**′ matrix to result from a mixed scheme in which constructions 2 and 4 each have internal nodes with assigned plant interaction potential.

Stage 3

After stages 1 and 2 have been achieved for each construction sub-matrix in turn, the modified zone matrix **R**′ can be processed to the end of the forward reduction, carrying through all control node coefficients relating to surface control

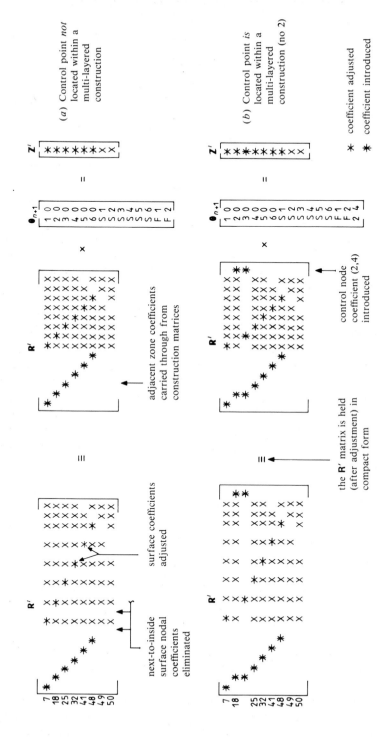

Figure 4.12 Partitioned zone energy balance matrix adjustment by construction CEs.

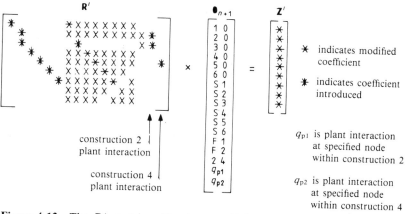

Figure 4.13 The **R′** matrix with plant coefficients introduced due to plant terms carried through from construction matrices **S₂** and **S₄**. The matrix indicates that a control point has been located within construction 2 and so two construction CEs are introduced.

node terms or intra-constructional node control coefficients introduced in stage 3. Any plant coefficient must also be carried through. Figure 4.14 demonstrates the range of possible reduction options, in each case assuming only one control node type. Note, however, that any multi-node control scheme is possible although matrix processing will become proportionally more complex.

The CE to emerge at this stage will relate the zone control node(s) to the required plant input or extract, at the user defined plant interaction node or nodes (if any are defined), in terms of modified **Z** matrix entries and all adjacency conditions.

Stage 4

The CE extracted at stage 3 is now solved for free floating conditions (no plant) or in terms of any user-imposed control statement concerning capacity, time or thermostatic constraints.

Stage 5

The zone surface temperatures are now obtained by simple back substitution operations performed on the modified zone matrix.

Stage 6

The intra-constructional node temperatures are obtained by similar back substitution in the appropriate constructional sub-matrices preserved from stage 1.

It is important to note that this solution procedure only addresses existing coefficients—as dictated by the system undergoing simulation—and so the single-zone matrix of figure 4.1 is never fully established as a complete two-dimensional

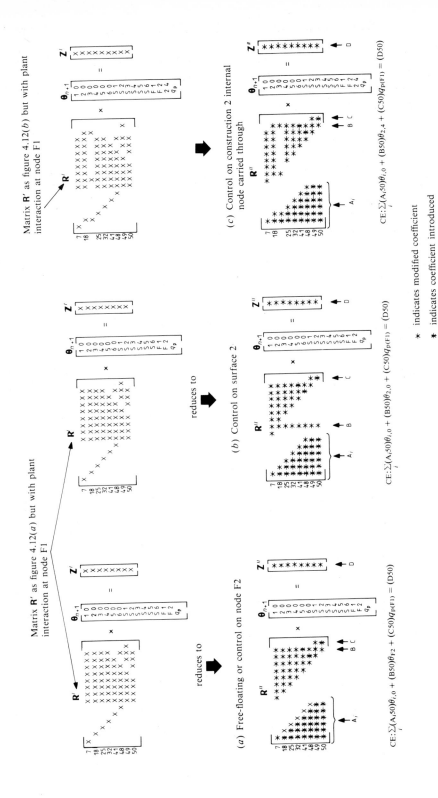

Figure 4.14 R matrix reduction options.

array. This gives the seed for fast and efficient matrix processing, permitting fast matrix reconstruction at each time-step and so model flexibility in use.

In a single-zone system, incorporating one control node and one plant interaction node, the CE to emerge from stage 4 will have the form

$$\sum_i A_i \theta_A + B\theta_c + Cq_p = D \qquad (4.3)$$

where A, B, C and D are the modified coefficients to emerge from the matrix reduction as shown in figure 4.14, θ_A are the adjacent zone temperatures at the future time-row (°C), θ_c is the temperature of zone nodes designated control points at simulation commencement (°C) and q_p is the plant interaction with nodes designated interaction nodes (W or W m^{-2} depending on node location).

To solve this equation, additional information must be harnessed which describes control expectations. In single-zone modelling, adjacent zone temperature information (at the future time-row) cannot be determined since no adjacent zone equation structure is present. With multi-zone simulation no such problem exists (see §4.2.2). Two possibilities exist for the single-zone case:

An adjacent zone may be external climatic in which case the corresponding θ_A temperature is known for all time.

Or, if internal, its temperature can be specified (by 'guesstimation' if not known), considered identical to the simulated zone or established algorithmically by some independent means (perhaps taken as the previous time-step value resulting from an independent single-zone treatment of the adjacent zone).

In any event, equation (4.3) will reduce to

$$B\theta_c + Cq_p = D$$

which can be solved, algorithmically, if the desired control temperature can be specified or if the current plant input or extract is known. It is at this point that controller characteristics are introduced—set point, control action (P, $P + I$ etc) and dead times etc—to allow determination of the actual flux input as a function of the sensed condition. Control system simulation is described in chapter 6 where selected plant systems are considered.

Multiple control nodes and/or plant interaction nodes can be accommodated by simply extracting additional CEs at stage 4 as appropriate. Note that iterative techniques, of the kind mooted in §4.2.2 for multi-zone matrix processing, may become necessary for such distributed control and plant interaction strategies.

4.2.2 Multi-zone solution

In multi-zone networks the procedure of stages 1–4 of §4.2.1 can be implemented independently for each zone to allow the stage 4 CE extraction to give, for each zone, a control/plant interaction equation—in the form of equation (4.3). Figure 4.15 gives the matrix representation of these collected zone equations.

If an approximate solution is sought, each zone equation can be solved independently by assuming all adjacent temperatures, at the future time-row, are set to the present time-row values as established as future time-row values at the previous time-step. This allows the future time-row control node temperature and plant interaction node flux to be evaluated for each zone. It is likely that this 'approximation' is adequate for most simulation applications so long as the time-stepping scheme is well adapted to the zone time constants.

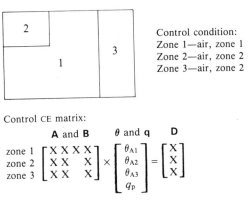

Control condition:
Zone 1—air, zone 1
Zone 2—air, zone 2
Zone 3—air, zone 2

Control CE matrix:

$$
\begin{array}{c} \text{zone 1} \\ \text{zone 2} \\ \text{zone 3} \end{array}
\begin{bmatrix} X & X & X & X \\ X & X & & X \\ X & X & & X \end{bmatrix}
\times
\begin{bmatrix} \theta_{A1} \\ \theta_{A2} \\ \theta_{A3} \\ q_p \end{bmatrix}
=
\begin{bmatrix} X \\ X \\ X \end{bmatrix}
$$

overhead: **A** and **B** θ and **q**   **D**

Figure 4.15 Matrix representation of zone CEs for an example three-zone problem.

If an exact solution is required, a mixed iterative/direct solution scheme is necessary. In such a scheme the adjacent temperatures, in each zone equation, are given some 'guesstimated' value (perhaps the present time-row value) to allow the evaluation of the θ_c and q_p terms for each zone. The backward substitution operation (stage 5) is then initiated and continued until each zone air temperature has been evaluated at the future time-row (or zone surface temperatures if the adjacency linkages are defined as proposed in §4.1.2). The new zone air (or surface) temperatures are then introduced as updated 'guesstimates' and the solution process iterates, at the same time-step, until the differences between successive predictions are minimised for all zones. In some cases, to encourage rapid iterative convergence, it will be necessary to apply a weighting factor to the temperature 'guesstimates' to cause over-relaxation as described in §2.4.3.

4.2.3 Solution on the basis of sensing element characteristics or comfort criteria

In real world applications zone temperature control is at the dictate of a thermostat which will be influenced by local air temperature and radiant exchanges with surrounding surfaces within its scene. Therefore the actual sensed temperature which should be assumed for simulation control purposes is given by

$$\theta_x = \alpha\theta_A + (1 - \alpha)\left(\sum_{i=1}^{N} \omega_{si}\theta_{si}\middle/\sum_{i=1}^{N} \omega_{si}\right) \qquad (4.4)$$

where θ_x is the sensed temperature, θ_A the local node air temperature, θ_{si} the surface temperature, N the number of surrounding surfaces in visual contact, ω_{si} the surface solid angle subtended at the controller, and α the controller convective weighting factor ($\alpha \to 1$ for aspirated sensors).

The CE (characteristic equation) to emerge from stage 3 of the matrix equation solution technique of §4.2.1 relates the control node temperature (as defined by the model user) to the required plant capacity injected at some interaction node(s) (again as defined by the user). Assuming that the control node is initially established as the air point alone, the solution stream can proceed as follows. In the zone CE, as given by equation (4.3), assume for the present that the adjacency conditions are evaluated at the present time-row as described in §4.2.2 (note that this is an exact treatment for a single-zone problem). The plant capacity term, q_p, is guessed to allow evaluation of the zone air node temperature θ_A ($\theta_c = \theta_A$ in equation (4.3)). Stage 5 of §4.2.1 is now implemented to establish the zone surface temperatures to allow assessment of the sensed temperature θ_x (from equation (4.4) above). The guessed plant capacity q_p can then be increased or reduced—within the constraints imposed by the plant capabilities—and the stage 5 procedure reiterated until the sensed temperature is adjusted to the desired value.

For exact multi-zone processing it is necessary to 'nest' this procedure within the overall iterative procedure detailed, for multi-zone systems, in §4.2.2.

If simulation control is to be achieved on the basis of comfort considerations it is possible to arrange that the sensed temperature, θ_x, is evaluated as a function of zone air and mean radiant temperatures *and* other factors, such as air velocity, humidity level and metabolic rate considerations, determined by independent means.

Consider, as an example, the use of dry resultant temperature as a comfort index. This is defined as the temperature recorded by a thermometer situated at the centre of a blackened sphere 100 mm in diameter and is given approximately by

$$\theta_{res} = \frac{\theta_{mrt} + 3.17\theta_A V^{\frac{1}{2}}}{1 + 3.17V^{\frac{1}{2}}}$$

where θ_{mrt} is the zone mean radiant temperature and V is the air velocity (m s^{-1}).

For sedentary occupation an air velocity of 0.1 m s^{-1} is not uncommon and so this equation reduces to the usual form:

$$\theta_{res} = 0.5(\theta_{mrt} + \theta_A)$$

and two conditions require to be satisfied (IHVE (now CIBS) 1970): $19 \leqslant \theta_{res} \leqslant 23$ and $-5 \leqslant (\theta_{mrt} - \theta_A) \leqslant 8$.

Note, however, that it would be incorrect to enforce such conditions rigorously, since resultant temperature is merely an indicative quantity. Indeed, more sophisticated comfort level predictors have been devised (Fanger 1972).

4.2.4 Treatment of non-linear processes

The prerequisite of the foregoing numerical method is that all governing partial differential equations be reduced to linear algebraic equations for matrix equation processing. It is nevertheless possible to achieve exact solution, when non-linear processes are present, by simple iteration applied to the entire matrix equation. Consider equation (3.11), which relates to building surface layers and includes the inter-surface longwave radiation exchanges. The h_r terms, as defined by equation (5.41), are determined, at the future time-row, as a function of surface temperatures as yet uncomputed. To permit future time-row coefficient matrix formulation (the **A** matrix of figure 4.1) it is usually acceptable, in terms of accuracy, to operate one time-step in arrears. An alternative mechanism is to guess the future time-row surface temperatures (perhaps by extrapolation from the previous temperature history) to allow the estimation of h_r. Steps 1 through 5 (of §4.2.1) are then implemented and the predicted surface temperatures to emerge used to re-establish the **A** matrix if the difference between the guessed and predicted values are outside some acceptable tolerance. In this way, iterative convergence is pursued. Such an approach can probably not be justified for daily model application to real design problems but can play some role in parametric based research study of the impact of linearisation on model predictions.

4.3 Mixed frequency inversion

When building and plant matrices are combined—as described in chapter 6—the overall matrix to result will exhibit a temporal mismatch between building-side components, such as multi-layered constructions and thermal stores, and plant-side components, such as coolers and boilers. In the former case, time constants will be measured in hours, in the latter case minutes. It is therefore attractive to consider the possibility of processing different partitioned matrices, representing regions with differing thermal capacities, at different frequencies depending on the rate at which region conditions change by a perceptible degree. This means that the entire matrix system need not be processed at a time-stepping interval deemed to be the lowest common denominator.

Stated briefly, each partitioned matrix is assigned an inversion frequency arrived at from consideration of its time constant. As the solution scheme of §4.2.1 is implemented (with plant matrices included, of course) each partitioned matrix is forward reduced according to its assigned frequency. This means that the CE to emerge is held constant, until the next inversion point is reached, independent of sub-interval inversions applied to other partitioned matrices possessing greater inversion frequencies.

The technique is obviously an approximation, but has the virtue of improving computational efficiency for marginal trade-off, in many applications, in accuracy.

This chapter has detailed the mechanisms by which large and sparse matrix structures, containing the transformed differential equations representing energy flow, can be solved rapidly and efficiently. Such techniques will form the cornerstone of cost effective simulation until greater small machine processing power becomes available—perhaps incorporating hardware matrix facilities.

References

Fanger P O 1972 *Thermal Comfort* (New York: McGraw-Hill)
IHVE (Now CIBS) 1970 *Guide Book A*

Bibliography

Churchill R V 1958 *Operational Mathematics* (2nd edn) (New York: McGraw-Hill)
Collatz L 1960 *The Numerical Treatment of Differential Equations* (Berlin: Springer)
George A and Liu J W 1981 *Computer Solution of Large Sparse Positive Definite Systems* (New Jersey: Prentice-Hall)
Hartree D R 1958 *Numerical Analysis* (2nd edn)(Oxford: Oxford University Press)
Miranker W L 1981 *Numerical Methods for Stiff Equations and Singular Perturbation Problems* (Boston: Reidel)
Stewart G W 1973 *Introduction to Matrix Computations* (New York: Academic)
Willoughby R A and Rose D J 1972 *Sparse Matrices and Their Applications* (New York: Plenum)

5

Energy related subsystems

Many of the derivations of chapter 3 and the matrix processing descriptions of chapter 4 assume that quantities such as areas, volumes, surface angles, solar gains, air flow, surface heat transfer coefficients, heat gains from casual sources and climate variables are each independently determinable for all time-rows. These are the energy subsystems which must be evaluated at each computational time-step in order that the mainstream matrix formulation and processing can proceed.

This chapter describes each of these subsystems and in most cases outlines the algorithmic treatment required to combine the background theory into a workable model. In this way it is envisaged that, in a working simulation model, a number of independent algorithms, one for each subsystem, would, at each time-step, pass information to a central matrix 'engine' concerned with matrix set-up and solution.

On closer inspection of subsystem descriptions it will become apparent that some share similar computational requirements—for example, point projection and containment testing in the case of insolation evaluation and view factor determination—and so computational groupings emerge which are best treated as common utilities for the entire system. Software development and organisation is the subject matter of chapter 7.

5.1 Geometrical considerations

A prerequisite of all user-friendly energy modelling systems is the ability to perform, by software, elementary operations on some geometrical data structure to produce derived quantities such as surface areas, contained volumes, angles of view and so on. These quantities are then used throughout any simulation processing to determine convection coefficients, solar injections, longwave radiation coefficients, air flows etc as detailed in §§5.2–5.8. These latter quantities are then utilised to establish the finite volume heat balance equation sets of chapter 3.

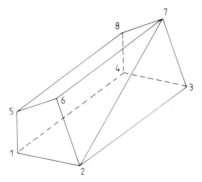

Figure 5.1 A general-shaped volume contained by seven planar polygons.

Many of the possible geometrical operations are discussed at length in the literature (for example, Foley and Van Dam 1982) and so only a brief summary, concerned with energy modelling requirements directly, is given here.

Consider the general-shaped volume of figure 5.1 as formed by the 7 bounding planar polygons. One convenient and well used geometrical representation technique is to specify the coordinates of each polygon vertex relative to some arbitrarily chosen Cartesian coordinate system. In addition, and separately, the body topology is held as an ordered vertex description for each bounding polygon. Thus, for the example given, each vertex is specified as some point (x_i, y_i, z_i) and the topology is given as (say):

Number of planar polygons = 7
Total number of vertices   = 8

Polygon	Number of vertices	Ordered description
1	4	1, 2, 6, 5
2	3	2, 3, 7
3	4	3, 4, 8, 7
4	4	4, 1, 5, 8
5	3	2, 7, 6
6	4	8, 5, 6, 7
7	4	1, 4, 3, 2

Note that the convention here is to describe each polygon as a collection of vertices specified anticlockwise when the polygon is viewed from 'outside'. Any intra-polygon hole is then specified by reversing the vertex order relative to the main body convention.

This data structure is convenient since it allows the application of single and composite transformations to perform basic scaling and/or rotation operations, producing modified vertex coordinates, but which require no corresponding topological modification. This means that the topology need only be held once for any shape classification since it is independent of scale and orientation Although suitable for building energy modelling applications more advanced techniques of surface representation exist (Rogers *et al* 1976) which are more suitable for other engineering fields.

A number of basic operations on this data set can now be defined. These allow the direct determination of the area of any plane polygon, the angles which define its direction and the overall contained volume of the body. Also of interest is the polygon perimeter often used to determine some characteristic dimension for use in surface convection assessment (see §5.5). Other operations such as translation and three-dimensional rotations are detailed in §5.2 where shading and insolation are considered.

For any planar polygon, p, consider the following summations applied to the polygon vertices:

$$\left. \begin{array}{l} \text{XSUM}_p = \sum_{i=1}^{NV} (y_i z_j + z_i y_j) \\[4mm] \text{YSUM}_p = \sum_{i=1}^{NV} (z_i x_j + x_i z_j) \\[4mm] \text{ZSUM}_p = \sum_{i=1}^{NV} (x_i y_j + y_i x_j) \end{array} \right\} \quad \begin{array}{l} j = i + 1 \\ \text{for } j > NV; \, j = 1 \end{array}$$

where NV is the total number of vertices in the polygon and x, y, z are corresponding coordinates. The area of this polygon is then given by

$$\text{AREA}_p = 0.5(\text{XSUM}_p^2 + \text{YSUM}_p^2 + \text{ZSUM}_p^2)^{1/2}.$$

If the vertex ordering convention is anticlockwise (when the polygon is viewed from outside) then the area is positive. And since clockwise conventions produce negative areas, algebraic summations give net polygon areas.

The perimeter length of any polygon is given by

$$\text{PERIM}_p = \sum_{i=1}^{NV} [(x_j - x_i)^2 + (y_j - y_i)^2 + (z_j - z_i)^2]^{1/2}$$

$$j = i + 1 \quad \text{for} \quad j > NV; \, j = 1.$$

It is usual to define the orientation of a polygon by azimuth and elevation angles as shown in figure 5.2. Here the azimuth is defined as the clockwise angle between the coordinate system Y axis (sometimes made to point north) and the projection of the polygon outward facing normal onto the XY plane (usually made to represent the horizontal plane). The plane elevation is defined as the angle between the outward facing normal and the projection of this normal onto a plane parallel to the XY plane. Adopting this convention gives, for the azimuth α_p

$$\alpha_p = \tan^{-1}(\text{XSUM}_p/\text{YSUM}_p)$$

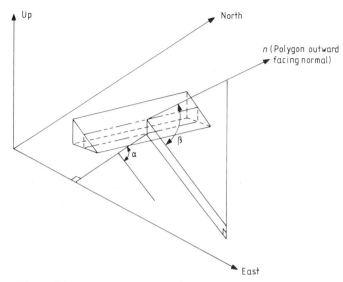

Figure 5.2 Polygon azimuth (α) and elevation (β) angles.

where, for $\text{YSUM}_p = 0$ $\alpha_p = -90°$ for $\text{XSUM}_p < 0$

$\alpha_p = \quad 0°$ for $\text{XSUM}_p = 0$

$\alpha_p = \quad 90°$ for $\text{XSUM}_p > 0$

and for the elevation β_p

$$\beta_p = \tan^{-1}[\text{ZSUM}_p/(\text{XSUM}_p^2 + \text{YSUM}_p^2)^{1/2}]$$

where, for $\text{XSUM}_p^2 + \text{YSUM}_p^2 = 0$ $\beta_p = -90°$ for $\text{ZSUM}_p < 0$

$\beta_p = \quad 0°$ for $\text{ZSUM}_p = 0$

$\beta_p = \quad 90°$ for $\text{ZSUM}_p > 0$.

The volume contained by all polygons is now found by algebraic summation of the volumes of the prism formed by connecting the vertices of each polygon in turn to some arbitrary point. Note that since polygons specified as holes (clockwise ordering) will still bound the volume, their vertex ordering must first be reversed. The contained volume is given by:

$$\text{VOL} = \frac{1}{6}\sum_{j=1}^{\text{NP}}(x_{j1}\text{XSUM}_j + y_{j1}\text{YSUM}_j + z_{j1}\text{ZSUM}_j)$$

where x_{j1}, y_{j1}, z_{j1} are the coordinates of the first mentioned vertex in polygon j, and NP is the total number of polygons.

5.2 Insolation of exposed building surfaces

With the liberal use of lightweight building materials and, in many 'passive' solar applications, large glazing areas, excessive solar gain can be an important design consideration even in countries with modest solar incidence. It is therefore necessary to have reliable and accurate means of predicting this gain, especially under elevated solar conditions. Section 5.3 covers solar prediction methodologies.

Since the solar energy absorbed by one of a group of buildings is often greatly influenced by the extent to which that particular building is shaded by other members of the group and by self-associated facade obstructions, a prerequisite of solar modelling is the ability to predict shading and insolation patches as a function of solar position and obstruction geometry. Indeed it is often an early design stage strategy to alter design parameters such as orientation, shape or obstruction geometry in an attempt so to modify the shading/insolation patterns at some critical time that environmental performance is improved without recourse to plant intervention.

Essential information for energy modelling purposes includes a time-series knowledge of the magnitude and position of the insolated portions determined separately for each external exposed opaque and transparent surface, and of the magnitude and position of internal exposed opaque and transparent insolation determined separately for each window/surface combination. These data are summarised in figure 5.3: the former data are required in the prediction of external opaque surface solar absorption/reflection and transparent surface absorption/transmission/reflection; and the latter data are required to track the

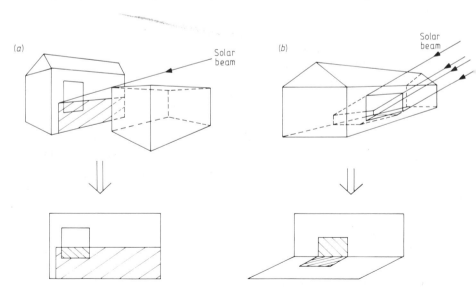

Figure 5.3 Shading/insolation data required at each time-row. (*a*) External opaque and transparent surface shading. (*b*) Internal opaque and transparent surface insolation.

window transmitted shortwave energy to its firstly incident internal surface (§5.3 discusses the treatment of the then reflected portion).

In essence there are two extant methods for the computation of insolation information on the basis of geometrical and solar position information: hidden surface methods and the technique of point projection.

The former method involves the 'clipping' of surface polygons—one against the other—until only those polygon portions remain that will be viewed from some chosen viewpoint (taken as the sun for insolation prediction). The technique is rigorous and can be designed to produce, in addition to insolation data, a polygon set which represents an entire scene. Indeed, techniques are now available which can apply colour and texture to these viewed polygons as a function of ambient or artificial lighting conditions. The computer models that result (Stearn 1982) allow the generation of realistic and extremely accurate (Purdie 1983) images of future reality at an early design stage. Sutherland *et al* (1976) have produced characterisations of several hidden surface algorithms and so detailed descriptions are omitted here.

The point projection technique is used to quantify the insolation patch and has less potential for visualisation. Given some collection of target and obstruction objects, each defined relative to some site Cartesian coordinate system, then the procedure is to relocate the *XZ* plane of the coordinate system in the plane of each face (in turn) of the target body. Each obstruction object can now be projected, parallel to the sun's rays, onto the face and the projected image expressed relative to the local face coordinate system. A simple grid can then be superimposed on each opaque and transparent surface allowing grid point containment testing against each individual shadow polygon. Prediction accuracy is controlled simply by adjusting the number of grid subdivisions of the homogeneous face in question.

5.2.1 Insolation transformation equations

Figure 5.4 shows a target and an obstruction body located in a right-hand coordinate system. Body *I* is a general-shaped object (defined as detailed in §5.1) for which insolation data are required. Body *J* is also a general-shaped object (an adjacent building perhaps or, if adjacent to the target body, some facade feature) and will be the cause of shading on body *I*. Body *J* has *N* vertices given by x_n, y_n, z_n.

The objective of the derivation which follows is to generate a set of transformation equations which allow any obstruction body vertex to be projected on to each target body face in turn with the projected coordinates expressed relative to a coordinate system normalised to the face in question.

The general axes relocation and vertex projection processes can be established as follows.

Translation: Move the coordinate system origin to the first defined vertex in the first face of the target body: body *I*, vertex 1 for example. This gives a new coordinate system in which—as shown in figure 5.4—any original point (xyz)

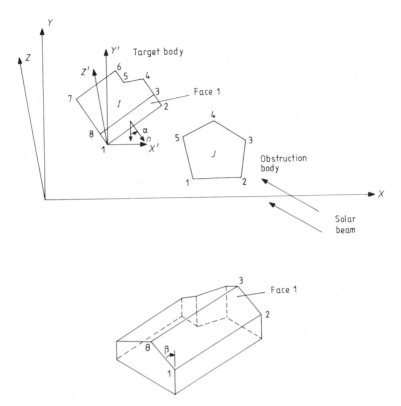

Figure 5.4 Geometry of a target and obstruction body: the coordinate system is right-handed.

translates to a new point $(x'y'z')$ according to the translation matrix:

$$(x'y'z') = (xyz1) \begin{bmatrix} 1 & 0 & 0 \\ 0 & 1 & 0 \\ 0 & 0 & 1 \\ -x0 & -y0 & -z0 \end{bmatrix} \qquad (5.1)$$

where $(x0y0z0)$ is the new origin in old coordinates; that is, the components of translation in the X, Y and Z directions.

Rotation: The translated axes must now be subjected to an X', Y' and Z' axis rotation to align the $X''Z''$ plane of the new coordinate system $X''Y''Z''$ in the plane of the target body face with the new Y'' axis pointing away from the sun. The rotation angles α, β and γ, as shown in figure 5.5, should be regarded as clockwise Z', X' and Y' axis rotations when viewed from the coordinate system origin. This three-axis rotation will result in a localised coordinate system allowing two-dimensional polygon manipulation after the insolation polygons have

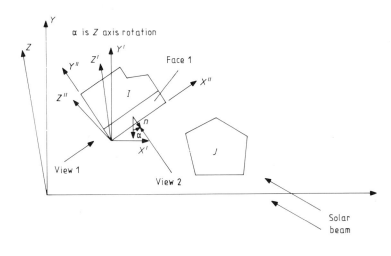

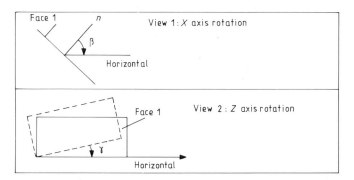

Figure 5.5 Axes transformation: axes rotations shown positive. Note that γ (in view 2) is the angle between the first two numbered vertices in the face and the horizontal, and is zero in this case.

been established by projection of the obstruction bodies. Any point $(x'y'z')$ therefore transforms to the point $(x''y''z'')$ according to the matrix relationship:

$$(x''y''z'') = (x'y'z') \begin{bmatrix} 1 & 0 & 0 \\ 0 & \cos\beta & \sin\beta \\ 0 & \sin\beta & \cos\beta \end{bmatrix} \begin{bmatrix} \cos\gamma & 0 & \sin\gamma \\ 0 & 1 & 0 \\ -\sin\gamma & 0 & \cos\gamma \end{bmatrix} \begin{bmatrix} \cos\alpha & -\sin\alpha & 0 \\ \sin\alpha & \cos\alpha & 0 \\ 0 & 0 & 1 \end{bmatrix}$$

$$(5.2)$$

where clockwise axes rotations (looking to the positive side of the origin) are positive.

Substitution of equation (5.1) in (5.2) gives the final axes transformation matrix:

$$(x''y''z'') = (xyz1) \begin{bmatrix} 1 & 0 & 0 \\ 0 & 1 & 0 \\ 0 & 0 & 1 \\ -x0 & -y0 & -z0 \end{bmatrix} \begin{bmatrix} 1 & 0 & 0 \\ 0 & \cos\beta & \sin\beta \\ 0 & \sin\beta & \cos\beta \end{bmatrix}$$

$$\times \begin{bmatrix} \cos\gamma & 0 & \sin\gamma \\ 0 & 1 & 0 \\ -\sin\gamma & 0 & \cos\gamma \end{bmatrix} \begin{bmatrix} \cos\alpha & -\sin\alpha & 0 \\ \sin\alpha & \cos\alpha & 0 \\ 0 & 0 & 1 \end{bmatrix}. \qquad (5.3)$$

The Z' axis rotation α is related to the face azimuth, a_f (see §5.1), as follows (units are degrees):

$$\alpha = a_f - 180 \qquad 0 < a_f \leqslant 180$$
$$\alpha = a_f \qquad 180 < a_f \leqslant 360.$$

The X' axis rotation β is related to the face elevation, e_f, as follows:

$$\beta = -e_f \qquad -90 \leqslant e_f \leqslant 90$$

and the Y' axis rotation γ is related to the angle, t_f, made by the line joining the first two vertices of the face with the horizontal:

$$\gamma = -t_f \qquad 0 \leqslant t_f < 360.$$

Projection: The vertices of each obstruction object can now be projected onto the $X''Z''$ plane of the $X''Y''Z''$ coordinate system to give the vertices of the projected shadow polygons relative to the local face coordinate system. Figure 5.6 shows the projection of some point $(x''y''z'')$ onto some target face with associated (and relocated) coordinate system $X''Y''Z''$ to give some projected point $(x_P y_P z_P)$. Now

$$\tan \xi_1 = BC/AB = (x'' - x_P)/y'' \qquad \text{for } \xi_1 \text{ positive as shown}$$
$$= (x_P - x'')/y'' \qquad \text{for } \xi_1 \text{ negative}$$

$$\Rightarrow x_P = x'' \pm y'' \frac{\sin \xi_1}{\cos \xi_1}.$$

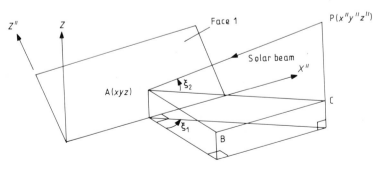

Figure 5.6 Projection of a point P on to the plane $X''Z''$.

Also,

$$\cos \xi_1 = AB/AC = y''/AC$$
$$\Rightarrow AC = y''/\cos \xi_1 \tag{5.4}$$

and

$$\tan \xi_2 = PC/AC$$
$$= (z'' - z_P)/AC \qquad \text{for } \xi_2 \text{ positive as shown}$$
$$= (z_P - z'')/AC \qquad \text{for } \xi_2 \text{ negative}$$
$$\Rightarrow z_P = z'' \pm AC \tan \xi_2. \tag{5.5}$$

Substituting equation (5.4) in (5.5) gives

$$z_P = z'' \pm y'' \tan \xi_2/\cos \xi_1 \qquad \text{and} \qquad y_P = 0.$$

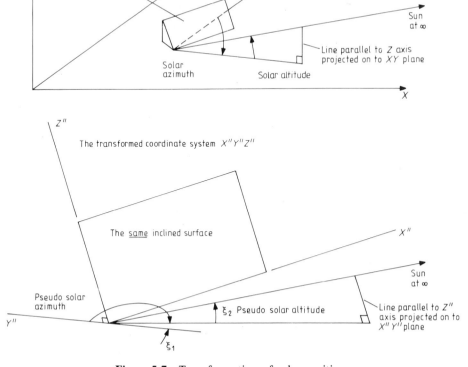

Figure 5.7 Transformation of solar position.

Note that if $y'' \geqslant 0$ then point P lies 'behind' the face relative to the sun and is therefore omitted from processing. This will have algorithmic implications in the case of bodies partially behind and partially in front of the face in question.

In matrix notation, projection is given by

$$(x_P y_P z_P) = (x'' y'' z'') \pm \begin{bmatrix} 1 & 0 & 0 \\ \dfrac{\sin \xi_1}{\cos \xi_1} & 1 & \pm \dfrac{\tan \xi_2}{\cos \xi_1} \\ 0 & 0 & 1 \end{bmatrix} \qquad (5.6)$$

where decisions on the addition and subtraction operators are made on the basis of the sign of ξ_1 and ξ_2. These angles are pseudo solar azimuth and elevation angles formed by direct transformation of the real solar angles in the XYZ coordinate system. This transformation is shown in figure 5.7 and the procedure for arriving at ξ_1 and ξ_2 is given in §5.2.3.

5.2.2 *The complete translation, rotation and projection equations*

Equations (5.3) and (5.6) can now be combined as follows:

$$(x_P y_P z_P) = (xyz1) \begin{bmatrix} 1 & 0 & 0 \\ 0 & 1 & 0 \\ 0 & 0 & 1 \\ -x0 & -y0 & -z0 \end{bmatrix} \begin{bmatrix} 1 & 0 & 0 \\ 0 & \cos \beta & -\sin \beta \\ 0 & \sin \beta & \cos \beta \end{bmatrix}$$

$$\times \begin{bmatrix} \cos \gamma & 0 & \sin \gamma \\ 0 & 1 & 0 \\ -\sin \gamma & 0 & \cos \gamma \end{bmatrix} \begin{bmatrix} \cos \alpha & -\sin \alpha & 0 \\ \sin \alpha & \cos \alpha & 0 \\ 0 & 0 & 1 \end{bmatrix} \begin{bmatrix} 1 & 0 & 0 \\ \pm \dfrac{\sin \xi_1}{\cos \xi_1} & 1 & \pm \dfrac{\tan \xi_2}{\cos \xi_1} \\ 0 & 0 & 1 \end{bmatrix}$$

where $x0, y0, z0$ define the new face origin expressed in old coordinates, β is the face elevation angle as defined in §5.1, γ is the local face x axis tilt as defined in §5.1, α is related to the face azimuth as defined in §5.1, ξ_1 is related to the solar azimuth when expressed relative to the face coordinate system, and ξ_2 is related to the solar altitude when expressed relative to the face coordinate system.

Expanding this transformation equation set gives

$$(x_P y_P z_P) = [(x - x0), (y - y0) \cos \beta + (z - z0) \sin \beta, -(y - y0) \sin \beta + (z - z0) \cos \beta]$$

$$\times \begin{bmatrix} \cos \gamma & 0 & \sin \gamma \\ 0 & 1 & 0 \\ -\sin \gamma & 0 & \cos \gamma \end{bmatrix} \begin{bmatrix} \cos \alpha & -\sin \alpha & 0 \\ \sin \alpha & \cos \alpha & 0 \\ 0 & 0 & 1 \end{bmatrix} \begin{bmatrix} 1 & 0 & 0 \\ \pm \sin \xi_1 / \cos \xi_1 & 1 & \pm \tan \xi_2 / \cos \xi_1 \\ 0 & 0 & 1 \end{bmatrix}.$$

Assuming $x - x0 = x_T$, $y - y0 = y_T$ and $z - z0 = z_T$, then

$$(x_P y_P z_P) = (x_T \cos \gamma \cos \alpha + y_T \sin \beta \sin \gamma \cos \alpha - z_T \cos \beta \sin \gamma \cos \alpha$$
$$+ y_T \cos \beta \sin \alpha + z_T \sin \beta \sin \alpha, \quad -x_T \cos \gamma \sin \alpha$$
$$- y_T \sin \beta \sin \gamma \sin \alpha + z_T \cos \beta \sin \gamma \sin \alpha + y_T \cos \beta \cos \alpha$$
$$+ z_T \sin \beta \cos \alpha, \quad x_T \sin \gamma - y_T \sin \beta \cos \gamma + z_T \cos \beta \cos \gamma)$$

$$\times \begin{bmatrix} 1 & 0 & 0 \\ \pm \sin \xi_1 / \cos \xi_1 & 1 & \pm \tan \xi_2 / \cos \xi_1 \\ 0 & 0 & 1 \end{bmatrix}$$

$$= (x_{TR} y_{TR} z_{TR}) \begin{bmatrix} 1 & 0 & 0 \\ \pm \sin \xi_1 / \cos \xi_1 & 1 & \pm \tan \xi_2 / \cos \xi_1 \\ 0 & 0 & 1 \end{bmatrix}$$

where the point $(x_{TR} y_{TR} z_{TR})$ is the original point (xyz) but expressed relative to some new coordinate system arrived at by axes translation and rotation only. That is,

$$x_{TR} = x_T \cos \gamma \cos \alpha + y_T \sin \beta \sin \gamma \cos \alpha - z_T \cos \beta \sin \gamma \cos \alpha$$
$$+ y_T \cos \beta \sin \alpha + z_T \sin \beta \sin \alpha$$
$$y_{TR} = -x_T \cos \gamma \sin \alpha - y_T \sin \beta \sin \gamma \sin \alpha + z_T \cos \beta \sin \gamma \sin \alpha$$
$$+ y_T \cos \beta \cos \alpha + z_T \sin \beta \cos \alpha$$
$$z_{TR} = x_T \sin \gamma - y_T \sin \beta \cos \gamma + z_T \cos \beta \cos \gamma. \tag{5.7}$$

Further matrix multiplication gives, for translation, rotation and projection,

$$x_P = x_T \cos \gamma \cos \alpha + y_T \sin \beta \sin \gamma \cos \alpha - z_T \cos \beta \sin \gamma \cos \alpha$$
$$+ y_T \cos \beta \sin \alpha + z_T \sin \beta \sin \alpha - x_T \cos \gamma \sin \alpha (\pm \tan \xi_1)$$
$$- y_T \sin \beta \sin \gamma \sin \alpha (\pm \tan \xi_1) + z_T \cos \beta \sin \gamma \sin \alpha (\pm \tan \xi_1)$$
$$+ y_T \cos \beta \cos \alpha (\pm \tan \xi_1) + z_T \sin \beta \cos \alpha (\pm \tan \xi_1)$$
$$y_P = -x_T \cos \gamma \sin \alpha - y_T \sin \beta \sin \gamma \sin \alpha + z_T \cos \beta \sin \gamma \sin \alpha$$
$$+ y_T \cos \beta \cos \alpha + z_T \sin \beta \cos \alpha$$
$$z_P = -x_T \cos \gamma \sin \alpha (\pm \tan \xi_2 / \cos \xi_1) - y_T \sin \beta \sin \gamma \sin \alpha (\pm \tan \xi_2 / \cos \xi_1)$$
$$+ z_T \cos \beta \sin \gamma \sin \alpha (\pm \tan \xi_2 / \cos \xi_1) + y_T \cos \beta \cos \alpha (\pm \tan \xi_2 / \cos \xi_1)$$
$$+ z_T \sin \beta \cos \alpha (\pm \tan \xi_2 / \cos \xi_1) + x_T \sin \gamma - y_T \sin \beta \cos \gamma$$
$$+ z_T \cos \beta \cos \gamma. \tag{5.8}$$

5.2.3 An insolation algorithm

These transformation equations can now be utilised as the basis of an algorithm to determine external and internal surface (opaque and transparent) insolation as a function of building geometry, obstructions geometry and solar position.

For external surfaces, one possible algorithm involves the following computational procedure. Each face of the target building is processed in turn and the surrounding obstructions, including other obstructing parts of the target building, projected onto the plane of each face. For any solar position and target face this is achieved by:

1 Transforming the current solar position, relative to the site coordinate system XYZ, to pseudo solar azimuth and elevation angles (ξ_1 and ξ_2) expressed relative to the local face coordinate system. This can be done by expressing the sun's position as a distant point (xyz) for insertion, along with the target face angles (α, β, γ), in equations (5.7). The point to emerge (x_{TR}, y_{TR}, z_{TR}) can then be re-expressed as angles ξ_1 and ξ_2 as shown in figure 5.7.
2 Each vertex of any obstruction body can now be projected on to the plane of the target face by application of equations (5.8) to give a shadow 'image'. This will appear as a two-dimensional representation of a three-dimensional object *but with all vertices of the image in the same local face $X''Z''$ plane* as shown in figure 5.6.
3 The next stage is to remove all internal line segments to leave only the shadow polygon which may or may not intersect with the target face.
4 After steps 2 and 3 have been applied to all participating obstruction bodies, the shadow polygons can be combined, by polygon union operations, to give the final shadow polygon set for display purposes. This can be done by applying the maxim that if two polygons overlap and an intersection point is located then union is obtained by retaining this point before proceeding clockwise to the next vertex (of either polygon) which does not incur an intersection on the way. The procedure continues clockwise around that polygon, retaining vertices, until the next intersection point is located. This point is retained and, changing polygons, the clockwise retention procedure is continued until either an intersection point is again reached (indicating polygon changeover) or the starting point is encountered to terminate the process.
5 The intersection between each distinct shadow polygon to emerge from step 4 with the target face is then determined by a similar procedure, but one which commences at a point which is a vertex of one polygon and contained by the other.
6 Final shading, and therefore insolation, estimation can be determined by simple point containment tests applied to a grid superimposed on the target face. Each grid point is assigned a value of zero to indicate that the point is insolated. Any point contained by a shadow polygon is then reassigned a value of one to indicate shading. This allows both magnitude and point-of-application estimation, with accuracy controlled by varying the grid size. Point containment can be established by radiating a line away from the point in any arbitrary direction. If the number of polygon intersections is odd the point is contained and, if even, it is not. Note that if shading pattern display is not an objective then the union and intersection operations of steps 4 and

5 can be discarded, with grid point shading determined on the basis of containment tests against each shading polygon, some of which may overlap. Although the number of point containment tests will rise, the overall computational effort may fall since polygon operations are usually more demanding.

For internal surfaces, insolation patch position and magnitude can be assessed by applying a grid to each window so that each grid point can be projected on to internal surfaces (by equations (5.8)) to establish, again by point containment tests, the grid point/receiving surface pairings.

The time-series data to emerge for each external and internal surface can now be utilised in the solar computations of the next section.

5.3 Shortwave radiation processes

Consider figure 5.8 which details the typical interaction between a group of building zones and the incident direct and diffuse beam. The direct and diffuse

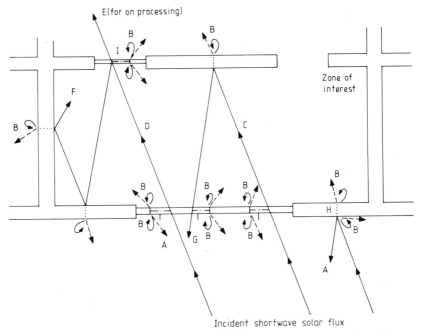

Figure 5.8 Building/solar interaction. A: reflected shortwave flux; B: ultimate flux emission by convection and longwave radiation; C: shortwave transmission to cause opaque surface insolation; D: shortwave transmission to cause transparent surface insolation; E: shortwave transmission to adjacent zone; F: enclosure reflections; G: shortwave loss; H: solar energy penetration by transient conduction; I: solar energy absorption prior to retransmission by the processes of B.

shortwave flux incident on external opaque surfaces will be partially absorbed and partially reflected and, subsequently, some portion of the absorbed component may be transmitted to the corresponding interior surface, by conduction processes, to elevate inside surface temperatures and so enter a building zone by the internal surface convective and radiative flowpaths. At the same time some portion of the absorbed flux will contribute to outside surface temperature elevation and so may be re-released to outside and lost from the system. If a multi-layered construction is opaque overall, but has transparent elements located at its outermost surface, then some portion of the incident direct and diffuse radiation flux will, in addition to the foregoing absorption and reflection, be transmitted inward until it strikes the intra-constructional opaque interface where absorption and reflection will again occur. In this case any incident shortwave energy will be partially absorbed, partially reflected and partially transmitted as it travels outwards; the process continuing, essentially instantaneously, until complete flux diminution.

With completely transparent multi-layered constructions such as windows, the direct and diffuse beam is partially reflected, partially absorbed and partially transmitted at each interface so that, if the construction is considered in total, the reflected component is lost from the system, the absorbed component is transmitted inward and outward by the conductive, convective and longwave radiative processes, and the transmitted component appears as shortwave flux within the corresponding zone. This directly transmitted beam will then continue on to cause internal surface insolation as a function of system geometry. This can be visualised as a time-dependent internal surface patch determined by the projection of the insolated portion of the transparent surface, at the prevailing solar incidence angle, onto the internal zone surfaces (by the theory of §5.2). Treatment of this directly incident flux will depend on the nature of the receiving surface(s): absorption and reflection for an opaque surface; or absorption, reflection and transmission (perhaps to another zone or back to outside) in the case of a transparent surface. If the internal surface is a specular reflector then the reflected beam can be tracked onwards by ray tracing techniques until diminished in magnitude below some predetermined small level. Alternatively, if the zone surfaces can be considered as diffuse or general reflectors, the apportioning of the reflected flux can be determined on the basis of view factor relationships if known (see §5.4) or, perhaps, by simple area weighting techniques if such an approximation is acceptable. The diffusely transmitted beam can be distributed over zone surfaces on the basis of view factor relationships between the sending transparent surface and the receiving surfaces. This treatment will introduce the essential bias away from surfaces in the same plane as the transmitter. As with the directly transmitted beam, the then reflected components can be apportioned between all zone surfaces according to some preferred distribution criteria.

The causal effect of these shortwave processes is implicit in the finite volume energy balance formulations of chapter 3, given that the shortwave flux injections at each discrete node can be established at each computational time-row. Simulation modelling requirements can therefore be summarised as time-series shortwave injection for nodes representing external opaque and transparent surfaces (see §5.3.2), intra-constructional elements, if these are part of window systems or other multi-layered constructions with transparent surface elements allowing shortwave penetration (see §5.3.3), and internal opaque and transparent surfaces (see §5.3.4).

The mathematical expressions which follow relate to these requirements and assume the existence of direct and diffuse radiation intensities on a horizontal plane and under real sky conditions. The availability of such data, their typicality and prediction options are the subject matter of §5.8.

5.3.1 Solar position

It is usual to express the position of the sun in terms of its altitude and azimuth angles which, in turn, depend on site latitude, solar declination and local solar time. Figure 5.9 illustrates these angles which are discussed in detail and established as mathematical expressions in a number of texts (for example IHVE 1973, ASHRAE 1981, Duffie and Beckman 1980). For this reason only cursory treatment is given here.

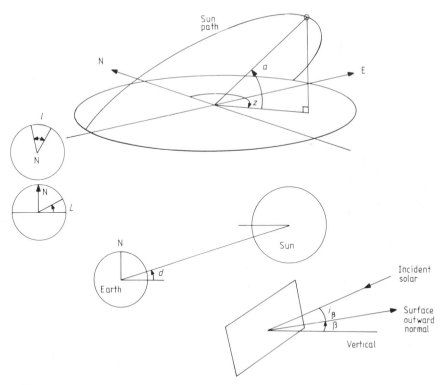

Figure 5.9 Solar position. The azimuth is denoted by z; a is the altitude, L latitude, l longitude, d declination and N is north.

The solar declination can be determined from

$$d = 23.45 \sin(280.1 + 0.9863\,Y)$$

where d is the solar declination in radians and Y is the year day number (January $1 = 1$, February $1 = 32$ etc). The solar altitude is then given by

$$a = \sin^{-1}[\cos(L)\cos(d)\cos(\theta_h) + \sin(L)\sin(d)]$$

where a is the solar altitude (radians), L the site latitude (north positive, south negative), and θ_h the hour angle (radians). The hour angle is the angular expression of solar time and is positive for times before solar noon and negative for times after. It is found from

$$\theta_h = 15(\,|\,12 - t_s\,|\,)$$

where t_s is the local apparent time or solar time. This is a time scale which relates to the apparent angular motion of the sun across the sky vault with solar noon corresponding to the point in time at which the sun traverses the meridian of the observer. It is important to note that solar time does not necessarily coincide with local mean (or clock) time, t_m, with the difference given by

$$t_s - t_m = \pm l/15 + e_t + \delta \tag{5.9}$$

where l is the longitude difference (degrees), e_t the equation of time (hours) and δ the possible correction for daylight saving (hours).

The longitude difference is the difference between an observer's actual longitude and the longitude of the mean or reference meridian for the local time zone. The difference is positive for locations to the west of the reference meridian and negative to the east. For Great Britain the reference meridian is at 0 degrees and local mean time is known as Greenwich Mean Time (GMT). Thus equation (5.9) becomes

$$t_s = \text{GMT} \pm l'/15 + e_t + \delta$$

where l' is the actual longitude of the observer (degrees). The equation of time makes allowance for the observed disturbances to the earth's rate of rotation and can be determined from

$$e_t = 9.87 \sin(1.978\,Y - 160.22) - 7.53 \cos(0.989\,Y - 80.11)$$
$$- 1.5 \sin(0.989\,Y - 80.11).$$

The solar azimuth angle is given by

$$z = \sin^{-1}(\cos d \sin \theta_h/\cos a).$$

In applying this equation it is necessary to distinguish between northern and southern latitudes and so the azimuthal corrections of table 5.1 should be applied.

The angular relationship between the direct solar beam and a surface of arbitrary inclination β, as shown in figure 5.9, is also of prime importance in solar modelling. The surface–solar incidence angle is given by

$$i_\beta = \cos^{-1}[\sin a \cos(1 - \beta) + \cos a \cos \omega \sin(1 - \beta)]$$

where β is the angle between the surface outward normal and the horizontal, upward-facing positive (see §6.1), ω is the surface–solar azimuth $= |a - \alpha|$, and α is the surface azimuth (see §5.1).

Note that negative values of $\cos(i_\beta)$ imply that the surface in question faces away from the sun and cannot be directly insolated.

Table 5.1 Azimuthal corrections (from IHVE 1973).

Condition	Time	Northern latitude	Southern latitude
$x < y$	AM	Z	$180 - Z$
	PM	$360 - Z$	$180 + Z$
$x = y$	AM	90	90
	PM	270	270
$x > y$	AM	$180 - Z$	Z
	PM	$180 + Z$	$360 - Z$

$x = \cos(15 |t - 12|)$ $\qquad y = \tan d \cot L$

5.3.2 Intensity of direct and diffuse radiation on inclined surfaces

The total radiation incident on an exposed opaque or transparent surface of arbitrary inclination β and with azimuth α has three potential components: direct beam, ground reflected and sky diffuse (it is usual to disregard reflections from surrounding buildings).

The direct component is relatively straightforward to assess since it involves only angular operations on the known horizontal direct intensity. To determine the ground reflected component it is normal practice to treat the ground as a diffuse reflector and so the combined horizontal direct and diffuse radiation is treated isotropically and reflected on to the inclined surface as a function of some representative view factor.

Estimation of the sky diffuse component is more problematic since the known horizontal diffuse intensity exhibits a mix of true diffuse behaviour (caused by general atmospheric scatter) and directional qualities; the mix varying between totally overcast conditions (fully isotropic sky) and completely clear skies (fully anisotropic sky). One possibility is to assume isotropic sky conditions at all times and so simplify computation since diffuse radiation is then independent of direction. It has been estimated that such an assumption leads to an underestimation of inclined surface intensities (Ma and Iqbal 1983). An alternative approach is to formulate an anisotropic model which accounts for the enhancement of diffuse radiation in the vicinity of the sun and at the horizon.

Direct beam component
This is given by

$$I_{d\beta} = I_{dh} \cos(i_\beta)/\sin a$$

where $I_{d\beta}$ is the direct intensity on an inclined surface (W m^{-2}), and I_{dh} is the direct intensity on the horizontal (W m^{-2}).

Ground reflected component
For an unobstructed vertical surface ($\beta = 0$) the view factor between the surface and the ground and between the surface and the sky is in each case 0.5, and so the radiation intensity at the surface due to isotropic ground reflection of the combined direct and diffuse radiation impinging on the horizontal is given by

$$I_{rv} = 0.5(I_{dh} + I_{fh})r_g$$

where I_{rv} is the ground reflected total radiation incident on a vertical surface (W m^{-2}), I_{fh} is the horizontal diffuse radiation (W m^{-2}) and r_g is the ground reflectivity (albedo).

For a surface of non-vertical inclination a simple view factor modification is introduced so that

$$I_{r\beta} = 0.5(1 - \cos \beta)(I_{dh} + I_{fh})r_g$$

where $I_{r\beta}$ is the ground reflected total radiation incident on a surface of inclination β (W m^{-2}).

Sky diffuse component
Two approaches are prominent in the treatment of anisotropic sky conditions.

In the first (Temps and Coulson 1977, Klucher 1979) the sky diffuse component on an inclined surface is determined by an expression which increases the intensity of the diffuse flux due to circumsolar activity and horizon brightening:

$$I_{s\beta} = I_{fh}[0.5(1 + \cos\beta)]\,\{1 + [1 - (I_{fh}^2/I_{Th}^2)]\sin^3(0.5\beta)\}$$
$$\times \{1 + [1 - (I_{fh}^2/I_{Th}^2)]\sin^3(90 - a)\}$$

where $I_{s\beta}$ is the sky diffuse radiation incident on a surface of inclination β(W m^{-2}) and I_{Th} is the total radiation on the horizontal, $I_{dh} + I_{fh}$ (W m^{-2}). Thus, when the sky is completely overcast, $I_{fh}/I_{Th} = 1$ and the expression reduces to the isotropic sky case.

In the second approach (Hay 1979) the known horizontal diffuse radiation is assumed to be composed of a uniform background diffuse component and a circumsolar component, with a weighting applied according to the degree of sky isotropy. The sky diffuse component is given by

$$I_{s\beta} = I_{fh}[(I_{dh}/I_{eh}) \cos i_\beta \cos(90 - a) + 0.5(1 - \cos \beta)(1 - I_{dh}/I_{eh})]$$

where I_{eh} is the extraterrestrial radiation incident on the horizontal (W m^{-2}). Again, as overcast sky conditions are approached so $I_{dh} \to 0$ and the isotropic expression is obtained.

The foregoing equations allow the computation of direct and diffuse shortwave radiation impinging on exposed external surfaces. These flux quantities, when modified by representative surface absorptivities, are the shortwave nodal heat generation terms, q_{SI}, of the simulation equations derived in chapter 3.

5.3.3 Reflection, absorption and transmission within transparent media

In the energy balance formulations of §3.2.1 relating to multi-layered composites such as windows and mixed transparent/opaque systems, heat generation terms were included. Some portion of this heat generation will be due to the absorption of shortwave energy as it passes through the transparent media. The objective of this section is two-fold: to describe techniques which allow the computation of the shortwave absorption within the various finite volumes representing the system and to formulate the whole system transmission and absorption in the case of window systems for subsequent use in §5.3.4.

The solar spectrum occupies that part of the electromagnetic spectrum extending from 0.3–5 μm. Thus the spectral reflectance (r) of a window system is given by

$$r = \int_{\lambda=0.3}^{\lambda=5.0} I_\lambda r_\lambda \, d\lambda \bigg/ \int_{\lambda=0.3}^{\lambda=5.0} I_\lambda \, d\lambda$$

where I_λ is the spectral irradiance (W m^{-2}). The objective of this section is to consider techniques for the determination of r (and transmittance and absorptance) for various window systems.

Consider figure 5.10 which shows a single transparent substrate subjected to some ambient beam or, alternatively, some beam transmitted from an adjacent layer. It is necessary to treat this radiation as two polarised vectors, one parallel ($\parallel$) and one perpendicular ($\perp$) to the substrate. Assuming solar radiation to be unpolarised, then the intensity of the incident radiation will consist of equal quantities of each component of polarisation so that

$$I_1 = 0.5(I_\parallel + I_\perp).$$

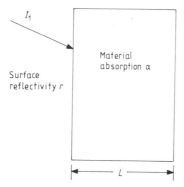

Figure 5.10 A transparent material subjected to shortwave flux (I_1).

The interface reflectivity is then given for each electric vector by Fresnel's specular reflection equations (Vasicek 1960):

$$r_\| = \frac{\tan^2(\theta_i - \theta_r)}{\tan^2(\theta_i + \theta_r)} \qquad r_\perp = \frac{\sin^2(\theta_i - \theta_r)}{\sin^2(\theta_i + \theta_r)}$$

where $r_\|$ is the reflectivity for the parallel vector, $r_\perp$ the reflectivity for the perpendicular vector, θ_i the beam angle of incidence and θ_r the corresponding angle of refraction. The interface transmissivity follows from

$$\tau_\xi = 1 - r_\xi$$

where ξ is $\|$ or $\perp$ as required.

As the separate radiation vectors pass through the transparent medium, absorption occurs and beam intensity can be assumed to diminish according to Bouguer's law (Duffie and Beckman 1980) which is based on the assumption that the absorbed radiation is proportional to the local intensity within the medium and the path length so that

$$dI/dx = -KI$$

where dx is the penetration path length, dI the intensity reduction, and K the absorption extinction coefficient. Table 5.2 gives some typical values for the extinction coefficient.

Table 5.2 Typical values for the extinction coefficient.

Material	$K(\text{cm}^{-1})$
Polyvinyl fluoride (Tedlar)	1.4
Teflon	0.59
Polyethylene (Mylar)	2.05
Ordinary window glass	0.3
White glass	0.04
Heat absorbing glass	1.3 (to 2.7)

Integrating over the total path length $0 \rightarrow L/\cos\theta_r$, where L is the medium thickness normal to the surface of the substrate, gives

$$\int_{I_1}^{I_2} dI/I = \int_0^{L/\cos\theta_r} - K\,dx$$

$$\Rightarrow \ln(I_2/I_1) = -KL$$

$$\Rightarrow I_2 = I_1 \exp(-KL/\cos\theta_r)$$

and so the overall absorption fraction is given by

$$\alpha = (I_1 - I_2)/I_1 = 1 - \exp(-KL/\cos\theta_r). \qquad (5.10)$$

The index of refraction μ is defined by Snell's Law as

$$\mu = \sin \theta_i / \sin \theta_r. \tag{5.11}$$

This index is subject to some variation depending on the particular wavelength being considered. This phenomenon is termed dispersion. In general the refractive index decreases as wavelength increases and the rate of decrease is greatest at the shorter wavelengths. Equations describing the dispersion behaviour of different materials can be found in the literature (for example Optical Society of America 1978). Now

$$\cos \theta_r = (1 - \sin^2 \theta_r)^{\frac{1}{2}}$$

and so, from equation (5.11),

$$\cos \theta_r = \left(1 - \frac{\sin^2 \theta_i}{\mu^2}\right)^{\frac{1}{2}}.$$

Substitution in equation (5.10) gives the absorptivity as a function of the prevailing incidence angle:

$$\alpha = 1 - \exp\left[-KL\bigg/\left(1 - \frac{\sin^2 \theta_i}{\mu^2}\right)^{\frac{1}{2}}\right]$$

and tables of KL values for different transparent materials can be found in the literature.

Consider now the superimposition (on figure 5.10) of the flowpath of one radiation vector as it undergoes multiple reflections at each interface and between-interface absorption as shown in figure 5.11. The total absorption within the transparent element is given by

$$A = \left(\sum^{\infty} F_{A2} - \sum^{\infty} F_{R1}\right) + \left(\sum^{\infty} F_{A1} - \sum^{\infty} F_{R2}\right) \tag{5.12}$$

where F_{A1} is the flux to *arrive* at interface 1 from interface 2, F_{A2} is the flux to *arrive* at interface 2 from interface 1, F_{R1} is the flux *reflected* at interface 1 towards interface 2 and F_{R2} is the flux *reflected* at interface 2 towards interface 1. Now

$$\sum^{\infty} F_{A2} = I_1(1 - r_{1\xi})\tau_a + I_1(1 - r_{1\xi})\tau_a^3 r_{2\xi} r_{1\xi} + I_1(1 - r_{1\xi})\tau_a^5 r_{2\xi}^2 r_{1\xi}^2 + \ldots$$

$$= I_1(1 - r_{1\xi})\tau_a(1 + \tau_a^2 r_{2\xi} r_{1\xi} + \tau_a^4 r_{2\xi}^2 r_{1\xi}^2 + \ldots)$$

where τ_a is $(1 - \alpha)$ and relates to the incidence angle at interface 1 and ξ is $\parallel$ or $\perp$, the parallel or perpendicular components corresponding to the incidence angle at interface 1. The terms within the square brackets constitute an infinite summation series with common ratio $\tau_a r_{2\xi} r_{1\xi}$ and so

$$\sum^{\infty} F_{A2} = \frac{I_1(1 - r_{1\xi})\tau_a}{1 - \tau_a^2 r_{2\xi} r_{1\xi}}.$$

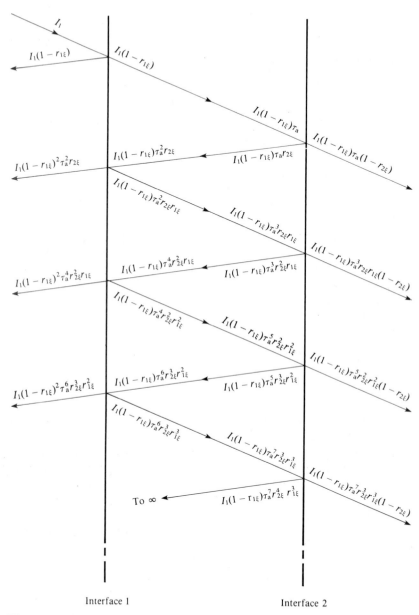

Figure 5.11 Multiple reflections and material absorption for a component of polarisation ξ.

Similarly,

$$\sum_{}^{\infty} F_{A1} = \frac{I_1(1 - r_{1\xi})\tau_a^2 r_{2\xi}}{1 - \tau_a^2 r_{2\xi} r_{1\xi}} \qquad \sum_{}^{\infty} F_{R1} = \frac{I_1(1 - r_{1\xi})}{1 - \tau_a^2 r_{2\xi} r_{1\xi}} \qquad \sum_{}^{\infty} F_{R2} = \frac{I_1(1 - r_{1\xi})\tau_a r_{2\xi}}{1 - \tau_a^2 r_{2\xi} r_{1\xi}}$$

Substitution in expression (5.12) gives

$$A = \frac{I_1(1 - r_{1\xi})\,[\tau_a - 1 + \tau_a^2 r_{2\xi} - \tau_a r_{2\xi}]}{1 - \tau_a^2 r_{2\xi} r_{1\xi}}. \tag{5.13}$$

If the transparent element exists in isolation with radiation incident only at interface 1 then this expression gives the absorption of the parallel and perpendicular components ($\xi = \parallel$; $\perp$) which can be summed to give the total absorption within the element. If shortwave radiation is simultaneously impinging on interface 2 (as with single glazing externally insolated and receiving internal reflections from surrounding surfaces) then further absorption of the second beam I_2 will occur. Expression (5.13) is then applied with $I_1 = I_2$ and the total absorption obtained by summing the component absorptions of each individual beam. If the transparent element is a member of a multi-layered group then two possibilities exist: some portion of the radiation impinging on interface 1 is transmitted on to an adjacent transparent element with some portion being subsequently retransmitted back across the common interface; or the second element is opaque. In the latter case the total absorption at the opaque surface is given by

$$I_1(1 - r_{1\xi})\tau_a(1 - r_{2\xi}) + I_1(1 - r_{1\xi})\tau_a^3 r_{2\xi} r_{1\xi}(1 - r_{2\xi}) + I_1(1 - r_{1\xi})\tau_a^5 r_{2\xi}^2 r_{1\xi}^2(1 - r_{2\xi}) + \ldots$$

$$= \frac{I_1(1 - r_{1\xi})\tau_a(1 - r_{2\xi})}{1 - \tau_a^2 r_{2\xi} r_{1\xi}}. \tag{5.14}$$

In the former case expression (5.14) (after both vector components have been added) defines the total onward flux transmitted to the adjacent element. If the interface reflectivities are known then it is possible to process the second element independently by treating the onward transmitted flux from element 1 as the initial incident flux on element 2. In this way, the process can continue for one forward pass through the multi-layered construction until all elements have been considered or until an opaque element is encountered. The flux transmitted from one element to the next will be partly polarised even if the original beam was not. It is therefore necessary to process each component of polarisation separately before combining the overall value for the entire system.

This is not the end of the process, however, since each element will reflect flux back to the element from which it initially received the shortwave energy. This is given by

$$I_i r_{i1\xi} + I_i(1 - r_{i1\xi})^2 \tau_{ai}^2 r_{i2\xi} + I_i(1 - r_{i1\xi})^2 \tau_{ai}^4 r_{i2\xi}^2 r_{i1\xi} + I_i(1 - r_{i1\xi})^2 \tau_{ai}^6 r_{i2\xi}^3 r_{i1\xi}^2 + \ldots$$

$$= I_i r_{i1\xi} + \frac{I_i(1 - r_{i1\xi})^2 \tau_{ai}^2 r_{i2\xi}}{1 - \tau_{ai}^2 r_{i2\xi} r_{i1\xi}}$$

where i is the element number.

It is therefore possible to establish an algorithm to transmit a single flux package from one transparent element to the other as calculation 'sweeps' are made in alternating directions until the flux quantities diminish in magnitude below some predetermined level. In this way the total absorption of each element can be determined as well as the total transmissivity of the combined system.

Let A_i be the absorption for element i and τ be the combined system transmissivity and consider the volumetric subdivision of chapter 3 applied to the transparent element of figure 5.10 with nodes situated at the element boundaries and at the centre plane. The fraction of the total absorption associated with each node is determined by a volumetric weighting. Given that the total path length of radiation passing from one boundary to the other is given by $L/\cos(\theta_r)$ then the fraction received by a boundary node is given by

$$q_{SI} = A_i L/4 \cos \theta_r$$

or

$$q_{SI} = A_i L/\{4[1 - (\sin^2\theta_i)/\mu^2]^{1/2}\} \tag{5.15}$$

and for a centre plane node

$$q_{SI} = A_i L/\{2[1 - (\sin^2 \theta_i)/\mu^2]^{1/2}\}.$$

If the boundary node separates two transparent elements then the nodal short-wave heat generation terms will have contributions from both elements. If the node separates a transparent element and an opaque element then the absorption of equation (5.15) must be added to that of equation (5.14) to give

$$q_{SI} = A_i L/\{4[1 - (\sin^2 \theta_i)/\mu^2]^{1/2}\} + \frac{I_i(1 - r_{i1\xi})(1 - r_{i2\xi})\tau_{ai}}{1 - \tau_{ai}^2 r_{i2\xi} r_{i1\xi}}.$$

As a low-energy feature some contemporary window systems incorporate thin films evaporated on to the outer glass surface to alter the reflectance of the glazing system. Figure 5.12 shows the measured spectra of various oxide films (from Howson *et al* 1984) which aim to reduce reflectance in the shortwave region and increase reflectance in the longwave region.

Vasicek (1960) suggested that a thin film is one for which

$$\mu_f t < 2.5\lambda_d$$

where μ_f is the refractive index of the film, t the film thickness (μm) and λ_d the spectral wavelength for which reflectance modification is required. Films which violate this condition can be treated in the same manner as glazing elements; that is, by the technique previously outlined. With thin films the change of phase of the radiation vector as it traverses the film must be included in the analysis. Vasicek has analysed the effects of the multiple reflections occurring within a thin film bounded by air and a transparent substrate. The film interface reflectance is given by

$$r_\xi = \left(\frac{am_1 + bm_2 - cm_3 - m_4}{am_1 + bm_2 + cm_3 + m_4}\right)^2$$

and table 5.3 gives the a, b, c and m values.

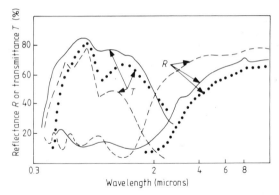

Figure 5.12 Measured spectra of various oxide films (from Howson *et al* 1984). Full curve cadmium 2:1 tin oxide; broken curve indium 10% tin oxide; dotted curve indium oxide.

Table 5.3 Values of a, b, c and m (from Rousseau and Mathieu 1973).

Coefficient	Normal incidence	Non-normal incidence $\parallel$ component	$\perp$ component
a	μ_0/μ	$\dfrac{\mu_0 \cos\theta_f}{\mu_s \cos\theta_i}$	$\dfrac{\mu_0 \cos\theta_i}{\mu_s \cos\theta_f}$
b	μ_0	$\dfrac{\mu_0}{\cos\theta_i}$	$\mu_0 \cos\theta_i$
c	$1/\mu_s$	$\dfrac{\cos\theta_f}{\mu_s}$	$1/\mu_s \cos\theta_f$
m_{11}	$\cos\delta_f$	$\cos\delta_f$	$\cos\delta_f$
m_{12}	$\dfrac{i\sin\delta_f}{\mu_f}$	$\dfrac{i\cos\theta_f \sin\delta_t}{\mu_f}$	$\dfrac{i\sin\delta_t}{\mu_f \cos\theta_f}$
m_{21}	$i\,\mu_f \sin\delta_t$	$\dfrac{i\,\mu_f \sin\delta_t}{\cos\theta_f}$	$i\,\mu_f \cos\theta_f \sin\delta_t$
m_{22}	$\cos\delta_f$	$\cos\delta_f$	$\cos\delta_f$

μ_0 = refractive index of air
μ_s = refractive index of substrate
θ_f = film angle of refraction
θ_i = beam angle of incidence
δ_f = change of phase
$\quad = 2\pi\mu_f\, t \cos\theta_f/\lambda_d$

A number of workers (Vasicek 1960, Heavens 1955, Born and Wolf 1965, Rousseau and Mathieu 1973) have produced formulations for multiple thin-film systems analysis based on matrix combination of individual films.

5.3.4 *Intra-zone shortwave distribution*

The formulations of §5.3.2 will permit calculation of the direct and diffuse intensities impinging on external building surfaces. For opaque surfaces, intensity modification by surface absorptivity and shading factors will give the shortwave heat injection to be applied to surface nodes via the excitation matrix (**C**) of chapters 3 and 4. For a transparent system, the formulations of §5.3.3 allow the assessment of internal element absorption for application, via the heat generation terms of equation (3.5), to intra-element nodes. The eventual heat exchange between surface nodes and the surrounding air (by convection) and other surfaces (by longwave radiation exchange) will then follow from the matrix equation solutions of chapter 4.

The formulations of §5.3.3 also allow the assessment of overall system properties such as transmissivity, absorptivity and reflectivity. This section addresses the use of these properties, in conjunction with the shading and insolation time-series information discussed in §5.2, to estimate the apportioning of shortwave energy between internal surfaces.

For a window system, the transmitted portion of the direct beam can be evaluated from

$$Q_{dt} = \frac{I_{dh}}{\sin a} \, \tau_{i_\beta}(1 - P_g)A_g \cos i_\beta \qquad (5.16)$$

where Q_{dt} is the transmitted direct beam flux (W), τ_{i_β} the overall transmissivity for the given incidence angle, P_g the window shading factor (proportion of 1) and $A_g \cos i_\beta$ the apparent window area (m^2). The τ value can be determined by the techniques of §5.3.3 or, alternatively, by reference to published data for different window arrangements and glass types (Pilkingtons 1973).

If, as is often the case, more than one internal surface will share this transmitted radiation, then the flux defined by equation (5.16) can be applied to those internal surfaces defined by the insolation data determined by the technique of §5.2. Any internal node will then receive a heat injection given by

$$q_{SI} = Q_{dt}P_I\alpha_I/A_s$$

where α_I is the surface absorptivity, A_s the surface area (m^2) and P_I the proportion of window direct beam transmission which strikes the surface in question (proportion of 1). The first reflected flux is given by

$$q_{RI} = q_{SI}(1 - \alpha_I)/\alpha_I.$$

The accumulated flux reflections from each surface can now be processed further to give the final apportioning between all internal surfaces. If the usual assumption of diffuse reflections is made, then apportionment can be decided on the basis of enclosure view factor information as described in the following section. For the case of specular reflections, a recursive ray tracing technique will be required.

It is important to remember that an internal surface may be composed of opaque and transparent portions with the latter allowing the onward transmission of incident shortwave flux to a connected zone or back to outside. Conservation of the integrity of multi-zone systems will be important in building systems incorporating passive solar features. Application of equation (5.16) with the $(I_{dh}/\sin a)$ term set to the incident flux value will then give the re-transmitted flux. The diffuse beam transmission can be determined from

$$Q_{ft} = (I_{s\beta} + I_{r\beta})A_g\tau_{51}\cos(51°)$$

where τ_{51} is the overall transmissivity corresponding to $51°$ incidence angle, representing the average approach angle for anisotropic sky conditions.

This flux quantity can now be processed by the technique described for the direct beam: internal surface smearing on the basis of specular or diffuse reflections as described in the following section concerned with the evaluation of longwave radiation exchanges.

5.4 Longwave radiation processes

Heat transfer by longwave radiation exchange between two surfaces in visual communication is an important area in building energy modelling but one which introduces mathematical complexity due to non-linear behaviour and the spatial relationship problems caused by complex geometries and inter-surface obstructions. In the following derivations internal and external exposed surfaces are treated separately since, in the latter case, detailed knowledge of surrounding surfaces (sky, buildings, ground) is usually unavailable; unless, of course, the modelling system has access to a detailed terrain and sky data structure.

The radiation flux emitted by a perfect 'black' body is defined by the relationship

$$q_b = \sigma A\theta^4 \tag{5.17}$$

where q_b is the black body radiation flux (W), σ is the Stefan–Boltzmann constant (W m^{-2} K^{-4}) and θ is the absolute temperature (K).

Real building materials do not behave as black bodies and deviate in their ability to absorb all incoming longwave energy. The radiant flux emitted by such a 'grey' body is given by a simple temperature-dependent modification to equation (5.17):

$$q = \epsilon\sigma A\theta^4 \tag{5.18}$$

where q is the grey body radiation flux (W) and ϵ is the temperature dependent surface emissivity.

Kirchhoff's law states that the emissivity of a surface is equal to its absorptivity. Although true for most contemporary building materials under normal temperature conditions, the law is violated by certain advanced surfaces—aiming to promote heat emission whilst minimising absorption—and by materials at high temperatures, where emission and absorption can take place at substantially different wavelengths. However, in the derivations which follow, Kirchhoff's law is accepted and so emissivity, and not absorptivity (α), appears throughout.

5.4.1 Exchange between internal surfaces

In any enclosure the radiation emitted by all surfaces will, after infinite reflections at all surfaces, be totally reabsorbed and in the process redistributed. Assuming, for the present purpose, that no energy is lost by longwave transmission directly to an adjacent enclosure and that the surface reflections are diffuse, then mathematically the solution of this problem is recursive, requiring that:

The initial flux emitted by all surfaces be tracked to first reflection at all other surfaces to give the first absorption at these surfaces. Thus if (say) four grey surfaces 1, 2, 3 and 4 are communicating to form an enclosure as shown in figure 5.13, then the flux emitted by each surface is given by

$$q_1 = \epsilon_1 \sigma A_1 \theta_1^4 \qquad q_2 = \epsilon_2 \sigma A_2 \theta_2^4$$
$$q_3 = \epsilon_3 \sigma A_3 \theta_3^4 \qquad q_4 = \epsilon_4 \sigma A_4 \theta_4^4$$

and, after each flux travels to all other surfaces (to arrive at the first reflection), the absorption at each surface will have contributions from each other surface such that:

$$a_1' = \qquad q_2 f_{2 \to 1} \epsilon_1 + q_3 f_{3 \to 1} \epsilon_1 + q_4 f_{4 \to 1} \epsilon_1$$
$$a_2' = q_1 f_{1 \to 2} \epsilon_2 \qquad\qquad + q_3 f_{3 \to 2} \epsilon_2 + q_4 f_{4 \to 2} \epsilon_2$$
$$a_3' = q_1 f_{1 \to 3} \epsilon_3 + q_2 f_{2 \to 3} \epsilon_3 \qquad\qquad + q_4 f_{4 \to 3} \epsilon_3$$
$$a_4' = q_1 f_{1 \to 4} \epsilon_4 + q_2 f_{2 \to 4} \epsilon_4 + q_3 f_{3 \to 4} \epsilon_4$$

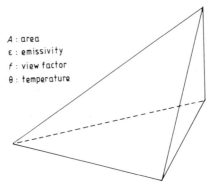

A : area
ϵ : emissivity
f : view factor
θ : temperature

Figure 5.13 Four grey bounding surfaces.

where a_i' is the total absorption of flux at grey surface i from all surfaces and at first reflection (W), $f_{j \to i}$ is the geometric view factor between surface j and i, and ϵ_i is the emissivity of surface i.

A single flux quantity can now be determined for each surface which represents the total flux apparently emanating from each surface and can be processed to next reflection at all other surfaces as before. This flux quantity is given by

$$r_i' = a_i'(1 - \epsilon_i)/\epsilon_i \qquad i = 1, 2, 3, 4$$

where r_i' is the total flux reflected at surface i after first reflection (W).

If this reflected flux is treated as the initial flux emission then after the second reflection the total absorption at each surface is given by:

$$a_1'' = a_1' \qquad\qquad + r_2' f_{2 \to 1} \epsilon_1 + r_3' f_{3 \to 1} \epsilon_1 + r_4' f_{4 \to 1} \epsilon_1$$
$$a_2'' = a_2' + r_1' f_{1 \to 2} \epsilon_2 \qquad\qquad + r_3' f_{3 \to 2} \epsilon_2 + r_4' f_{4 \to 2} \epsilon_2$$
$$a_3'' = a_3' + r_1' f_{1 \to 3} \epsilon_3 + r_2' f_{2 \to 3} \epsilon_3 \qquad\qquad + r_4' f_{4 \to 3} \epsilon_3$$
$$a_4'' = a_4' + r_1' f_{1 \to 4} \epsilon_4 + r_2' f_{2 \to 4} \epsilon_4 + r_3' f_{3 \to 4} \epsilon_4$$

where a_i'' is the total absorption of flux at grey surface i from all surfaces and after two reflections; and the reflections are given by:

$$r_1'' = a_1''(1 - \epsilon_1)/\epsilon_1 \qquad r_2'' = a_2''(1 - \epsilon_2)/\epsilon_2$$
$$r_3'' = a_3''(1 - \epsilon_3)/\epsilon_3 \qquad r_4'' = a_4''(1 - \epsilon_4)/\epsilon_4.$$

In general terms the absorptions and reflections at each recursive step are found from:

$$a_i^n = a_i^{n-1} + \sum_{j=1}^{N} r_i^{n-1} f_{j \to i} \epsilon_i \qquad 1 \leqslant n \leqslant \infty$$
$$a_i^0 = 0$$
$$r_i^0 = q_i$$

and in practice the recursive process continues until the reflected flux is reduced to insignificance. In most applications incorporating conventional building materials with high (~ 0.9) values of longwave emissivity and absorptivity this situation is reached after about three recursive steps. Many more recursions will be required in the case of enclosures incorporating low emissivity surfaces for which absorption rate is low.

It is important to note that if the objective of the recursive algorithm is to determine the h_r values between each zone surface pair for incorporation within the system matrix (see chapters 3 and 4) to allow future time-row longwave radiation exchanges to be evaluated, in part, on the basis of future time-row surface temperatures (to improve accuracy), then the relevant inter-surface energy exchange at each recursive step must be extracted and separately summed.

On the other hand, if it is deemed acceptable to determine surface heat flux values for future time-rows on the basis of present time-row temperature values (since future time-row values are as yet uncomputed) for injection to the various surfaces via the q heat generation terms of the energy balance equations, then it is possible to establish, for simultaneous solution, a set of surface radiosity balance equations. Note, however, that this will not directly yield net radiative exchanges between each surface pairing from which an effective h_r value can be computed to increase future time-row temperature prediction accuracy. Consider the spectral surface radiosity, which is equal to the sum of the emitted and subsequently reflected intensities:

$$R_i = \epsilon_i \sigma \theta_i^4 + \varrho_i \sum_{j=1}^{N} R_j f_{j \to i} A_j / A_i$$

$$= \epsilon_i \sigma \theta_i^4 + \varrho_i \sum_{j=1}^{N} R_j f_{i \to j} \qquad (5.19)$$

where R_i is the radiosity of surface i (W m^{-2}), ϱ_i is the reflectivity of surface i, and N is the total number of participating surfaces. Note that $f_{i \to i} = 0$.

Equation (5.19) can be re-expressed as

$$\sum_{j=1}^{N} \frac{(\delta_{ij} - \varrho_i f_{i \to j})}{\epsilon_i} R_j = \sigma \theta_i^4 \qquad (5.20)$$

where

$$\delta_{ij} = \begin{cases} 1 \text{ for } i = j \\ 0 \text{ for } i \neq j. \end{cases}$$

For all participating surfaces, equation (5.20) can be written in matrix form as

$$\mathbf{AR} = \mathbf{\Phi}$$

where

$$\mathbf{A} = \begin{bmatrix} m_{11} m_{12} & \cdot & \cdot & \cdot & m_{1N} \\ m_{21} & & & & \vdots \\ \vdots & & & & \vdots \\ m_{N1} \cdot & \cdot & \cdot & \cdot & m_{NN} \end{bmatrix} \qquad \mathbf{R} = \begin{bmatrix} R_1 \\ R_2 \\ \vdots \\ R_N \end{bmatrix} \qquad \mathbf{\Phi} = \begin{bmatrix} \sigma \theta_1^4 \\ \sigma \theta_2^4 \\ \vdots \\ \sigma \theta_N^4 \end{bmatrix}$$

$$m = (\delta_{ij} - \varrho_i f_{i \to j}) / \epsilon_i$$

and the solution for all surface radiosities is given by

$$\mathbf{R} = \mathbf{A}^{-1} \mathbf{\Phi}$$

which is equivalent to

$$R_i = \sum_{j=1}^{N} m'_{ij} \sigma \theta_j^4 \qquad i = 1, 2, 3, 4, \ldots, N$$

where m'_{ij} are elements of the $\mathbf{A}^{-1}$ matrix.

Given the radiosity of each surface, the net surface flux, due to all surface interaction, is given by

$$q_i = [\epsilon_i \sigma \theta_i^4 - (1 - \varrho_i) R_i] / \varrho_i.$$

Whilst the coding effort involved in the computer implementation of the recursive or simultaneous matrix method is relatively trivial it must be recalled that, in general, it is desirable to operate in terms of h_r values which are partly expressed as a function of future time-row values if an accurate solution is sought; and, since radiation exchange must be determined for both time-rows of any time-step and for all zones considered in a simulation, a severe computational penalty may accrue. For these reasons simplified analytical methods can become attractive which, in many circumstances, are no less accurate.

Consider the worked example of table 5.4 arrived at by assuming the stated conditions imposed on the geometry of figure 5.13 and by applying the recursive method. Note that only the flux emitted by surface 1 is detailed and that all subsequent reflections are shown. The three flux quantities arriving at any surface are not combined to give a single reflective quantity for further onward processing but, instead, are processed as distinct quantities. This is done simply to allow the identification of principal components as the original flux decays and it should be noted that combination will greatly improve computational efficiency.

By inspection, it can be seen that by the third reflection some 98.7% of the original flux emission from surface 1 has been reabsorbed and that this rises to approximately 99.7% at the fourth reflection; these values are not untypical for most building enclosures at or near comfort conditions. Recognising that the absorptions at surfaces 2, 3 and 4, at first reflection, are equivalent to the first term of all two-surface interactions $1\to2\to1\to2\to1\to2$ ... $\to\infty$, $1\to3\to1\to3\to1\to3$ ... $\to\infty$, $1\to4\to1\to4\to1\to4$ ... $\to\infty$ then it is possible to devise an expression for these infinite series which will account for the terms at first reflection and selected terms at all subsequent odd numbered reflections thereafter. The terms accounted for by all two-surface interactions are shown by subscript 2 in table 5.4. Further inspection reveals that three-surface interactions $1\to2\to3\to1\to2\to3\to1\to2\to3$ ... $\to\infty$, $1\to2\to4\to1\to2\to4\to1\to2\to4$ ... $\to\infty$, etc (note that $1\to2\to1\to(1)\to2\to1\to(1)\to2\to1$ ... $\to\infty$ is a special case) give rise to expressions which account for all terms identified by subscript 3. Similar reasoning for four-surface interactions can be applied to account for terms shown by subscript 4 and so on, depending on the accuracy required. Of course, the greater the number of interacting surfaces, the nearer will this technique come to the original recursive or matrix method.

The following derivation applies to this analytical method and gives a simple expression for $h_{ri\to j}$ such that when all possible pairings are treated the near total flux field has been processed.

Consider any two grey surfaces (planar polygons) oriented arbitrarily in 3-dimensional space and consider the flux emitted by surface 1 as it is reflected between each surface to infinity as shown in figure 5.14. The radiant flux emitted by surface 1 is given by $\epsilon_1\sigma A_1\theta_1^4$, and of this some portion $f_{1\to2}$ will arrive at surface 2. Of this amount, $\epsilon_1\sigma A_1\theta_1^4 f_{1\to2}$, a portion ϵ_2 will be absorbed and $(1-\epsilon_2)$ will be radiated diffusely. Accepting $f_{2\to1}$ as the geometric view factor between surface 2 and 1, then the flux to arrive back at surface 1 is given by $(1-\epsilon_2)\epsilon_1\sigma A_1\theta_1^4 f_{1\to2}f_{2\to1}$. Some portion of this is then reflected from surface 1,

Table 5.4(a)   Application of the recursive technique to the problem of figure 5.13.

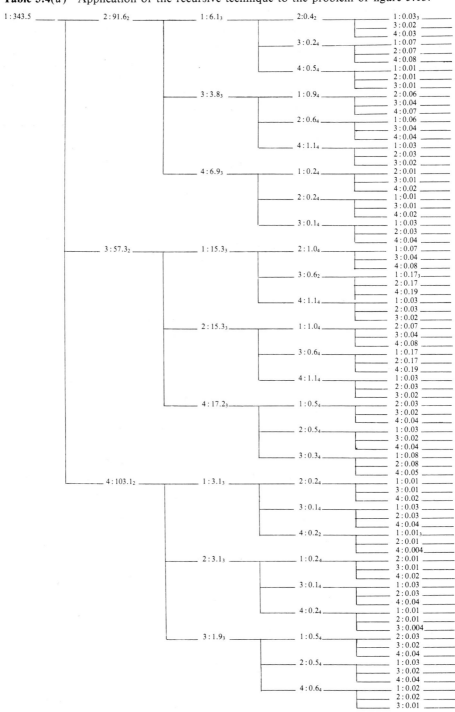

Table 5.4(b) Data used in deriving values in part (a).

Surface (i)	θ (K)	A (m^2)	ϵ	$f_{i \to j}$ for $j =$			
				1	2	3	4
1	295	1	0.8	0	0.33	0.33	0.33
2	297	1	0.8	0.33	0	0.33	0.33
3	293	1	0.5	0.33	0.33	0	0.33
4	295	1	0.9	0.33	0.33	0.33	0

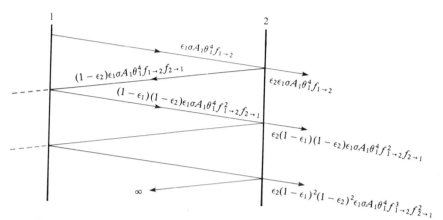

Figure 5.14 Flux absorbed at surface 2 due to two-surface interaction.

$(1 - \epsilon_2)(1 - \epsilon_1)\epsilon_1\sigma A_1\theta_1^4 f_{1 \to 2} f_{2 \to 1}$, and the interreflection process continues to infinity.

Focusing on surface 2, then the radiant flux intercepted and absorbed due to infinite two-surface interactions (between surfaces 1 and 2 in this case) is given by

$$q_{(2)1 \to 2} = \epsilon_2\epsilon_1\sigma A_1\theta_1^4 f_{1 \to 2} + \epsilon_2(1 - \epsilon_1)(1 - \epsilon_2)\epsilon_1\sigma A_1\theta_1^4 f_{1 \to 2}^2 f_{2 \to 1}$$
$$+ \epsilon_2(1 - \epsilon_1)^2(1 - \epsilon_2)^2\epsilon_1\sigma A_1\theta_1^4 f_{1 \to 2}^3 f_{2 \to 1}^2 + \ldots\infty$$
$$= \epsilon_2\epsilon_1\sigma A_1\theta_1^4 f_{1 \to 2}[1 + (1 - \epsilon_1)(1 - \epsilon_2)f_{1 \to 2}f_{2 \to 1}$$
$$+ (1 - \epsilon_1)^2(1 - \epsilon_2)^2 f_{1 \to 2}^2 f_{2 \to 1}^2 + \ldots\infty]$$

where the subscript (2) indicates two-surface interaction.

Since the terms contained in the square bracket comprise a geometrical progression with common ratio $CR = (1 - \epsilon_1)(1 - \epsilon_2)f_{1 \to 2}f_{2 \to 1}$, then the series when summed to infinity gives $1/(1 - CR)$, and so

$$q_{(2)1 \to 2} = \frac{\epsilon_2\epsilon_1\sigma A_1\theta_1^4 f_{1 \to 2}}{1 - (1 - \epsilon_1)(1 - \epsilon_2)f_{1 \to 2}f_{2 \to 1}}. \tag{5.21}$$

In a similar manner, three plane interactions can be considered as shown in

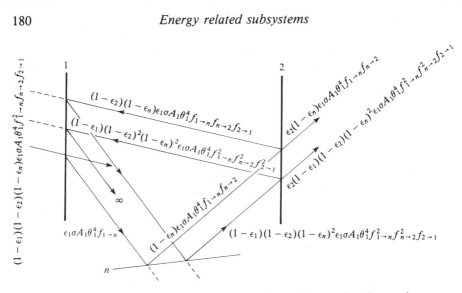

Figure 5.15   Flux absorbed at surface 2 due to three-surface interaction.

figure 5.15. For this case the flux absorbed at surface 2 is given by

$$
\begin{aligned}
q_{(3)1\to 2} &= \epsilon_2(1-\epsilon_n)\epsilon_1\sigma A_1\theta_1^4 f_{1\to n}f_{n\to 2} \\
&+ \epsilon_2(1-\epsilon_1)(1-\epsilon_2)(1-\epsilon_n)^2\epsilon_1\sigma A_1\theta_1^4 f_{1\to n}^2 f_{n\to 2}^2 f_{2\to 1} \\
&+ \epsilon_2(1-\epsilon_1)^2(1-\epsilon_2)^2(1-\epsilon_n)^3\epsilon_1\sigma A_1\theta_1^4 f_{1\to n}^3 f_{n\to 2}^3 f_{2\to 1}^2 + \ldots \infty \\
&= \frac{\epsilon_2(1-\epsilon_n)\epsilon_1\sigma A_1\theta_1^4 f_{1\to n}f_{n\to 2}}{1-(1-\epsilon_1)(1-\epsilon_2)(1-\epsilon_n)f_{1\to n}f_{n\to 2}f_{2\to 1}}
\end{aligned}
$$

and so the flux arriving at surface 2 from 1 via all other surfaces directly is given by

$$
q_{(3)1\to 2} = \sum_{i=1}^{N}\left(\frac{\epsilon_2(1-\epsilon_i)\epsilon_1\sigma A_1\theta_1^4 f_{1\to i}f_{i\to 2}}{1-(1-\epsilon_1)(1-\epsilon_2)(1-\epsilon_i)f_{1\to i}f_{i\to 2}f_{2\to 1}}\right). \tag{5.22}
$$

The special case of three-surface interaction is obtained when flux emitted at 1 finds its way back to 1 from another enclosure surface as shown in figure 5.16 for the case $1\to 2\to 1$. The flux absorbed at *1* is given by

$$
\begin{aligned}
q_{(3)1\to 1} &= \epsilon_1(1-\epsilon_2)\epsilon_1\sigma A_1\theta_1^4 f_{1\to 2}f_{2\to 1} + \epsilon_1(1-\epsilon_1)(1-\epsilon_2)^2\epsilon_1\sigma A_1\theta_1^4 f_{1\to 2}^2 f_{2\to 1}^2 \\
&+ \epsilon_1(1-\epsilon_1)^2(1-\epsilon_2)^3\epsilon_1\sigma A_1\theta_1^4 f_{1\to 2}^3 f_{2\to 1}^3 + \ldots \infty \\
&= \frac{\epsilon_1(1-\epsilon_2)\epsilon_1\sigma A_1\theta_1^4 f_{1\to 2}f_{2\to 1}}{1-(1-\epsilon_1)(1-\epsilon_2)f_{1\to 2}f_{2\to 1}}. \tag{5.23}
\end{aligned}
$$

This process can be continued indefinitely to consider four interacting surfaces and so on; each addition representing the next significant contribution in the recursive stages of table 5.4.

The addition of equations (5.21) and (5.22) gives the approximation to the total

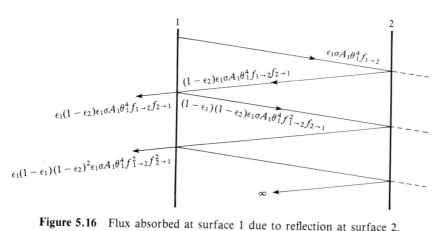

Figure 5.16 Flux absorbed at surface 1 due to reflection at surface 2.

flux absorbed by surface 2 (because of two- and three-surface diffuse reflections to infinity) due to the emission at surface 1:

$$q_{1 \to 2} = \frac{\epsilon_2 \epsilon_1 \sigma A_1 \theta_1^4 f_{1 \to 2}}{1 - (1 - \epsilon_1)(1 - \epsilon_2) f_{1 \to 2} f_{2 \to 1}}$$

$$+ \sum_{i=1}^{N} \frac{\epsilon_2 (1 - \epsilon_i) \epsilon_1 \sigma A_1 \theta_1^4 f_{1 \to i} f_{i \to 2}}{1 - (1 - \epsilon_1)(1 - \epsilon_2)(1 - \epsilon_i) f_{1 \to i} f_{i \to 2} f_{2 \to 1}} \qquad i \neq 1, 2 \qquad (5.24)$$

and, by transposing signs, the approximate flux absorbed by surface 1 due to emission at 2 is:

$$q_{2 \to 1} = \frac{\epsilon_1 \epsilon_2 \sigma A_2 \theta_2^4 f_{2 \to 1}}{1 - (1 - \epsilon_1)(1 - \epsilon_2) f_{1 \to 2} f_{2 \to 1}}$$

$$+ \sum_{i=1}^{N} \frac{\epsilon_1 (1 - \epsilon_i) \epsilon_2 \sigma A_2 \theta_2^4 f_{2 \to i} f_{i \to 1}}{1 - (1 - \epsilon_1)(1 - \epsilon_2)(1 - \epsilon_i) f_{2 \to i} f_{i \to 1} f_{1 \to 2}} \qquad i \neq 1, 2. \qquad (5.25)$$

The net flux exchange between surface 2 and 1 is found from

$q_{2,1} = [\text{flux absorbed at 1 from 2 } (q_{2 \to 1})] - [\text{flux absorbed at 2 from 1 } (q_{1 \to 2})]$

$= (5.25) - (5.24)$

$$= \frac{\epsilon_1 \epsilon_2 \sigma (A_2 \theta_2^4 f_{2 \to 1} - A_1 \theta_1^4 f_{1 \to 2})}{1 - (1 - \epsilon_1)(1 - \epsilon_2) f_{1 \to 2} f_{2 \to 1}}$$

$$+ \sum_{i=1}^{N} \frac{\epsilon_1 (1 - \epsilon_i) \epsilon_2 \sigma A_2 \theta_2^4 f_{2 \to i} f_{i \to 2}}{1 - (1 - \epsilon_1)(1 - \epsilon_2)(1 - \epsilon_i) f_{2 \to i} f_{i \to 1} f_{1 \to 2}}$$

$$- \sum_{i=1}^{N} \frac{\epsilon_2 (1 - \epsilon_i) \epsilon_1 \sigma A_1 \theta_1^4 f_{1 \to i} f_{i \to 2}}{1 - (1 - \epsilon_1)(1 - \epsilon_2)(1 - \epsilon_i) f_{1 \to i} f_{i \to 2} f_{2 \to 1}} \qquad i \neq 1, 2. \qquad (5.26)$$

Note that for the special case of infinite parallel surfaces $(f_{i \to j} = 1)$, the net flux exchange is also given by a flux balance conducted at *either* surface so that

$$q_{2,1} = q_1 - q_{2 \to 1} - q_{1 \to 1}$$
$$= (5.20) - (5.25) - (5.23)$$

or

$$= q_2 - q_{1 \to 2} - q_{2 \to 2}$$
$$= (5.20) - (5.24) - [(5.23) \text{ with signs transposed}].$$

Equation (5.26) defines the net transfer of energy between two surfaces (separated by a non-absorbing media) and due to one other participating surface. For the example of table 5.4 accuracy is reasonably good—in excess of 94.9% of the energy emitted at surface 1 is accounted for if all two- and three-surface combinatorial pairs are considered. Since the flux magnitude decays rapidly it is reasonable to account for the missing energy (5.1% in this case) by determining the difference between the computed value and the initial emission and then to apportion this difference between surfaces on the basis of some simple criterion such as area or view factor weighting. This will become inaccurate in the case of low emissivity surfaces for which the full recursive or matrix treatment is recommended. Alternatively, an analytical solution can be preserved by adding four interacting surface terms and so on. Figure 5.17 details for each level (4 participating surfaces, 5 participating surfaces, etc) example flowpaths then considered by the foregoing analytical technique, and table 5.5 shows the recursive technique applied in a low surface emissivity context. The reader is left to reconcile the analytical rigour required to satisfy this atypical problem.

The determination of the view factor relationship is a non-trivial task in the case of real enclosures with the possibility of inter-surface obstructions, re-entrant angles, surface openings representing windows and doors and specular reflections (Moore and Numan 1983). A number of approaches to view factor assessment are possible, ranging from rigorous numerical methods to simple approximations. These are the subject matter of the next section.

Table 5.5(b) Data used in deriving values in part (a).

Surface (i)	θ (K)	A (m^2)	ϵ	$f_{i \to j}$ for $j =$ 1	2	3	4
1	300	1	0.8	0	0.33	0.33	0.33
2	290	1	0.1	0.33	0	0.33	0.33
3	290	1	0.2	0.33	0.33	0	0.33
4	290	1	0.1	0.33	0.33	0.33	0

Table 5.5(a) Application of the recursive technique to low emissivity surfaces.

1:367.4
- 2:12.3
  - 1:29.4
    - 2:0.25
      - 1:0.59
      - 3:0.15
      - 4:0.07
    - 3:0.49
      - 1:0.52
      - 2:0.07
      - 4:0.07
    - 4:0.25
      - 1:0.59
      - 2:0.07
      - 3:0.15
  - 3:7.4
    - 1:7.8
      - 2:0.07
      - 3:0.13
      - 4:0.07
    - 2:0.98
      - 1:2.4
      - 3:0.59
      - 4:0.29
    - 4:0.98
      - 1:2.4
      - 2:0.29
      - 3:0.59
  - 4:3.7
    - 1:8.8
      - 2:0.07
      - 3:0.15
      - 4:0.07
    - 2:1.1
      - 1:2.7
      - 3:0.66
      - 4:0.33
    - 3:2.2
      - 1:2.4
      - 2:0.29
      - 4:0.29
- 3:24.5
  - 1:26.1
    - 2:0.22
      - 1:0.46
      - 3:0.06
      - 4:0.12
    - 3:0.44
      - 1:0.52
      - 2:0.07
      - 4:0.07
    - 4:0.22
      - 1:0.46
      - 2:0.12
      - 3:0.06
  - 2:3.3
    - 1:7.8
      - 2:0.09
      - 3:0.13
      - 4:0.09
    - 3:2.0
      - 1:2.1
      - 2:0.26
      - 4:0.26
    - 4:0.98
      - 1:2.4
      - 2:0.59
      - 3:0.29
  - 4:3.3
    - 1:7.8
      - 2:0.07
      - 3:0.13
      - 4:0.07
    - 3:2.0
      - 1:2.1
      - 2:0.26
      - 4:0.26
    - 4:0.98
      - 1:2.4
      - 2:0.59
      - 3:0.29
- 4:12.2
  - 1:29.4
    - 2:0.25
      - 1:0.59
      - 3:0.15
      - 4:0.07
    - 3:0.49
      - 1:0.52
      - 2:0.07
      - 4:0.07
    - 4:0.25
      - 1:0.59
      - 2:0.07
      - 3:0.15
  - 2:3.7
    - 1:8.8
      - 2:0.07
      - 3:0.15
      - 4:0.07
    - 3:2.2
      - 1:2.4
      - 2:0.29
      - 4:0.29
    - 4:1.1
      - 1:2.7
      - 2:0.33
      - 3:0.66
  - 3:7.4
    - 1:7.8
      - 2:0.07
      - 3:0.13
      - 4:0.07
    - 2:0.95
      - 1:2.4
      - 3:0.59
      - 4:0.29
    - 4:0.95
      - 1:2.4
      - 2:0.29
      - 3:0.59

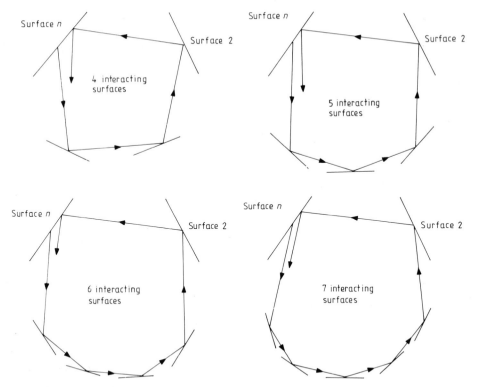

Figure 5.17 Flux absorbed at surface 2 from surface n and due to four or more interacting surfaces.

5.4.2 View factor determination

Consider figure 5.18 which shows a ray of intensity I contained within a solid angle $d\Omega$ to an elemental area dA and propagated in a direction Ω making an angle θ with the normal n to the surface element. The emitted flux contained in $d\Omega$ is

$$dq = I\cos\theta\,d\Omega \tag{5.27}$$

and the radiative flux contained in a solid angle over an entire hemisphere is obtained by integration such that

$$q = \int I\cos\theta\,d\Omega. \tag{5.28}$$

Now

$$d\Omega = \sin\theta\,d\theta\,d\phi$$

where ϕ is an azimuthal angle, as illustrated in figure 5.18. Substitution in

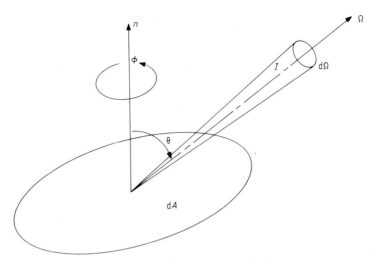

Figure 5.18 View factor prediction.

equation (5.28) gives

$$q = \int_{\phi=0}^{2\pi} \int_{\theta=0}^{\pi/2} I \cos \theta \sin \theta \, d\theta \, d\phi = \pi I. \tag{5.29}$$

Consider now figure 5.19 which shows two planar elemental areas dA_i and dA_j separated by some distance r and with polar angles θ_i and θ_j between r and normals n_i and n_j respectively. The radiative flux leaving dA_i that arrives at dA_j, as given by equation (5.27), is

$$dq_i = I_i \cos \theta_i \, d\Omega_{ij}$$

and

$$d\Omega_{ij} = dA_j \cos \theta_j / r^2$$
$$\Rightarrow dq_i = I_i \cos \theta_i \, dA_j \cos \theta_j / r^2. \tag{5.30}$$

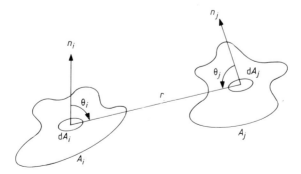

Figure 5.19 Two communicating elemental areas.

Introducing the elemental view factor of $\mathrm{d}f_{\mathrm{d}A_i \to \mathrm{d}A_j}$, which by definition is the ratio of the radiative flux leaving $\mathrm{d}A_i$ that arrives directly at $\mathrm{d}A_j$ to the total radiative flux leaving $\mathrm{d}A_i$,

$$\mathrm{d}f_{\mathrm{d}A_i \to \mathrm{d}A_j} = \mathrm{d}q_i/q$$
$$= \cos\theta_i \, \mathrm{d}A_j \cos\theta_j / \pi r^2.$$

In a similar manner, the point view factor $f_{\mathrm{d}A_i \to A_j}$ can be defined as the fraction of the radiative energy leaving $\mathrm{d}A_i$ which is received by a surface A_j:

$$f_{\mathrm{d}A_i \to A_j} = \int^{A_i} \mathrm{d}f_{\mathrm{d}A_i \to \mathrm{d}A_j}$$

$$= \int^{A_j} \frac{\cos\theta_i \cos\theta_j \, \mathrm{d}A_j}{\pi r^2}. \tag{5.31}$$

Now obviously the flux emitted by A_j is $I_j A_j$ and the fraction received by A_i is $I_j \int^{A_j} \mathrm{d}f_{\mathrm{d}A_j \to \mathrm{d}A_i} \, \mathrm{d}A_j$ and therefore $f_{A_j \to \mathrm{d}A_i}$ is given by

$$f_{A_j \to \mathrm{d}A_i} = I_j \int^{A_j} \frac{\mathrm{d}f_{\mathrm{d}A_j \to \mathrm{d}A_i}}{I_j A_j} \, \mathrm{d}A_j$$

$$= \frac{1}{A_j} \int^{A_j} \mathrm{d}f_{\mathrm{d}A_j \to \mathrm{d}A_i} \, \mathrm{d}A_j$$

and so, from equation (5.30),

$$f_{A_j \to \mathrm{d}A_i} = \frac{\mathrm{d}A_i}{A_j} \int^{A_j} \frac{\cos\theta_i \cos\theta_j}{\pi r^2} \, \mathrm{d}A_j. \tag{5.32}$$

The reciprocity relationship follows by equating equations (5.31) and (5.32):

$$\mathrm{d}A_i f_{\mathrm{d}A_i \to A_j} = A_j f_{A_j \to \mathrm{d}A_i}. \tag{5.33}$$

Lastly, the flux emitted by A_i is $I_i A_i$ and the fraction received by A_j is $I_i \int^{A_i} f_{\mathrm{d}A_i \to A_j} \, \mathrm{d}A_i$ and therefore the area-to-area view factor $f_{A_i \to A_j}$ becomes

$$f_{A_i \to A_j} = \frac{1}{A_i} \int^{A_i} f_{\mathrm{d}A_i \to A_j} \, \mathrm{d}A_i$$

and from equation (5.31)

$$f_{A_i \to A_j} = \frac{1}{A_i} \int^{A_i} \int^{A_j} \frac{\cos\theta_i \cos\theta_j}{\pi r^2} \, \mathrm{d}A_i \, \mathrm{d}A_j \tag{5.34}$$

with, as before,

$$A_i f_{A_i \to A_j} = A_j f_{A_j \to A_i}.$$

The determination of the view factor relationship by double integration over the surfaces of two directly communicating bodies can be achieved by a finite difference representation of equations (5.31) and (5.34). One algorithmic approach (Moore and Numan 1982) starts by subdividing some surface polygon into a number of small elemental areas and for each establishing a unit hemisphere above the centre point representing dA_i as shown in figure 5.20. Each unit hemisphere is then subdivided into a number of equal solid angles by 'strip and patch' subdivision as shown. Every solid angle can then be projected until intersection with another polygon occurs so that a 'viewed' polygon is associated with each hemispherical patch to account for all radiation emitted by dA_i. If full or non-specular reflection occurs, then only those patches are considered which receive the directional flux reflected by dA_i. In this case, a specular reflection model will be required to determine the angle of reflection of any incident ray. Returning to the diffusing case, the point view factor of equation (5.31) is now determined by simply summing the contributions of each viewed polygon and the view factor between two finite surfaces A_i and A_j found from

$$f_{A_i \to A_j} = \frac{1}{A_i} \int^{A_i} f_{dA_i \to A_j} \, dA_i$$

to give the equivalent of equation (5.34). Such an algorithm therefore guarantees that $\sum^j f_{i \to j} = 1$ and that reciprocity will prevail. As proof of this, consider equation (5.31) which can be rewritten

$$f_{dA_i \to A_j} = \int^{\Omega_i} \frac{\cos \theta_i}{\pi} \, d\Omega .$$

Now, for $f_{dA_i \to A_j} = 1$, then the integration must give

$$\int^{\Omega_i} \cos \theta_i \, d\Omega = \pi .$$

If this equation is approximated as a summation and if all $d\Omega_n$ are the same then

$$\sum^n \cos \theta_n \, d\Omega_n = d\Omega_n \sum^n \cos \theta_n. \tag{5.35}$$

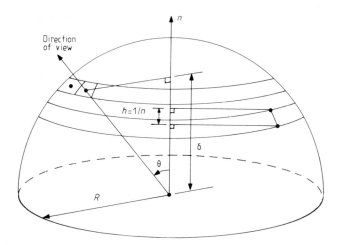

Figure 5.20 Strip and patch subdivision of a unit hemisphere.

To have identical $d\Omega_n$, by the definition of solid angle, the areas subtended on the surface of a unit hemisphere (centred on dA_i) must be the same. Consider figure 5.20; the area of a strip of the hemisphere is given by $A_s = 2\pi Rh$. Therefore, to divide the hemisphere into strips of equal area requires having equal h. Having achieved this, patches of equal area can be created by implementing equal subdivisions of each strip. These patches must then subtend equal solid angles at the centre node at the base of the hemisphere. In each case, the width of the patch is defined by some angle ϕ in a plane parallel to the hemisphere equator.

Note that for a unit hemisphere

$$\cos\theta = \delta$$

where δ is the projection of the radius onto the normal. If the hemisphere is divided into n strips, then the cosines will be

$$1/2n, \ 1/2n + 1/n, \ 1/2n + 2/n, \ \ldots, \ 1/2n + (n-1)/n$$

and summing by the usual formula for arithmetic series gives

$$S_n = (n/2)[2a + (n-1)b]$$

where, in this case, $a = 1/2n$ and $b = 1/n$ and so it follows that

$$S_n = n/2.$$

If each strip is further subdivided into m equal patches then the cosine summation will be $mn/2$. Since each patch subtends a solid angle of $2\pi/mn$ steradians, equation (5.35) becomes

$$(2\pi/mn)(mn/2) = \pi$$

and so $\sum^j f_{i \to j} = 1$ and reciprocity is achieved.

In the implemented algorithm, the point view factors are calculated by finding which viewed polygon is 'seen' through each of the equal area patches and then summing the contributions of each viewed polygon. A polygon is seen if the axis of the solid angle intersects that polygon at a point closer than it intersects any other polygon. Although all radiation is accounted for in theory, care should be taken in practice since numerous implementation inexactitudes (mesh generation, polygon clipping, point containment testing, application of point view factors to finite mesh areas etc) may introduce errors. These can be minimised by increasing the number of strip and patch divisions applied to the unit hemisphere. Note, however, that since the problem is quadrate any accuracy improvement will be hard fought.

For many simplified geometries such computational rigour may not be justified and simplifying assumptions are often acceptable.

View factor between two small surfaces

If two communicating surfaces can be considered small compared with the remaining surfaces comprising the enclosure then an approximation to $f_{i \to j}$ is to use the elemental view factor defined by equation (5.30).

View factor between large parallel surfaces

In this case most of the radiation from one surface will be intercepted by the other and so $f_{i \to j} = f_{j \to i} \to 1$.

View factor in the case of one surface enclosing another

Again $f_{i \to j} \to 1$, where surface 1 is contained, and from the reciprocity theorem $f_{j \to i} = f_{i \to j} A_i / A_j$.

View factors in the absence of geometric information

In many modelling applications—especially at an early design stage—detailed geometrical decisions are, perhaps, not yet taken and so it is impossible to determine view factors on the basis of expressions involving relative angles. In such cases it is convenient to utilise one of the following approaches:

It is possible to accept as representative one of the standard geometrical arrangements found in the literature (Hottel 1930, Chung and Samitra 1972). Many such arrangements have been evaluated and view factor graphs established for various dimensionally representative ratios.

Simple area weighting techniques can be used often without detectable loss of accuracy. For example approximate view factors are given by

$$f_{i \to j} = A_j / (\sum A - A_i)$$
$$f_{j \to i} = A_i / (\sum A - A_i)$$

where the area summation is over all participating surfaces. Note that these expressions satisfy the reciprocity theorem but cannot ensure that $\sum^j f_{i \to j} = 1$ since they have no basis in angular considerations. They give exact answers only in the case of a perfect cube.

Exact analytical solutions

It is often possible to subdivide two communicating surfaces in such a way that by the application of the appropriate view factor algebra (Welty 1974) the desired view factors can be directly determined from the two basic analytical formulations for parallel and perpendicular rectangular planes.

With reference to figure 5.21, the view factor for two parallel surfaces *of equal area* is given as a function of dimensions x, y and z:

$$f_{(\parallel)1 \to 2} = \frac{2}{xy\pi} \left[x(y^2 + z^2)^{\frac{1}{2}} \tan\left(\frac{x}{(y^2 + z^2)^{\frac{1}{2}}}\right) \right.$$

$$+ y(x^2 + z^2)^{\frac{1}{2}} \tan\left(\frac{y}{(x^2 + z^2)^{\frac{1}{2}}}\right) - xz\tan^{-1}\left(\frac{x}{z}\right)$$

$$\left. - yz\tan^{-1}\left(\frac{y}{z}\right) + \frac{z^2}{2} \ln\left(\frac{(x^2 + z^2)(y_2 + z^2)}{(x^2 + y^2 + z^2)z^2}\right) \right] = f_{(\parallel)}[x, y, z] \qquad (5.36)$$

and for two perpendicular surfaces, *with common edge a* and 'height' dimensions

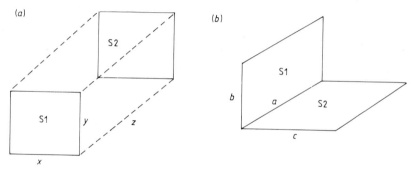

Figure 5.21 Basic view factor configurations. (*a*) Basic parallel case. (*b*) Basic perpendicular case.

b and c:

$$f_{(\perp)1 \to 2} = \frac{1}{\pi}\left[ab\tan^{-1}\left(\frac{a}{b}\right) + ac\tan^{-1}\left(\frac{a}{c}\right) - a(b^2 + c^2)^{\frac{1}{2}}\tan^{-1}\left(\frac{a}{(b^2 + c^2)^{\frac{1}{2}}}\right)\right.$$

$$+ (a^2 + b^2)\ln(a^2 + b^2) + (a^2 - c^2)\ln(a^2 + c^2)$$

$$- \frac{(a^2 - b^2 - c^2)}{4}\ln(a^2 + b^2 + c^2) - \frac{a^2}{4}\ln a^2 + \frac{b^2}{4}\ln b^2$$

$$\left. + \frac{c^2}{4}\ln c^2 - \frac{(b^2 + c^2)}{4}\ln(b^2 + c^2)\right] = f_{(\perp)}[a, b, c]. \tag{5.37}$$

These two basic expressions can now be combined, by shape factor algebra, to give the generalised parallel and perpendicular formulations which do not include the equal area or common edge restrictions of equations (5.36) and (5.37).

With reference to figure 5.22, the general parallel plane view factor ($\Phi_{(\|)}$) is given by

$$\Phi_{(\|)1 \to 2} = 1/4\{f_{(\|)}[(x_{1,2} - x_{2,1}), (y_{1,2} - y_{2,1}), z_1]$$

$$+ f_{(\|)}[(x_{2,2} - x_{1,1}), (y_{1,2} - y_{2,1}), z_1] + f_{(\|)}[(x_{1,2} - x_{2,1}), (y_{2,2} - y_{1,1}), z_1]$$

$$+ f_{(\|)}[(x_{2,2} - x_{1,1}), (y_{2,2} - y_{1,1}), z_1] - f_{(\|)}[(x_{1,2} - x_{2,1}), (y_{1,1} - y_{2,1}), z_1]$$

$$- f_{(\|)}[(x_{1,1} - x_{2,1}), (y_{1,2} - y_{2,1}), z_1] - f_{(\|)}[(x_{2,2} - x_{1,1}), (y_{1,1} - y_{2,1}), z_1]$$

$$- f_{(\|)}[(x_{2,2} - x_{1,2}), (y_{1,2} - y_{2,1}), z_1] - f_{(\|)}[(x_{1,1} - x_{2,1}), (y_{2,2} - y_{1,1}), z_1]$$

$$- f_{(\|)}[(x_{1,2} - x_{2,1}), (y_{2,2} - y_{1,2}), z_1] - f_{(\|)}[(x_{2,2} - x_{1,2}), (y_{2,2} - y_{1,1}), z_1]$$

$$- f_{(\|)}[(x_{2,2} - x_{1,1}), (y_{2,2} - y_{1,2}), z_1] + f_{(\|)}[(x_{1,1} - x_{2,1}), (y_{1,1} - y_{2,1}), z_1]$$

$$+ f_{(\|)}[(x_{2,2} - x_{1,2}), (y_{1,1} - y_{2,1}), z_1] + f_{(\|)}[(x_{1,1} - x_{2,1}), (y_{2,2} - y_{1,2}), z_1]$$

$$+ f_{(\|)}[(x_{2,2} - x_{1,2}), (y_{2,2} - y_{1,2}), z_1]\} \tag{5.38}$$

and the general perpendicular plane value ($\Phi_{(\perp)}$) by

$$\Phi_{(\perp)1 \to 2} = 1/2\{f_{(\perp)}[(a_{2,1} - a_{1,2}), (b_2 - b_0), (c_2 - c_0)]$$

$$- f_{(\perp)}[(a_{2,1} - a_{1,2}), (b_2 - b_0), (c_1 - c_0)]$$

$$+ f_{(\perp)}[(a_{2,2} - a_{1,1}), (b_2 - b_0), (c_2 - c_0)]$$

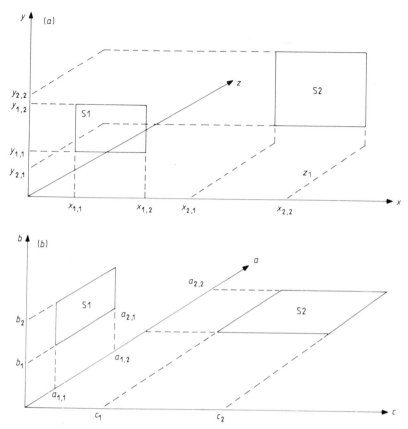

Figure 5.22 General view factor configurations. (*a*) General parallel case. (*b*) General perpendicular case.

$$-f_{(\perp)}[(a_{2,2} - a_{1,1}), (b_2 - b_0), (c_1 - c_0)]$$
$$+f_{(\perp)}[(a_{2,1} - a_{1,2}), (b_1 - b_0), (c_2 - c_0)]$$
$$+f_{(\perp)}[(a_{2,1} - a_{1,2}), (b_1 - b_0), (c_1 - c_0)]$$
$$-f_{(\perp)}[(a_{2,2} - a_{1,1}), (b_1 - b_0), (c_2 - c_0)]$$
$$+f_{(\perp)}[(a_{2,2} - a_{1,1}), (b_1 - b_0), (c_1 - c_0)]$$
$$-f_{(\perp)}[(a_{2,1} - a_{1,1}), (b_2 - b_0), (c_2 - c_0)]$$
$$+f_{(\perp)}[(a_{2,1} - a_{1,1}), (b_2 - b_0), (c_1 - c_0)]$$
$$-f_{(\perp)}[(a_{2,2} - a_{1,2}), (b_2 - b_0), (c_2 - c_0)]$$
$$+f_{(\perp)}[(a_{2,2} - a_{1,2}), (b_2 - b_0), (c_1 - c_0)]$$
$$+f_{(\perp)}[(a_{2,1} - a_{1,1}), (b_1 - b_0), (c_2 - c_0)]$$
$$-f_{(\perp)}[(a_{2,1} - a_{1,1}), (b_1 - b_0), (c_1 - c_0)]$$
$$+f_{(\perp)}[(a_{2,2} - a_{1,2}), (b_1 - b_0), (c_2 - c_0)]$$
$$-f_{(\perp)}[(a_{2,2} - a_{1,2}), (b_1 - b_0), (c_1 - c_0)]. \qquad (5.39)$$

Equations (5.38) and (5.39) can now be usefully employed, by straightforward addition and subtraction operations, to evaluate the view factors for other arrangements built up from the elementary cases. The view factor between the surfaces 1 and 2 of figure 5.23, for example, is given by

$$f_{1 \to 2} = f_{1 \to 2,3,4} - f_{1 \to 3} - f_{1 \to 4}$$

with each of the factors on the right-hand side determined from equation (5.39).

Once determined, these view factors can be used in the methods of §5.4.1 to assess the longwave flux exchanges, exactly or to an approximation. These exchanges can be applied to surface nodes via the heat generation terms of the nodal simulation equations derived in chapter 3. Alternatively, a linearised h_r value can be determined for incorporation in the equation coefficients and hence insertion in the system future time-row coefficients matrix. The next task of §5.4 is then to establish this linearised coefficient to permit the latter option.

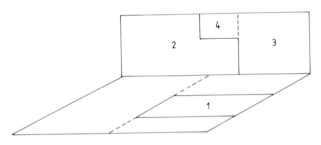

Figure 5.23 A view factor relationship established from elementary surface pairings.

5.4.3 Linearised longwave radiation coefficients

In most building energy applications it is usual to linearise longwave radiation exchanges. This is also the prerequisite of the simultaneous equation solution techniques described in chapter 4, although methods are outlined there for solving, by combined direct/iterative techniques, the non-linear process exactly.

We assume that the net longwave radiation gain at surface 1 from 2 can be written as

$$q_{2,1} = h_{r2,1} A_1 (\theta_2 - \theta_1) \tag{5.40}$$

where $h_{r2,1}$ is the linearised radiative heat transfer coefficient for surface 1 from 2 (W m^{-2}K^{-1}).

Note that the heat flux is expressed in terms of the receiving surface area as required by the energy balance formulations of §3.2.2. This is in contrast to the usual practice of expressing radiative transfer as a function of the emitter surface area. Equating equations (5.26) and (5.40) yields the formulation for the linearised longwave radiation coefficient:

$$h_{r2,1} = \frac{\epsilon_1 \epsilon_2 \sigma (A_2 \theta_2^4 f_{2\to1} - A_1 \theta_1^4 f_{1\to2})}{A_1(\theta_2 - \theta_1)[1 - (1 - \epsilon_1)(1 - \epsilon_2)f_{1\to2}f_{2\to1}}$$

$$+ \sum_{i=1}^{N} \frac{\epsilon_1(1 - \epsilon_i)\epsilon_2 \sigma A_2 \theta_2^4 f_{2\to i}f_{i\to1}}{A_1(\theta_2 - \theta_1)[1 - (1 - \epsilon_1)(1 - \epsilon_2)(1 - \epsilon_i)f_{2\to i}f_{i\to1}f_{1\to2}}$$

$$- \sum_{i=1}^{N} \frac{\epsilon_2(1 - \epsilon_i)\epsilon_1 \sigma A_1 \theta_1^4 f_{1\to i}f_{i\to2}}{A_1(\theta_2 - \theta_1)[1 - (1 - \epsilon_1)(1 - \epsilon_2)(1 - \epsilon_i)f_{1\to i}f_{i\to2}f_{2\to1}} \qquad i \neq 1, 2.$$

Now reciprocity states that $f_{2\to1} = f_{1\to2}A_1/A_2$, and since $(\theta_2^4 - \theta_1^4) = (\theta_2^2 + \theta_1^2)(\theta_2 + \theta_1)(\theta_2 - \theta_1)$, then

$$h_{r2,1} = \frac{\epsilon_1 \epsilon_2 \sigma f_{1\to2}(\theta_2^2 + \theta_1^2)(\theta_2 + \theta_1)}{[1 - (1 - \epsilon_1)(1 - \epsilon_2)f_{1\to2}^2 A_1/A_2]}$$

$$+ \epsilon_1 \epsilon_2 \sigma A_2(\theta_2^2 + \theta_1^2)(\theta_2 + \theta_1) \sum_{i=1}^{N} \frac{(1 - \epsilon_i)f_{1\to i}f_{2\to i}}{A_i[1 - (1 - \epsilon_1)(1 - \epsilon_2)(1 - \epsilon_i)f_{1\to i}f_{i\to2}f_{2\to1}]}.$$

$$(5.41)$$

Accepting compliance with the first law of thermodynamics as the prime consideration in energy modelling it follows that

$$q_{1,2} = q_{2,1} \text{ (ignoring sign)}$$

and so

$$h_{r1,2} = h_{r2,1}A_1/A_2. \qquad (5.42)$$

Thus, when computing the radiative heat transfer coefficient for inclusion within the matrix structures of chapter 4 (at each computational time-step), it is only necessary to evaluate explicitly equation (5.41) for half the total number of combinatorial surface pairings (that is, $1 \to 2$, $1 \to 3$, $1 \to 4$, ... etc; $2 \to 3$, $2 \to 4$, ... etc; $3 \to 4$, $3 \to 5$, ... etc; etc) since all self-coupling is zero and the remaining half can be determined directly from the relationship of equation (5.42). This, of course, implies that view factor information need only be established for the corresponding surface pairings.

5.4.4 Exchange between external surfaces

The net longwave radiation exchange at some exposed external building surface is given by the simultaneous difference between the emitted and received flux. If the surroundings are represented by some mean black body equivalent temperature, θ_e, then the net exchange can be expressed by the Boltzmann equation as

$$q = A_s \epsilon \sigma (\theta_e^4 - \theta_s^4)$$

Energy related subsystems

where A_s is the surface area (m^2), ϵ the surface emissivity, σ the Stefan–Boltzmann constant (W m^{-2}K^{-1}), and θ_s the surface absolute temperature (K).

The equivalent temperature is a function of sky, ground and surrounding obstruction surface temperatures and is given by

$$\theta_e^4 = \alpha\theta_{sky}^4 + \beta\theta_{grd}^4 + \gamma\theta_{obs}^4$$

where α, β and γ are the sky, ground and obstructions view factors relative to the surface in question. Table 5.6 gives some representative values. The task in hand, then, is to estimate these temperatures from the known climatic data.

Table 5.6 Representative values of sky, ground and obstruction view factors for an external surface.

Location	α	β	γ
1 City centre: surrounding buildings at same height, vertical surface	0.36	0.36	0.28
2 City centre: surrounding buildings higher, vertical surface	0.15	0.33	0.52
3 Urban site: vertical surface	0.41	0.41	0.18
4 Rural site: vertical surface	0.45	0.45	0.10
5 City centre: sloping roof	0.50	0.20	0.30
6 Urban site: sloping roof	0.50	0.30	0.20
7 Rural site: isolated	0.50	0.50	0.00

Sky temperature estimation
The sky temperature under non-cloudy conditions can be determined from

$$R_s = 5.31 \times 10^{-13}\theta_{sc}^6$$

where θ_{sc} is the screen air temperature (K). This expression has been compared with measured data from different global locations (Swinbank 1963) and was found to be valid. If the assumption is made that the clear sky behaves as a black body then

$$R_s = \sigma\theta_{sky}^4 \qquad (5.43)$$

and so

$$\theta_{sky} = 0.05532\,\theta_{sc}^{1.5}.$$

In the presence of clouds, the mean sky temperature increases and an alternative expression has been proposed (Cole 1976):

$$R_s' = (1 - CC)R_s + CC\epsilon_c\sigma\theta_{sc}^4 \qquad (5.44)$$

where R_s' is the cloudy sky radiation (W), CC the cloud cover factor (0–1), and ϵ_c the emissivity of the cloud base. The cloud base emissivity is given by

$$\epsilon_c = (1 - 0.84CC)\{0.527 + 0.161\exp[8.45(1 - 273/\theta_{sc})] + 0.84CC\}. \qquad (5.45)$$

Substitution of equation (5.45) in (5.44) and using equation (5.43) gives the final expression for the effective sky temperature under cloudy conditions:

$$\theta_{sky} = [\![9.365574 \times 10^{-6}(1 - CC)\theta_{sc}^6$$
$$+ \theta_{sc}^4 CC(1 - 0.84CC)(\{0.527 + 0.161 \ \exp[8.45(1 - 273/\theta_{sc})]\}$$
$$+ 0.84CC)]\!]^{0.25}.$$

Ground temperature estimation
The simplest method of estimation is to use the concept of sol–air temperature so that

$$\theta_{grd} = \theta_A + (\alpha_g I_h - q_{LW})/R_{so}$$

where θ_A is the air temperature (°C), α_g the ground absorptivity, I_h the total solar irradiance (W m^{-2}), q_{LW} the net longwave radiation exchange (W m^{-2}) and R_{so} the combined convective/radiative ground surface layer resistance (m^2 °C W^{-1}).

Application of this expression will require, firstly, that the longwave exchange term be evaluated. This in turn will require knowledge of the sky and surrounding building temperatures and, of course, the objective here is to compute one surface of the latter. It is entirely possible to accept the immediate past surface temperature for this surface to allow the calculation to proceed. Alternatively, a ground nodal scheme can be introduced to the system matrix equation (of chapter 4) to allow explicit modelling of the ground exchange processes and so the removal of ground temperature from the present calculations.

Obstruction surface temperature estimation
In the absence of detailed information concerning the nature of such surfaces it is not possible to calculate their temperature. However, it is perhaps acceptable to assume that surrounding buildings have surface temperatures not far removed from those of the corresponding surfaces (north face, south face, etc) of the building being simulated.

5.5 Surface convection

In building heat transfer applications, the convective heat flux at surface layers can be evaluated in a manner analagous to that implied by equation (5.40):

$$q_c = h_c A_s(\theta_A - \theta_s) \tag{5.46}$$

where A_s is the surface area and the remaining terms are as defined for equation (2.25). In this form the convective heat transfer coefficient h_c represents some surface-averaged value and so the apparent simplicity of equation (5.40) is misleading since, in reality, its numerical value is a complex function of the characteristics of the fluid flow field at the spatial position in question, the temperature-dependent thermophysical properties and system geometrical features. It follows, then, that an assessment of a surface-averaged value of h_c

will require a detailed investigation of the dynamics of fluid flow and many
standard texts exist (for example, McAdams 1954) which describe the fundamentals of boundary layer theory underlying convection coefficient assessment.

Kreith (1973) has described the four general methods for the determination of
h_c values: dimensional analysis combined with experimental data; mathematical
solutions of the continuity, momentum and energy equations; approximate
boundary layer analysis based on integral techniques; and an approach which
relies on the analogy between heat, mass and momentum transfer. Many workers
have employed the first method to correlate experimental data in terms of dimensionless groupings to provide empirical relationships for h_c of much use in
simulation modelling.

In building simulation applications it is usual to consider convection
coefficients separately for internal and external building surfaces. It is also
necessary to distinguish between natural and forced convection.

5.5.1 Natural convection at internal surfaces

When a fluid comes into contact with a heated surface, heat transfer takes place
by conduction and fluid temperature variations are established which give rise to
density variations. Buoyancy forces then establish fluid motion to carry away the
conducted heat. This process is known as natural (or buoyancy-driven) convection. The fluid flow to result will be either laminar, in which each fluid particle
follows a smooth streamline and does not interfere with adjacent streamlines, or
turbulent, in which fluid particles can cross the streamlines to increase the potential for heat transfer.

Three dimensionless groupings are of importance in natural convection
estimation:

Nusselt number (Nu)

$$Nu = h_c d / k$$

where d is some characteristic dimension (m), and k is the fluid thermal conductivity (W m^{-1} °C^{-1}).

Prandtl number (Pr)

$$Pr = C_p \mu / k$$

where C_p is the specific heat at constant pressure (J kg^{-1} °C^{-1}) and μ is the fluid
viscosity (kg m^{-1} s^{-1}).

Grashof number (Gr)

$$Gr = \varrho^2 g \beta (\theta_s - \theta_A) d^3 / \mu^2$$

where g is the gravitational constant (m s^{-2}), β the coefficient of expansion
(K^{-1}), θ_s the surface temperature (°C), θ_A the bulk fluid temperature (°C), and
ϱ the fluid density (kg m^{-3}).

Most data correlations are obtained from the experimental evaluation of the natural convection heat transfer from heated plates under experimental conditions. For example, Fujii and Imura (1970), working with a 30×15 cm^2 plate of arbitrary inclination, produced the following correlations between the foregoing dimensionless groupings:

Vertical plate and inclined plate facing upward or downward; laminar region.

$$Nu = 0.56(GrPr \cos \theta)^{\frac{1}{4}} \qquad 10^5 < GrPr \cos \theta < 10''$$

where, in the case of the inclined plate, the gravitational constant of the Grashof number is adjusted to the component parallel to the surface and θ is the angle with the vertical, positive downward.

Horizontal plate facing upward, vertical plate and inclined plate facing upward; turbulent region.

$$Nu = 0.13(GrPr)^{\frac{1}{3}} \qquad 5 \times 10^8 < GrPr.$$

Horizontal plate facing downward; laminar region.

$$Nu = 0.58(GrPr)^{\frac{1}{3}} \qquad 10^6 < GrPr < 10^{11}.$$

Inclined plate facing upwards; transition region.

$$Nu = 0.13[(GrPr)^{\frac{1}{3}} - (Gr_cPr)^{\frac{1}{3}}] + 0.56(Gr_cPr \cos \theta)^{\frac{1}{4}}$$

where Gr_c is the Grashof number corresponding to the transition region from laminar to turbulent flow.

Many other empirical formulations can be found in the literature (Jacob 1949, Wong 1977, ASHRAE 1981). In recent years a number of workers have used empirical equations to generate expressions which give natural convection coefficients for building surfaces as a function of surface–fluid temperature difference, characteristic dimensions and the direction of heat flow. To achieve this, a set of assumed conditions are imposed on the empirical equations, further widening the error margin associated with the h_c estimate.

Alamdari and Hammond (1983) have produced a general expression for both vertical and horizontal surfaces (with heat flow upward) valid over the range $10^4 < GrPr < 10^{12}$, which encompasses most of the flow conditions found within buildings:

$$h_c = \{[a(\Delta\theta/d)^p]^m + [b(\Delta\theta)^q]^m\}^{\frac{1}{m}} \qquad (5.47)$$

where a, b, p, q and m are given in table 5.7.

For horizontal surfaces undergoing downward heat flow the natural convective heat transfer coefficient is given by:

$$h_c = 0.6(\Delta\theta/d)^{\frac{1}{5}}.$$

For vertical surfaces, the characteristic dimension d is given by the surface height, whereas for horizontal surfaces the characteristic dimension is the

Table 5.7 Empirical coefficients for equation (5.47) from Alamdari and Hammond (1983).

Surface aspect	a	b	p	q	m
Vertical	1.5	1.23	1/4	1/3	6
Horizontal	1.4	1.63	1/4	1/3	6

hydraulic diameter, found from

$$d = 4A/P$$

where A is the surface area (m^2) and P the perimeter length (m).

Equivalent formulations are given elsewhere (for example, ASHRAE 1981, Min *et al* 1956).

5.5.2 Forced convection at internal and external surfaces

When fluid motion is caused by some external force such as fan or wind power, convection is termed forced. In this case an additional dimensionless grouping is important:

Reynolds number (Re)

$$Re = V \, d\varrho/\mu$$

where V is the fluid velocity (m s^{-1}).

As with natural convection estimation, dimensional analysis techniques can be employed to determine some generalised expression. This is found to be similar to the expression for natural convection with the Reynolds number replacing the Grashof number giving

$$Nu = CRe^n Pr^m$$

with the coefficient and exponent terms being assessed from experimental observation or theoretical considerations. For example, based on a copper plate experiment using a parallel flow of air at a reference temperature of 21.1 °C, McAdams (1954) gives the following expression:

$$h_c = 5.678 \left[a + b \left(\frac{V}{0.3048} \right)^n \right] \tag{5.48}$$

where h_c is the forced convection coefficient (W m^{-2} °C^{-1}), a, b and n are empirical values given in table 5.8, and V is the parallel flow velocity (m s^{-1}). This is equivalent to the adopted ASHRAE expression. For non-reference temperatures a simple adjustment to the velocity term is recommended as given by $284.26 \, V/(273.16 + \theta_n)$ where θ_n is the non-reference temperature (°C).

In any modelling application, two issues are of some importance: the resolution of the prevailing wind direction into a surface parallel component in

Table 5.8 Empirical coefficients and exponents for equation (5.48) from McAdams (1954).

Nature of surface	$v < 4.88\,\text{ms}^{-1}$			$4.88 \leqslant v < 30.48\,\text{ms}^{-1}$		
	a	b	n	a	b	n
Smooth	0.99	0.21	1	0	0.50	0.78
Rough	1.09	0.23	1	0	0.53	0.78

the case of external surfaces and an assessment of system generated velocities in the case of internal surface application.

The latter issue is a difficult one which, in most cases, does not lend itself to mathematical treatment and it is necessary to determine the velocity vectors by measurement (or by detailed three-dimensional air flow simulation).

The former issue, wind velocity resolution, is a simpler one. Based on experimental data, Ito *et al* (1972) have developed empirical relationships for the assessment of V—as required in equation (5.48)—from the free stream velocity V_f. The wind direction relative to some building surface is given by

$$d_r = \alpha_p + 180 - d_f \qquad \text{(if } |d_r| > 180, d_r = 360 - |d_r|)$$

where d_r is the wind direction relative to surface, α_p the surface azimuth (see §5.1), and d_f the free stream wind direction (degrees from north, clockwise positive).

The local surface velocity is dependent on whether the surface is windward or leeward and on the magnitude of the free stream velocity. Its value is approximated by:

Windward $(|d_r| < 90)$

$$\text{If } V_f > 2 \text{ m s}^{-1} \qquad V = 0.25 V_f$$
$$\text{If } V_f < 2 \text{ m s}^{-1} \qquad V = 0.5.$$

Table 5.9 Example convection coefficients.

	Natural convection equation (5.47)			Forced convection equation (5.48)	
$\Delta\theta$	h_{c1}	h_{c2}	V	h_{c3}	h_{c4}
0.1	0.81	0.77	0.5	7.58	8.33
0.5	1.24	1.31	1	9.53	10.47
1	1.49	1.64	5	25.17	26.68
5	2.36	2.80	10	43.21	45.81

h_{c1} vertical surface
h_{c2} horizontal surface
h_{c3} smooth surface
h_{c4} rough surface

Leeward ($|d_r| > 90$)

$$V = 0.3 + 0.05 V_f.$$

Table 5.9 gives example values of the internal and external convection coefficients—as evaluated from equations (5.47) and (5.48)—against a number of $\Delta\theta$ and V conditions.

5.6 Air flow

Whilst it is true that many energy modelling systems are *capable* of processing the energy flowpaths with a high degree of precision, it is also true that many of these systems continue to treat air flowpaths in a rudimentary manner relying, in some cases, on the specification of some globally relevant, but often 'guesstimated', design air change rate. For dynamic modelling purposes air flow information must be evaluated at each time-step and separately for each conceivable flowpath connecting a zone to ambient conditions or to other zones participating in the network. Intra- or inter-zone buoyancy modelling will also be required in cases where a vertical volume subdivision has been established for that purpose.

The objective of this section is to consider the techniques which can be used to establish the v terms of the fluid node simulation expression given by equation (3.12). In essence there are three methods of predicting time-dependent (or climate-dependent) air flow. In increasing order of complexity, these are:

Simplified expressions—usually concerned only with infiltration air flow— which relate the air permeability to such factors as temperature difference (between inside and outside) and the velocity of the free or undisturbed air stream. Such expressions are often produced by application of regression techniques to data collected from field studies or extracted from laboratory tests. For example, ASHRAE give a simplified equation for domestic building air change rates:

$$\text{ACR} = A + B\,\Delta\theta + Cv$$

where ACR represents the volume changes per hour, $\Delta\theta$ is the inside/outside temperature difference (°F), v is the wind speed (mph) and A, B, C are empirical coefficients. Based on the entries of table 7.11B and 7.11C from the ASHRAE fundamentals volume (ASHRAE 1981), values for A, B and C can be evaluated to give, for a 'tight' construction

$$\text{ACR} = 0.11 + 0.0066\,\Delta\theta + 0.012v$$

and for a construction of average leakage

$$\text{ACR} = 0.15 + 0.0087\,\Delta\theta + 0.02v.$$

An approach of intermediate complexity is to apply mass balance techniques to some nodal network in which nodes represent discrete volumes of air (zones, portions of zones and ambient conditions of various aspects) and inter-node connections are the resistances representing the distributed leakage paths through which air will flow if subjected to some pressure difference caused by wind or stack effect due to density variations.

The third approach is to establish a detailed model of the flow field by developing a numerical solution of the governing partial differential equations relating to the conservation of mass, energy, momentum and turbulence when integrated over finite control volumes as was done in chapter 3 for energy flow. Such a method (Patankar and Spalding 1972, Patankar 1980) is appropriate to the study of fluid flow under complex mixing conditions, geometries or boundary conditions. A number of computer models exist for two- and three-dimensional fluid flow analysis, although they often relate to the steady state.

In building energy modelling applications the former two methods are often most useful, with the mass balance approach being most commensurate with the techniques of chapters 2 and 3. What follows then is a summary description of this approach.

5.6.1 The mass balance approach

The principal factors which combine to produce air movement are the pressure and temperature differences that exist between inside and outside (and between adjacent internal air volumes) and the distribution and magnitude of the leakage paths. There are several distinct aspects to the overall prediction problem and these are considered briefly in turn.

Surface pressure distribution
The distribution of surface pressure can be regarded as the boundary condition of the air flow network for solution. Assessment of this distribution will depend on a knowledge of prevailing wind conditions—speed, direction and vertical velocity profile—and, more problematic, on the influence of local obstructions, facade and terrain features, and on other microclimatic effects such as wind hollows and shelter belts.

One way of establishing data on surface pressure distribution is by wind tunnel testing applied to scale models and, if precision is the main objective, this will be necessary even when obstruction geometries and terrain features are fairly simple. In recent years computer models have emerged (Launder and Spalding 1972, Hanson *et al* 1982) which may eventually, after considerable refinement, replace the experimentally difficult and often expensive wind tunnel approach.

Regardless of the method of assessment, it is usual practice to express surface pressure gradients as a set of dimensionless coefficients, one set for each representative surface location i, so that for any wind direction d (figure 5.24)

$$C_{id} = \frac{P_{id}}{\frac{1}{2}\varrho V_f^2}$$

where C_{id} is the pressure coefficient for surface i and corresponding to wind from direction d, P_{id} is the surface pressure (N m^{-2}), ϱ the air density (kg m^{-3}) and V_f the free stream wind speed for direction d (m s^{-1}).

Typically a coefficient set is comprised of 16 values at $22\frac{1}{2}°$ intervals and so the coefficient for any particular wind direction can reflect site obstruction features. Coefficients can be negative, as well as positive, to reflect leeward and windward exposures, and pressures at non-reference heights can be simply determined by pressure corrections found from

$$\Delta P_{id} = \varrho g (h - h_i) \qquad (5.49)$$

where ΔP_{id} is the pressure correction (N m^{-2}), h the wind reference height (m), h_i the height of point i (m) and g the gravitational constant (m s^{-2}).

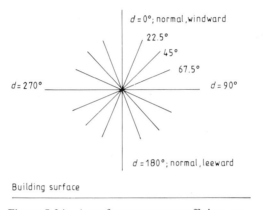

Figure 5.24   A surface pressure coefficient set.

Leakage distribution and behaviour
In modelling applications, the flow of air is often (for convenience) resolved into two components: the flow resulting from some steady (or mean) pressure difference over some relatively small time interval (<1 hour) and the flow caused by random pressure fluctuations superimposed on this mean condition.

Many attempts have been made to quantify crack flow behaviour and the true leakage characteristics of building components such as windows. Dick (1950), for example, has produced equivalent area data for a variety of building components such as window cracks, timber floors and so on. Other workers (Malinowski 1971, Harris-Bass *et al* 1974) have experimentally examined the induction of fluctuating flows across an opening such as a window due to wind turbulence acting to compress the air in the internal volume and so cause infiltration. Unfortunately leakage flowpaths are subject to large size variations and uncertainties, are difficult to measure and, to add to these problems, might well be controlled by occupants. It has also been observed in a number of field studies that the flow field is often more strongly related to hidden or background leakage paths than to the more obvious openings such as windows.

Table 5.10 Some example leakage dimensions.

	Typical crack width (mm/m)	Equivalent area (m²)	Comments
Window:			
not weather-stripped	1.0†	N/A	{Windows of
weather-stripped	0.25†	N/A	{average fitting.
deliberate ventilation opening	~2.5‡ (1 m of window perimeter)	0.02	Based on a 2 m × 1 m window.
Doors:			
not weather-stripped	1.5†–5.0‡	0.0084–0.028	For standard
weather-stripped	0.8†–3.0‡	0.0045–0.168	door of
deliberate ventilation opening	10‡ (1 m of door perimeter)	0.056	0.8 m × 2.0 m average fitting
Walls:			
unplastered	N/A	0.00002†	Based on a
air brick	N/A	0.005–0.033†	10 m² wall
Roof: eaves:			
attic ventilation	N/A	0.15‡	Based on 20 m × 10 m (plan area) roof
Floor:			
skirting	1.6†	N/A	
tongue and groove	0.8†	N/A	Floor uncovered
sub-floor ventilation	1.3–3.2† (1 m of external wall length)	N/A	
Flues:			
open fire	N/A	0.033†	
gas fire	N/A	0.013–0.033†	
Ventilators:			
fixed louvres constant flow	N/A	0.015†	
@0.9 m s⁻¹	N/A	0.0084†	Based on face
9 m s⁻¹	N/A	0.0022†	0.3 m × 0.075 m

†data from Dick (1950) ‡data from Cockcroft (1979)

Table 5.10 gives some example opening and crack dimensions for the flowpaths occurring in buildings.

In general, the equations which represent the mean air flow through simple restrictions are conveniently expressed by an equation of the form

$$v = ka(\Delta P)^x \tag{5.50}$$

where ΔP is the pressure difference across the restriction (N m^{-2}), k is an empirical constant dependent on the nature of the flow restriction, a is a characteristic dimension such as length or area, and x is an empirical exponent.

For a simple orifice (such as a partially open window) at high Reynolds number x is close to 0.5 (ASHRAE 1981). For cracks and similar restrictions with a large aspect ratio x is close to 0.65, rising to unity for completely laminar flow. For this class of flow restriction empirical relationships have been established to determine k and x as a function of crack width. For example

$$x = 0.5 + 0.5 \exp(-W/2)$$
$$k = 9.7(0.0092)^x$$

where W is the crack width in mm (and therefore a of equation (5.50) becomes the crack length).

With large vertical openings, such as doorways, more complex flow patterns are found (Brown 1962a). If a temperature difference exists across such an opening, then air flow can occur in both directions due to the action of small density variations over the door height causing a positive pressure difference at the bottom (or top) of the opening with a corresponding negative pressure difference at the top (or bottom). This situation is illustrated in figure 5.25.

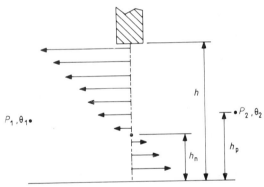

Figure 5.25 Bi-directional air flow across a doorway.

Cockcroft (1979) has studied this effect and produced an expression for the air flow through such an opening:

$$v = (3/2)[C_D Wh(2/\varrho)^{1/2}(C_a^{3/2} - C_b^{3/2})/C_t] \tag{5.51}$$

where C_D is the coefficient of discharge, W the opening width and h the opening height (m), $C_a = (1 - r_p)C_t + (P_1 - P_2)$, $C_b = (P_1 - P_2) - r_p C_t$, $C_t = g\varrho h/R$ $\times (1/\theta_2 - 1/\theta_1)$, θ_1, θ_2 are the absolute temperatures on either side of the opening (K), P_1, P_2 the pressures on either side (N m^{-2}), R is the gas constant, $r_p = h_p/h$ and h_p is the height of the reference nodes on either side (m).

On evaluation this equation yields a sum of real and imaginary parts. Real parts indicate a flow in the positive direction and imaginary parts indicate a flow in the reverse direction. The equation has a singularity at infinity for $C_t = 0$. In this case no buoyancy effects exist, and so an equation of the form of equation (5.50) can be used. Other formulations can be found in the literature (e.g. Etheridge 1977).

The neutral height h_n—the height at which no net pressure difference can be measured across an opening—is found from

$$h_n = h\left(r_p - \frac{(P_1 - P_2)}{C_t}\right).$$

Similar expressions can also be derived for large horizontal openings (Brown 1962b).

Temperature differences between internal and external air conditions also give rise to pressure differences according to

$$\Delta P = g \, \Delta \varrho \, (h - h_n)$$

where $\Delta \varrho$ is the density difference (kg m^{-3}) and h_n is the height at which the pressure difference is zero, termed the neutral height (m). This is the stack effect which predominates in high rise buildings (Tamura and Wilson 1967).

Influence of occupants
The actions of occupants can have a major influence on air movement. In general, these actions can only be handled by statistical methods which are inappropriate to small time-step simulation modelling. This unpredictability is, paradoxically, a justification rather than a condemnation of a detailed air flow simulation approach, since it is only when a large number of simulations are undertaken, against different behavioural assumptions, that robust design solutions can be formulated which will accommodate any anticipated behaviour.

System characteristics
It is a trivial matter to include gross air input or extract to any zone (network node) if this is known. On the other hand, modelling of the specific idiosyncracies of mechanical system flow fields is a complex problem and the subject of much current research activity (for example, Nielsen 1982). However, if system characteristics are known (for example duct flow resistances, fan curves, etc) then these can be included as part of the nodal network for simultaneous solution as described in the next section.

5.6.2 *Iterative solution procedure*

The characteristic leakage equations can now be harnessed as appropriate to represent some air flow network in which branches represent windows, doors, cracks etc connecting nodes representing internal and external zones into and out of which air can flow. Each node will therefore have an associated pressure and temperature with a difference dictating air flow in any connecting branch.

External zone nodes are assigned pressure and temperature values at any time-step from the known surface pressure coefficients and prevailing wind data. Since the flow equations are non-linear, the non-fixed pressures associated with internal zone nodes must be determined by iterative means. One appropriate technique is the Newton–Raphson method (Kreyszig 1979). Consider a differentiable function of pressure $f(P) = 0$. The solution of this equation is found by the successive (or iterative) application of the general formula

$$P_{n+1} = P_n - f(P_n)/f'(P_n) \qquad n = 0, 1, 2, \ldots \qquad (5.52)$$

Thus, for any internal node i, the flow equations can be combined to give mass balance

$$\sum^{j} \varrho_j v_{ij} = 0 \qquad (5.53)$$

and the solution of the entire equation set can be found from a four-step procedure:

1 An arbitrary initial pressure is assigned to each internal node.
2 For any internal node defined by equation (5.53), with the assigned pressures assumed for surrounding node pressures, equation (5.52) is iteratively applied to establish the nodal pressure which achieves mass balance.
3 The next internal node is then processed as in step 2 but with the newly computed pressure (if appropriate) for the previously processed internal nodes.
4 Network iteration then proceeds until the change in nodal pressure (and hence air flow) is less than some small amount.

Some problems will occur where networks include a mixture of large and small openings, or where small pressure changes produce relatively large changes in air flow. In this case some over- or under-relaxation on the pressure adjustments will be required to produce fast and stable solution schemes. Cockcroft (1979), for example, has developed a solution procedure in which the air flow through each connection is determined on the basis of the initial guessed pressure field. The summation of flows for each internal node then gives the related residual error defining a departure from the desired mass balance. The objective then is to force the largest residual error to less than some predefined small value (say 1 litre/second or some percentage of the sum of air flows into each node) whilst not increasing residual errors elsewhere. This is done iteratively by finding the largest residual error and then adjusting the associated nodal pressure, by the Newton–Raphson method, to reduce the error to the predetermined level. The flows affected by the pressure change are recalculated and a new worst residual determined. The procedure then repeats until the worst residual is acceptable. Solution instabilities are avoided by tracking the effect of pressure change on residual error evolution and by limiting the maximum nodal pressure correction to say $50 \, \mathrm{N \, m^{-2}}$.

Another convergence device is to restrict the area of large openings to some fraction of that specified. When a solution is found for this fictitious network, the areas can be increased (usually doubled) and a new solution found on the basis of initial pressure values carried over from the first solution. This doubling process then continues until a solution is found for the specified network, or until successive solutions produce the same result. Walton (1982) has proposed a simpler solution technique which, it is claimed, exhibits better convergence than the Newton–Raphson method, although convergence is slower.

5.6.3 Simultaneous energy/air flow simulation

It is entirely possible, and in most cases desirable, to combine the direct solution methods for energy flow as expounded in chapter 4 with the iterative techniques appropriate to air flow evaluation. Thus at each simulation time-step inter-nodal air flow determination, based on the known temperature conditions for the time-step, can proceed in tandem with the other evaluations (solar gain, heat transfer coefficients etc) required to establish finally the overall system energy balance matrix for simultaneous solution.

It is important to note that this requires that the boundary conditions for the energy and air flow subsystems are matched. As an example, it is possible to conduct a meaningful energy analysis of an internally located zone which is surrounded by rooms of specified temperature and flux conditions. On the other hand, a detailed air flow prediction will usually require a more extensive building description expanded to include all leakages distribution up to and including the real building envelope.

5.7 Casual heat sources

A substantial portion of the heat injection at zone surface and air nodes—equations (3.10) and (3.12)—will often be due to the radiant and convective components of the so-called 'free' gains from lights, occupants and heat generating equipment. If each casual heat source is specified, then the total radiant and convective heat fluxes are given by

$$Q_{RI}(\xi) = \sum_{i=1}^{M} Q_{Si}(\xi)R_i$$

$$Q_{CI}(\xi) = \sum_{i=1}^{M} Q_{Si}(\xi)C_i$$

where $Q_{RI}(\xi)$ is the total radiant flux at time ξ (W), $Q_{CI}(\xi)$ the total convective flux (W), $Q_{Si}(\xi)$ the sensible flux output for gain i (W), R_i the radiant portion of radiant/convective split for casual gain i, C_i the convective portion and M the total number of casual sources.

The apportioning of the total radiant flux between zone surfaces can be achieved by simple area weighting or, for more accuracy, on the basis of view factor data generated from a knowledge of source position and zone geometry. The convective flux can be applied to the node representing the air volume which contains the heat source(s). Note that it may be necessary to process casual gains individually if different node point injections are required.

Of the normal heat sources found in buildings, perhaps lighting systems offer the most potential for energy savings (Cooper and Crisp 1983). Critical sizing of fittings and switching on the basis of prevailing lighting levels can result in significant electrical energy reduction and, in the case of air-conditioned buildings, reduced cooling loads. It is therefore a common simulation exercise to moderate lighting levels, and therefore casual heat gain, between lower and upper design levels at the dictate of penetrating daylight.

5.7.1 Lighting heat gain control on the basis of natural lighting levels

What is required, then, is the ability to predict zone daylight levels, to allow the setting of artificial lighting levels, to generate, in turn, heat gain values for insertion in the excitation matrix (**C**) of chapter 4. Such a model will allow the appraisal of alternative lighting control strategies and window designs.

Comprehensive monochromatic lighting models already exist (McLean 1982) to perform the task of natural lighting level prediction, and multi-chromatic models are under development (Stockmar 1983, ABACUS 1983) which, in addition, can supply spectral data suitable for colour display systems. However, these are large systems, difficult to integrate with a dynamic energy model such as the one derived in this book. The following simplified method is therefore included to allow simple switching operations in the absence of a comprehensive lighting simulation system.

The total amount of natural light which enters an enclosure depends on the size and position of windows relative to the overall enclosure area and shape, the luminance distribution (brightness) of the sky vault and the extent and surface characteristics of external obstructions. The amount of light at any point within the enclosure is, in addition to the above, dependent on the reflectance of internal surfaces.

Whilst random cloudiness will determine sky luminance distribution (Tragenza 1983), for the purpose of most winter design calculations it is usual to assume that the sky is overcast. Such an assumption eliminates cloud movement and the direct lighting effect of the sun, an important factor in tropical and semi-tropical regions, but offers a design convention acceptable in most temperate regions. Note that cloud patterns and solar lighting can be included by simulation systems of the kind referred to previously.

In general terms, an overcast sky is brightest at the zenith and approximately one-third as bright at the horizon. This describes the luminance distribution of the internationally agreed CIE standard overcast sky, that is:

$$L_\alpha = L_z(1 + 2 \sin \alpha)/3 \qquad (5.54)$$

where L_α is the luminance of the sky at an elevation of α above the horizontal (asb or cd m^{-2}) and L_z is the luminance of the sky at the zenith (asb or cd m^{-2}).

In reality the luminance distribution is constantly changing and, as a result, it is difficult to determine, absolutely, the quantity of light available within an enclosure. Instead, it has become common practice to specify the ratio of the illuminance at various points within the enclosure to that available outside. This ratio is termed the *daylight factor*, and its use offers the advantage that it remains approximately constant as the sky luminance distribution changes. The daylight factor consists of three components:

1 The *sky component* is the ratio of the illumination of a point by light arriving *directly* from the sky to the illumination caused by the complete and unobstructed sky. For a completely unobstructed window, the projected solid angle of the window, subtended at the point in question, determines the sky component. Any window obstruction (window frame, surrounding buildings, etc) will therefore act to reduce the sky component.
2 The *external reflected component* is the ratio of the light arriving at the point after initial reflection from external surfaces to that available from the complete and unobstructed sky.
3 The *internal reflected component* is the ratio of the light arriving at the point after reflection from the various enclosure surfaces to that available simultaneously from the complete and unobstructed sky.

Consider the arbitrary point P of figure 5.26. For a CIE sky, the luminous intensity normal to the elemental area of window ($\mathrm{d}x\,\mathrm{d}y$) is given by

$$I_N = L_\alpha \, \mathrm{d}x \, \mathrm{d}y \tag{5.55}$$

and the luminous intensity due to the elemental area in the direction of the point

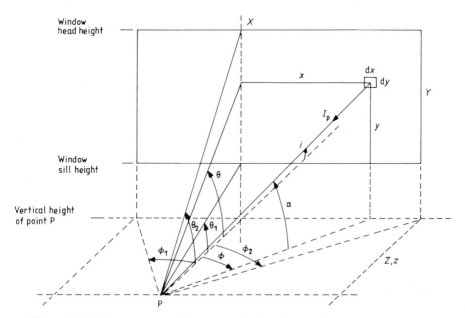

Figure 5.26 Sky component due to a vertical unobstructed rectangular window.

Energy related subsystems

P is given by

$$I_P = I_N \cos i \tag{5.56}$$

where i is the angle of incidence as shown in figure 5.26.

The illumination on a horizontal plane at P, due to the elemental area, is found from the inverse square and cosine laws of illumination:

$$dE_P = (I_P \cos \gamma)/R^2$$
$$= (I_P \sin \alpha)/R^2 \tag{5.57}$$

where γ is the angle between the line joining the elemental area and the point P and the normal at P, R is the distance between the elemental area and the point P (m), and $\alpha = (90 - \gamma)$.

Combining equations (5.54)–(5.57) gives

$$dE_P = \frac{L_z(1 + 2\sin\alpha)}{3R^2} \cos i \sin \alpha \, dx \, dy$$

$$= \frac{L_z z y (1 + 2y/R)}{3R^4} dx \, dy$$

where z is the perpendicular distance from point P to the plane of the window (m).

Now $z \tan\theta = y$ where θ is the altitude angle, and $z \tan\phi = x$ where ϕ is the azimuth angle, and so it follows that

$$\frac{d\theta}{dy} = \frac{1}{z \sec^2\theta} \qquad \text{and} \qquad \frac{d\phi}{dx} = \frac{1}{z \sec^2\phi}$$

$$E_P = \frac{L_z}{3} \int_{y_1}^{y_2} \int_{x_1}^{x_2} \frac{z^2 \tan\theta \, [(1 + 2z\tan\theta)/(x^2 + y^2 + z^2)^{1/2}]}{(x^2 + y^2 + z^2)^2} dx \, dy$$

$$= \frac{L_z}{3} \int_{\theta_1}^{\theta_2} \int_{\phi_1}^{\phi_2} \frac{\tan\theta \, [(1 + 2\tan\theta)/(\tan^2\theta + \tan^2\phi + 1)^{1/2}] \sec^2\theta \sec^2\phi}{(\tan^2\theta + \tan^2\phi + 1)^2} d\theta \, d\phi$$

$$= \frac{L_z}{3} \int_{\theta_1}^{\theta_2} \int_{\phi_1}^{\phi_2} \left[\frac{\tan\theta \sec^2\theta \sec^2\phi}{(\tan^2\phi + \sec^2\theta)^2} + \frac{2\tan^2\theta \sec^2\theta \sec^2\phi}{(\tan^2\phi + \sec^2\theta)^{5/2}} \right] d\theta \, d\phi. \tag{5.58}$$

Consider now figure 5.27, which shows a point P under the unobstructed hemispherical sky vault. The horizontal plane illumination at P due to the elemental ring located at any altitude α is given by

$$dE = (L_\alpha \, dA \sin\alpha)/R^2$$

$$= \frac{L_z(1 + 2\sin\alpha)}{3R^2} \sin\alpha \, (2\pi R^2 \cos\alpha \, d\alpha)$$

$$= \frac{2\pi L_z}{3} (1 + 2\sin\alpha) \sin\alpha \cos\alpha \, d\alpha.$$

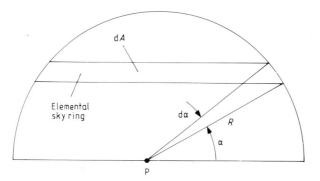

Figure 5.27 The unobstructed hemispherical sky vault.

The illumination due to the total sky vault is therefore given by

$$E = \int_0^{\pi/2} \frac{\pi L}{3}(1 + 2\sin\alpha)\sin 2\alpha \, d\alpha \qquad = 7\pi L_z/9.$$

The sky component of the daylight factor then follows from

$$SC = \frac{E_P}{E}$$

$$= \frac{3}{7\pi} \int_{\theta_1}^{\theta_2} \int_{\phi_1}^{\phi_2} \left[\frac{\tan\theta \sec^2\theta \sec^2\phi}{(\tan^2\phi + \sec^2\theta)^2} + \frac{2\tan^2\theta \sec^2\theta \sec^2\phi}{(\tan^2\phi + \sec^2\theta)^{5/2}} \right] d\theta \, d\phi.$$

The exact analytical solution of this double integral (Clarke 1977) is given by

$$\begin{aligned}
SC = \frac{3}{14\pi} \Bigg\{ &\frac{1}{A_1^{0.5}} \left[\tan^{-1}\left(\frac{B_2}{A_1}\right)^{0.5} - \tan^{-1}\left(\frac{B_1}{A_1}\right)^{0.5} \right] \\
&+ \frac{1}{A_2^{0.5}} \left[\tan^{-1}\left(\frac{B_1}{A_2}\right)^{0.5} - \tan^{-1}\left(\frac{B_2}{A_2}\right)^{0.5} \right] \\
&+ \frac{4}{3A_1}(A_1 - 1)^{0.5} \left[\left(\frac{B_2}{A_1 + B_2}\right)^{0.5} - \left(\frac{B_1}{A_1 + B_1}\right)^{0.5} \right] \\
&+ \frac{4}{3A_2}(A_2 - 1)^{0.5} \left[\left(\frac{B_1}{A_2 + B_2}\right)^{0.5} - \left(\frac{B_2}{A_2 + B_2}\right)^{0.5} \right] \\
&+ \frac{4}{3}\tan^{-1}\left[\frac{(A_2 - 1)B_2}{A_2 + B_2}\right]^{0.5} - \frac{4}{3}\tan^{-1}\left[\frac{(A_2 - 1)B_1}{A_2 + B_1}\right]^{0.5} + \frac{4}{3}\tan^{-1}\left[\frac{(A_1 - 1)B_1}{A_1 + B_1}\right]^{0.5} \\
&- \frac{4}{3}\tan^{-1}\left[\frac{(A_1 - 1)B_2}{A_1 + B_2}\right]^{0.5} \Bigg\}
\end{aligned} \qquad (5.59)$$

where $A_1 = \sec^2\theta_1$, $A_2 = \sec^2\theta_2$, $B_1 = \tan^2\phi_1$ and $B_2 = \tan^2\phi_2$.

This equation gives the sky component at any point due to a vertical window with sill, head and edges defined by altitude angles θ_1 and θ_2 and azimuth angles ϕ_1 and ϕ_2. These angles can become negative if point P is to one side, above or below the window. In this case the sky component must be evaluated by the 'shape factor algebra method' as detailed in §5.4. By the same technique, the sky component for a partially obstructed window can be determined by subtracting the value for the obstructed portion from that obtained for the whole window.

External reflected component
It is common practice to assume that the relative luminance of external obstructions is one-tenth that of the sky. This allows the external reflected component to be found by dividing the sky component of the obstructed portion, when treated as unobstructed, by ten.

Internal reflected component
This component is a function of the available direct and external reflected light, the reflectance of internal surface finishes and the geometric configuration of the exclosure. There are many methods for assessing this component (Hopkinson 1963) and one of the most popular is the 'split flux' method (AJ 1954) applied to the theory of inter-reflection for an integrating sphere. This method is an empirical formulation, based on formulae proposed by Pleijel (1952) and Dresler (1954), which considers the flux above and below a horizontal plane passing through the mid-height of the window as separate quantities. Each flux component is then modified only by receiving surfaces as illustrated in figure 5.28.

The integrating sphere theory is derived from consideration of the inter-reflection of the flux within a sphere as shown in figure 5.29. At the end of the first reflection of some initial flux entering the sphere, the average internal reflected component of the illuminance is given by

$$\text{IRC}_1 = FR/4\pi r^2$$

where IRC_1 is the average internal reflected component after one reflection, F is the flux entering the sphere (lm), R the average reflectance of internal surfaces, and r the radius of the sphere (m). At the end of the second reflection, the average internal reflected component is

$$\text{IRC}_2 = FR^2/4\pi r^2$$

and so on to infinity:

$$\text{IRC} = \frac{FR}{4\pi r^2} (1 + R + R^2 + \ldots)$$

$$= \frac{FR}{4\pi r^2} \left(\frac{1}{1 - R}\right)$$

$$= \frac{F_\text{T}}{A(1 - R)} \tag{5.60}$$

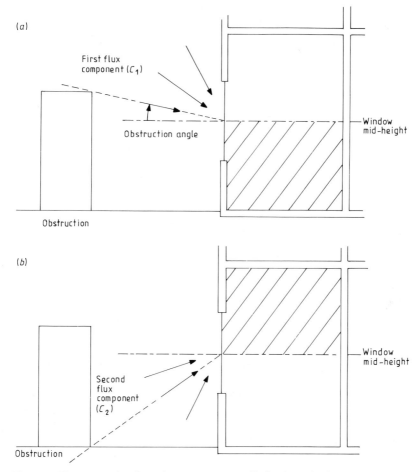

(a)

First flux
component (C_1)

Obstruction angle

Window
mid-height

Obstruction

(b)

Window
mid-height

Second
flux
component
(C_2)

Obstruction

Figure 5.28　Internal reflected component: 'split flux' method. (*a*) Upper flux component, (*b*) lower flux component.

where IRC is the total average internal reflected component, F_T the total first reflected flux (lm) and A the total internal surface area (m^2).

In applying the approach of equation (5.60) to an actual enclosure two main assumptions are made: the interior surfaces of the enclosure are diffuse reflectors;

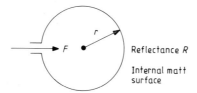

F

r

Reflectance R

Internal matt
surface

Figure 5.29　An integrating sphere.

and the enclosure shape is not significantly removed from a sphere. The latter assumption is likely to contribute most to observed differences between actual and calculated values. However, for design purposes such errors are often disregarded for all but extreme departures from the ideal sphere. If accuracy is required then the more computationally sophisticated models, relying on ray tracing techniques, will be required.

The total first reflected flux is given by

$$F_T = W(C_1 R_{fw} + C_2 R_{CW})$$

where W is the window area (m^2); R_{fw} is the average reflectance of the working plane and those wall surfaces below the horizontal plane passing through the mid-height of the window but excluding the window wall; and R_{CW} is the average reflectance of the ceiling and those wall surfaces above the horizontal plane but excluding the window wall.

C_1 and C_2 are functions of the flux entering the enclosure from above and below the horizontal plane. C_1 is dependent on the luminance distribution of the sky and an angle defining external obstructions. C_2 is dependent on the luminance of the ground and external obstructions below the horizontal plane. The ground luminance distribution is usually assumed to be 10% of the sky distribution. This gives a value for C_2 which is not significantly removed from 5 and, for convenience, C_2 is set equal to 5 by adjusting the value of C_1 accordingly. Equation (5.60) can now be expressed in final form:

$$IRC = W(C R_{fw} + 5 R_{CW})/A(1 - R)$$

where C is the adjusted function of C_1 as given in table 5.11.

Daylight factor
This is now obtained by adding the individual components for each window when

Table 5.11 Values of C.

Obstruction angle ($^\circ$)	C
0	39
10	35
20	31
30	25
40	20
50	14
60	10
70	7
80	5

each is modified by glass transmission and maintenance factors:

$$DF = \sum_{}^{N} TM(\text{SC} + \text{ERC} + \text{IRC})$$

where T is the window transmittance, M the maintenance factor and N the total number of windows. Within an energy simulation, this equation can be used to assess daylight factor magnitude and distribution on the basis of which lighting levels can be established from some predefined control statement. Simulation results will then include the effects of variable heat gains from the lighting installation.

5.8 Climate

In most energy modelling exercises the *modus operandi* is to test alternative design possibilities against short period climatic data considered representative of typical or extreme (hot, cold and moderate) weather influences. By such comparative means the favoured design elements are selected and the final (or near final) scheme subjected to long term simulation, usually annual, to determine energy consumption trends. In some cases atypical weather patterns will be required to test component response under extreme loading. The selection of climatic time-series collections, to provide meaningful simulation boundary conditions, will require great care. Of fundamental importance is the typicality of the collection in relation to the elements of the design problem in hand. Generally two conditions require to be met:

1 The data should represent the conditions under which, at some time, the building will be required to function (winter design, winter typical, summer peak, long term average, and so on).
2 The data should have some quantifiable severity measure which establishes their suitability for selection.

The former requirement is concerned with the availability of relevant, and site specific, climatic data and the latter requirement with the rank ordering of these available data according to severity criteria which include building-specific factors.

Section 5.8.1 deals with the availability of climatic data, in computer readable form, suitable for energy simulation applications. Section 5.8.2 outlines the various 'Test Reference Year' selection procedures and proposes an index of climatic severity. And, finally, §5.8.3 briefly considers solar radiation prediction often necessary in the absence of measured data.

5.8.1 Availability of climatic collections

Table 5.12 summarises the main climatic variables required for energy simulation modelling. Many weather observation centres exist world-wide to collect these

Table 5.12 Climatic data requirements of energy models.

Dry bulb temperature
Wet bulb temperature
Global (or direct) solar radiation
Diffuse solar radiation
Wind speed
Wind direction
Atmospheric pressure

and, optionally, one or more of:

Longwave radiation balance
Precipitation
Cloud cover and type
Sunshine hours

(and other) data at frequencies of one hour and greater. Until quite recently few of these centres recorded direct and diffuse (or global and diffuse) solar radiation intensities—as required by energy modelling systems—although cloud data (amount and type) are often available to permit intensity assessment by prediction.

In Britain the Meteorological Office has archived data relating to various country-wide stations as well as a small number of stations overseas. Table 5.13 gives the various solar radiation stations in the UK and indicates the climatic

Table 5.13 UK solar radiation stations and climatic elements measured.

Station	Latitude	Longitude	Elevation (m)	Element measured
Lerwick	60° 08′ N	01° 11′ W	82	T, D, L, B, SS
Eskdalemuir	55° 19′ N	03° 12′ W	242	T, D, L, B, SS
Aldergrove	54° 39′ N	06° 13′ W	68	T, D, L, B, SS
Aberporth	52° 08′ N	04° 34′ W	133	T, D, SS
Cardington	52° 06′ N	00° 25′ W	29	T, D, SS
London	51° 31′ N	00° 07′ W	77	T, D, L, SS
Kew	51° 28′ N	00° 19′ W	5	T, D, L, B, SS, I, F
Bracknell	51° 23′ N	00° 47′ W	73	T, D, L, SS, I, F, N, S, E, W
Jersey	49° 13′ N	02° 12′ W	83	T, D, L, B, SS
Aberdeen	57° 10′ N	02° 05′ W	35	T
Dunstaffnage	56° 28′ N	05° 26′ W	3	T
Dundee	56° 27′ N	03° 04′ W	30	T, B
Hurley	51° 32′ N	00° 49′ W	43	T

T = total horizontal solar radiation. D = diffuse horizontal solar radiation. L = total illumination on horizontal. B = radiation balance. SS = sunshine hours. I = direct normal solar radiation. F = diffuse illumination on horizontal. N, S, E, W = total solar radiation on vertical surfaces facing north, south, east and west respectively.

parameters recorded at each. As part of the Science and Engineering Research Council's Specially Promoted Programme, 'Energy in Buildings', a UK climatological database has been established, initially for non-commercial use (Page *et al* 1983). Table 5.14 gives the content of this database.

In the US extensive climatic data are available—for example 'Test Reference Year' collections (see §5.8.2) have been compiled for more than sixty cities—although solar values are often missing and replaced by cloud cover observations.

Table 5.14 Contents of the SERC meteorological database (hourly data only).

Station	Period of data	Hourly parameters stored for *each* station
Aberporth	1/ 1/74–30/12/81	Global solar radiation, diffuse solar radiation,
Aldergrove	1/ 1/74–31/12/81	sunshine duration, dry bulb temperature, wet
Cardington	1/ 1/74–30/05/80	bulb temperature, atmospheric pressure, wind
Eskdalemuir	1/ 1/74–29/11/81	speed, wind direction, rainfall amount, rainfall
Kew	1/ 1/62–29/12/80	duration, total cloud amount, low cloud
Lerwick	1/ 1/74–30/12/81	amount.
London	1/10/74–31/12/81	

5.8.2 *Climatic severity assessment*

The need to simulate accurately building energy exchanges has lead to the necessity of establishing techniques by which climatic data, of a given severity, can be extracted from the years of historical data available for any given location.

In 1974 the NATO committee on 'Challenges of Modern Society' initiated, as part of a project on the rational use of energy, a subproject dealing with 'Climatic Conditions and Reference Years'. The objective of this project was to recommend methods for producing a 'Test Reference Year' (TRY) for any locality for which extensive climatic data were available. A TRY is a weather collection consisting of 8760 hourly sets of mandatory and optional climatic data which, against some criteria, can be declared representative. At present a number of countries have TRY data ready for commercial or research use, although the procedures used to establish a TRY will vary by country.

In the USA (Stamper 1977) the procedure is to eliminate those years, in a period of record, containing months with extremely high or low mean temperatures until only one year—the TRY—remains. This is achieved by marking those months within the period which can be described, in terms of mean monthly temperature, as shown in table 5.15. The procedure then continues by marking those months which can be described as the next-to-hottest July, the next-to-coldest January and so on until one year remains without any marked months. The method is extremely simple to apply and the resulting TRY is considered useful for comparative studies, but not for the estimation of long term energy consumption.

Table 5.15 US 'Test Reference Year'
selection method.

hottest	July
coldest	January
hottest	August
coldest	February
hottest	June
coldest	December
hottest	September
coldest	March
warmest	May
coolest	November
warmest	October
coolest	April
coolest	July
mildest	January
coolest	August
mildest	February
coolest	June
mildest	December
coolest	September
mildest	March
coolest	May
warmest	November
coolest	October
warmest	April

In Japan, an alternative selection procedure is used (Saito and Matsuo 1974) based, not on the climatic variables, but on the cooling and heating loads to result from the application of the climate to a standardised design problem. These loads are computed hour-by-hour over a ten year period, for two different enclosures and four different orientations. This gives eight different ten-year profiles. The procedure is then repeated for each individual year in the period of record and the yearly profile considered 'nearest' to the ten-year profile (over all eight cases) is declared the TRY.

The Danish selection procedure (Lund 1976, Anderson 1974) is based on a rigorous statistical analysis applied to eleven years of climatic data. TRY selection is based on the daily mean dry bulb temperature, the daily maximum dry bulb temperature and the daily total of solar radiation according to three criteria:

1   Months with abnormal weather conditions are excluded at the outset.
2   Months with typical mean values (of the three parameters given above) are selected by comparing mean values for each month with the mean value for the same month but established from the whole period data.

3 Months with typical variations of the three parameters are selected. This is done by comparing the deviation of the three parameters, from the previously selected monthly mean values, with the corresponding deviations for the whole period.

Each month is rank ordered according to these criteria and the TRY selected. A South African procedure involves the selection of typical hot days on the basis of either the daily maximum sol–air temperature or daily maximum dry bulb temperature occurring on 10, 5 and 2.5% of the days in the period considered. Typical cold days are selected on the basis of daily minimum temperatures. A study initiated by the UK Building Research Establishment and the Meteorological Office proposed a similar method (IHVE 1973).

In the UK the Chartered Institute of Building Services has produced an 'Example Year' based on a selection method proposed by Holmes and Hitchin (1978). Based on global and diffuse solar radiation, wind speed, dry bulb temperature and degree days, the method eliminates any year containing a monthly mean which varies by more than two standard deviations from the long-term mean for this month. The recommended 'Example Year' is October 1964–September 1965 from the Kew observatory or, if a calendar year is required, the 1967 Kew collection. These data are considered adequate for predicting energy demands but not for peak load estimation.

Excluding the Japanese procedure, each of these methods suffers from two main defects:

1 Only simple synoptic data are used to discriminate between different sets of climatic data.
2 No real attempt is made to include the characteristics of the building in the selection procedure; the intention is to rank order the climatic data in terms of climatic factors alone.

In one recently completed research project (Markus *et al* 1984) a technique was developed which allows the assessment of an index which defines the stress any given climate would place on a house design of known characteristics. This Climatic Severity Index (CSI), which is house-type-specific, indicates, by a single number on a dimensionless scale, the stress placed on the building's energy system by any given climatic collection. Of course a CSI already exists—the simple degree day total—but this is inadequate because it excludes the effects of important climatic parameters and only includes building characteristics in the most rudimentary manner (via the base temperature). The CSI has a number of anticipated uses:

Determination of TRY *collections*
The index—by allowing the assessment of the stress caused by any declared climatic collection—readily lends itself to the classification of climatological data and therefore the selection of reference years of varying severity.

Design location decisions
Many national, regional and urban agencies have a range of alternative sites which can be developed for housing. Of course, many factors will affect their decision, but one should be the energy consumption consequences. For this purpose the sites must be compared on a standard energy-related index.

Determination of starting design hypothesis
Because (as will be shown) the CSI allows a rapid appraisal of the interplay between insulation level, capacity level, wind permeability and solar utilisation or exclusion, it can be used to establish the approximate level of each parameter best suited to the house type, design location, financial limits etc. This then forms the basis of a proposal which can be carried forward for refinement by, for example, dynamic simulation. Many other uses have been identified and these are elaborated in the foregoing reference.

Figure 5.30 details the main tasks involved in CSI development. Considering each task briefly in turn:

1   The ESP system (Clarke 1982) was used to predict the dynamic energy behaviour of selected house types when each were subjected to a range of UK climates.
2   Monthly mean values of each of the main climatic parameters were extracted from the climate collections to be used in the project. These data were

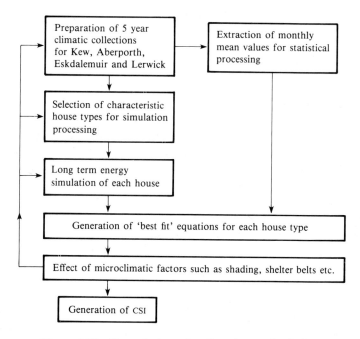

Figure 5.30   Formulation of a climatic severity index.

required for later use in the statistical part of the project. Selected data included five year collections for each of Kew (South England), Aberporth (Wales), Eskdalemuir (Scottish Borders) and Lerwick (Shetland).

3 With regard to the range of house types processed, the assumption was made that house energy characteristics can be adequately represented by the five properties of 'window size', 'insulation level', 'capacity level', 'capacity position' and 'wind permeability'. If three levels of each property are sufficiently representative (high, medium and low, or inside, middle and outside in the case of capacity position) then this gives a combinatorial possibility of 3^5 or 243 potential house types. That is, if the five design variables are accepted as adequate determinants of the energy performance of a house and if, further, the reasonable assumption is made that they are completely independent of each other, then any specific house can be regarded as a unique combination of the five variables. To computer process each house type against the 20 years of climatic data was prohibited by available resources and so an approximate 1/9th replicate scheme was established by randomly selecting 27 house types.

4 At this stage long term simulations were conducted to process effectively each of the selected house types against the twenty year climatic record.

5 Monthly energy requirements were extracted from the simulation results database and subjected, along with the monthly climatic statistics extracted earlier, to curve fitting techniques to establish, for each house type, a best fit relationship of the form:

$$E = a\bar{\theta}^2 + b\bar{R}_d^2 + c\bar{R}_f^2 + d\bar{V}^2 + e\bar{\theta} + f\bar{R}_d + g\bar{R}_f + h\bar{V} + i\bar{\theta}\bar{R}_d$$
$$+ j\bar{\theta}\bar{R}_f + k\bar{\theta}\bar{V} + l\bar{R}_d\bar{R}_f + m\bar{R}_d\bar{V} + n\bar{R}_f\bar{V} + o$$

where E is the monthly energy (kW h), $\bar{\theta}$ the monthly mean temperature ($^\circ$C), $\bar{R}_d$ the monthly mean direct normal solar radiation (W m^{-2}), $\bar{R}_f$ the monthly mean diffuse horizontal solar radiation (W m^{-2}), $\bar{V}$ the monthly mean wind speed (m s^{-1}), and o is a constant term (or y intercept); a–n are the least squares coefficients. Table 5.16 gives the equations to result for each house type.

6 At this stage the effect, on the foregoing equations (one for each house type), of a variety of microclimatic modifiers is determined by modifying the climatic data (to simulate shelter belts, for example) or by superimposing such factors as shading on the energy simulations.

7 The CSIs are now generated by a four stage process:

By entering the maximum and minimum climatic conditions—to which the house types were subjected during simulation—into each regression equation, an energy difference between maximum and minimum loading can be established for each house type and for each climatic variable. An example of this is given in table 5.17 for the house type 16 (average house) and 26 (passive solar house) equations.

By then computing the ratio of the individual energy differences to the overall totals, the contribution of each particular climatic variable is ascertained.

Table 5.16 Energy regression equations.

Term	$\bar{\theta}^2$	$\bar{R}_d^2$	$\bar{R}_f^2$	$\bar{V}^2$	$\bar{\theta}$	$\bar{R}_d$	$\bar{R}_f$	$\bar{V}$	$\bar{\theta}\bar{R}_d$	$\bar{\theta}\bar{R}_f$	$\bar{\theta}\bar{V}$	$\bar{R}_d\bar{R}_f$	$\bar{R}_d\bar{V}$	$\bar{R}_f\bar{V}$	constant
1	0.9100	0.0076	0.0276	0.1630	−64.6656	−0.0896	−1.4820	16.5864	0.0472	−0.0096	−1.1807	−0.0272	−0.0492	−0.0295	850.0271
2	0.6844	0.0116	0.0276	0.0934	−50.3507	−1.4727	−3.5503	19.2228	−0.0204	0.2580	−1.3318	−0.0256	0.0130	−0.0659	590.2528
3	0.9803	0.0220	0.0413	0.0513	−93.3554	−4.3369	−3.5748	26.0130	0.0713	0.0780	−1.8874	−0.0418	0.0353	−0.1049	1308.1035
4	1.2703	0.0295	0.0409	−1.0818	−99.8170	−5.2093	−2.5492	40.8774	0.0694	0.1374	−1.3193	−0.0495	0.0449	−0.2570	1253.9896
5	1.0045	0.0067	0.0274	0.1509	−59.9448	−0.5894	−1.7462	25.1951	0.0372	0.0296	−1.5546	−0.0220	−0.0039	−0.0727	724.4446
6	1.2734	0.0137	0.0446	0.1706	−76.3315	−1.9088	−4.6295	26.8021	0.0217	0.1322	−1.7218	−0.0361	0.0152	−0.0678	996.7990
7	0.3256	0.0229	0.0287	−0.2266	−87.8016	−3.8537	−3.1527	30.3013	−0.0423	0.2072	−1.9627	−0.0389	0.0285	−0.1022	1269.7730
8	1.2973	0.0231	0.0517	0.2335	−83.8376	−3.4787	−4.3549	29.2712	0.0570	0.1344	−1.9535	−0.0479	0.0008	−0.1304	1100.1423
9	1.9297	0.0370	0.0664	−0.0897	−130.1218	−9.1346	−5.9252	41.9583	0.2313	0.2819	−2.5699	−0.0502	0.0855	−0.3076	1701.7621
10	1.0291	0.0094	0.0386	0.1051	−69.9616	−0.7531	−3.6782	15.4652	0.0376	0.0810	−0.8655	−0.0280	−0.0148	−0.0607	926.7016
11	1.4308	0.0235	0.0493	−0.0346	−99.1432	−4.6952	−4.7566	22.5083	0.0976	0.1527	−1.3431	−0.0429	0.0291	−0.1259	1324.1257
12	1.2771	0.0173	0.0467	0.0231	−68.5724	−2.2832	−5.0732	30.5679	0.0351	0.2211	−1.8449	−0.0394	0.0213	−0.1063	819.5991
13	1.0605	0.0112	0.0233	−0.5264	−76.8607	−2.2238	−2.3820	38.4960	0.0193	0.1039	−1.7694	−0.0272	0.0651	−0.0897	936.4577
14	0.9605	0.0105	0.0315	0.1391	−54.2414	−1.3047	−3.0575	18.6869	0.0192	0.1033	−1.4864	−0.0271	−0.0037	−0.0384	667.8197
15	0.9630	0.0081	0.0303	0.1799	−69.0387	0.0668	−1.5254	25.5187	0.0433	−0.0093	−1.6541	−0.0307	−0.0562	−0.0265	905.2008
16	1.0952	0.0110	0.0261	0.0491	−83.9506	−2.4921	−2.5211	31.0448	0.0248	0.0553	−1.9939	−0.0253	0.0449	−0.0693	1114.5220
17	0.2538	0.0165	0.0209	0.0221	−51.6724	−1.8704	−2.5772	21.8033	−0.0575	0.2840	−1.3268	−0.0290	−0.0117	−0.0879	621.1768
18	0.9906	0.0055	0.0246	0.1172	−55.2493	−0.7056	−1.7556	16.5954	0.0406	0.0289	−1.1329	−0.0172	−0.0018	−0.0742	665.1710
19	1.0004	0.0092	0.0291	0.1121	−72.1607	−1.6364	−3.1017	28.0738	0.0172	0.0803	−1.8924	−0.0266	0.0426	−0.0266	951.8901
20	1.1712	0.0098	0.0396	0.1618	−58.2987	−0.8520	−3.9856	16.6714	0.0188	0.1146	−1.1084	−0.0270	−0.0047	−0.0823	714.2599
21	0.8543	0.0128	0.0283	0.0179	−61.0690	−1.9737	−2.7151	21.3190	0.0111	0.1359	−1.4775	−0.0274	0.0003	−0.0895	759.9731
22	1.0812	0.0235	0.0420	0.1279	−86.0188	−4.1079	−3.9742	40.7104	0.0380	0.2388	−2.8644	−0.0426	0.0001	−0.1410	1099.4317
23	1.4335	0.0235	0.0497	−0.0359	−98.9667	−4.6606	−4.7857	22.4904	0.0960	0.1532	−1.3453	−0.0432	0.0282	−0.1240	1322.3968
24	0.8388	0.0044	0.0245	0.1201	−51.3552	−0.3429	−2.2132	26.1886	0.0137	0.0930	−1.7062	−0.0176	−0.0127	−0.0655	604.8176
25	1.2439	0.0173	0.0423	−0.0268	−86.5663	−2.9538	−4.9234	21.4916	0.0413	0.1606	−1.3686	−0.0391	0.0445	−0.0659	1163.1544
26	0.2439	0.0576	0.0259	0.0739	−96.4461	−10.2289	−6.4945	33.1380	0.2211	0.9374	−1.9247	−0.0208	−0.0182	−0.3696	1087.7877
27	0.9905	0.0100	0.0386	0.1294	−78.1842	0.0041	−3.1972	16.0948	0.0379	0.0383	−0.9732	−0.0348	−0.0558	−0.0312	1092.4584
28	0.6281	0.0415	0.0515	0.0205	−117.0153	−8.2064	−5.7879	43.1737	0.0303	0.5098	−2.9572	−0.0533	0.0436	−0.2821	1656.9116
29	1.4444	0.0540	0.0628	−0.3205	−111.9623	−10.0009	−7.4048	33.4261	0.2210	0.6378	−1.9180	−0.0529	0.0085	−0.2701	1380.3015
30	1.0526	0.0106	0.0407	0.1167	−82.6358	0.1481	−3.1904	25.5825	0.0354	0.0391	−1.4604	−0.0383	−0.0635	−0.0291	1145.3155
Coeff.	a	b	c	d	e	f	g	h	i	j	k	l	m	n	o

Energy equation: $E = a\bar{\theta}^2 + b\bar{R}_d^2 + c\bar{R}_f^2 + d\bar{V}^2 + e\bar{\theta} + f\bar{R}_d + g\bar{R}_f + h\bar{V} + i\bar{\theta}\bar{R}_d + j\bar{\theta}\bar{R}_f + k\bar{\theta}\bar{V} + l\bar{R}_d\bar{R}_f + m\bar{R}_d\bar{V} + n\bar{R}_f\bar{V} + o$

Table 5.17 Contribution of the climatic parameters to the energy difference between maximum and minimum climatic loadings.

Climatic parameter	Max/min condition	Energy difference (kW h)	
		House 16	House 26
$\bar{\theta}$	15.3/−1.9	1186	1596
$\bar{R}_d$	155.2/0.6	120	194
$\bar{R}_f$	94.4/3.3	2	361
$\bar{V}$	10.6/2.7	250	269
All	as above	1559	2420

If the best combination of climatic variables (lowest energy case) is made, arbitrarily, to correspond to a CSI of 2 (to allow later extension of the range if colder climates are added) and the worst combination a CSI of (say) 10, the contribution of each individual climatic parameter can be determined by multiplying these upper and lower CSI values by the previously determined ratios. Figure 5.31 shows one way of expressing the result: the contribution of each climatic parameter to the CSI can be shown graphically, separately and for each house type. Alternatively, a CSI equation can be formulated for each house type. Each equation will then give the contribution of each climatic parameter to the total CSI. Table 5.18 gives the final CSI equations for each house type. It is important to note that although the climatic contributions will vary for different house types, the normalisation technique ensures that the total CSI for each house type, when subjected to the same climate, will be approximately the same. As proof of this, consider the entries of table 5.19 for three radically different house types. This means that CSI maps (equivalent to degree day maps) can be constructed, for all locations with available monthly mean climatic data, by the application of any equation from table 5.18, and that each climate collection can be assigned a unique CSI value.

A family of curves can now be produced which relate climatic severity to energy consumption for the 30 house types with the remaining house types assessed by interpolation. Figure 5.32 gives example curves for three house types.

For any country for which a CSI map exists, the CSI for any locality can be read off and used, in conjunction with these curves (as exemplified by the curves of figure 5.32), to determine the relative energy consequences of adopting alternative housing mixes or locations. Alternatively, the curves of figure 5.32 can be redrawn for the case of constant energy as shown in figure 5.33. This is achieved by determining (from the equations) the altered CSI condition which would ensure energy equity against some standard house type: house type 16 in this case. This allows, for any CSI, the determination of an altered CSI which defines a region within which the house type of interest can be placed without incurring an energy penalty, relative to the base house type.

Table 5.18　Final CSI equations. $\text{CSI} = (a\bar{\theta} + e_a) + (b\bar{R}_d + e_b) + (c\bar{R}_f + e_c) + (d\bar{V} + e_d)$
$= a\bar{\theta} + b\bar{R}_d + c\bar{R}_f + d\bar{V} + e_t.$

House type	Design combination†	a	b	c	d	e_a	e_b	e_c	e_d	e_t
1	0/1/0/1/0	−0.356	−0.007	−0.008	0.127	6.118	1.281	0.860	−0.234	8.024
2	2/2/0/1/1*	−0.372	−0.003	−0.008	0.185	6.391	0.521	0.808	−0.339	7.381
3	1/0/0/1/0	−0.409	−0.005	−0.002	0.136	7.022	0.799	0.243	−0.250	7.814
4	2/1/1/1/0	−0.402	−0.003	−0.007	0.129	6.908	0.513	0.733	−0.236	7.918
5	0/2/0/0/1	−0.359	−0.004	−0.007	0.210	6.167	0.605	0.750	−0.386	7.136
6	1/2/1/0/1	−0.409	−0.002	−0.002	0.201	7.026	0.269	0.199	−0.368	7.125
7	2/0/0/1/0	−0.436	−0.001	−0.002	0.143	7.500	0.258	0.192	−0.262	7.688
8	1/1/1/0/1	−0.405	−0.001	−0.004	0.198	6.966	0.127	0.441	−0.363	7.171
9	1/0/1/0/0	−0.350	−0.011	−0.002	0.137	6.018	1.952	0.196	−0.251	7.914
10	0/2/1/1/1	−0.417	−0.005	−0.001	0.125	7.159	0.910	0.070	−0.230	7.909
11	1/2/1/1/0	−0.421	−0.005	−0.000	0.116	7.238	0.939	0.031	−0.213	7.995
12	2/2/1/0/1*	−0.375	−0.003	−0.004	0.225	6.441	0.518	0.392	−0.413	6.937
13	2/2/0/0/0	−0.401	−0.003	−0.001	0.202	6.894	0.529	0.072	−0.370	7.125
14	1/2/0/1/1	−0.403	−0.003	−0.000	0.200	6.928	0.557	0.018	−0.367	7.136
15	0/0/0/0/0	−0.333	−0.008	−0.008	0.166	5.720	1.359	0.888	−0.304	7.663
16	1/1/0/0/0	−0.400	−0.004	−0.000	0.183	6.870	0.775	0.016	−0.366	7.325
17	2/1/0/1/1*	−0.375	−0.005	−0.004	0.172	6.445	0.938	0.436	−0.315	7.504
18	0/0/0/1/1	−0.400	−0.001	−0.006	0.173	6.873	0.249	0.644	−0.318	7.448
19	1/2/0/0/0	−0.408	−0.001	−0.002	0.206	7.108	0.243	0.190	−0.377	7.047
20	0/1/1/1/1*	−0.386	−0.006	−0.001	0.171	6.631	1.054	0.114	−0.313	7.486
21	1/1/0/1/1	−0.435	−0.000	−0.000	0.189	7.469	0.031	0.044	−0.347	7.197
22	1/0/0/0/1	−0.392	−0.002	−0.001	0.233	6.744	0.423	0.072	−0.427	6.812
23	1/1/1/1/0	−0.422	−0.005	−0.000	0.117	7.254	0.912	0.037	−0.214	7.989
24	0/1/0/0/1*	−0.370	−0.003	−0.002	0.257	6.359	0.545	0.172	−0.471	6.606
25	2/2/1/1/0	−0.426	−0.002	−0.005	0.129	7.322	0.272	0.506	−0.236	7.863
26	2/0/1/1/1*	−0.346	−0.005	−0.015	0.127	5.954	0.804	1.541	−0.233	8.066
27	0/0/1/1/0	−0.157	−0.038	−0.001	0.043	2.703	6.511	0.145	−0.079	9.280
28	2/0/1/0/0	−0.385	−0.006	−0.003	0.154	6.609	1.063	0.281	−0.282	7.761
29	1/0/1/1/1	−0.383	−0.007	−0.005	0.121	6.576	1.131	0.550	−0.221	8.036
30	0/1/1/0/0	−0.355	−0.009	−0.004	0.141	6.100	1.607	0.424	−0.258	7.873

†A/B/C/D/E*:　A = capacity level　　　　　0 low; 1 medium; 2 high
　　　　　　　　B = capacity position　　　　1 inside; 2 middle; 3 outside
　　　　　　　　C = windows　　　　　　　　0 standard; 1 large
　　　　　　　　D = infiltration　　　　　　　0 standard; 1 tight
　　　　　　　　E = insulation　　　　　　　0 standard; 1 high
　　　　　　　　* indicates blind control active

5.8.3　Solar radiation prediction

The intensity of extraterrestrial solar radiation, evaluated at the mean sun–earth distance and integrated over all wavelengths, is termed the solar constant. Because of the elliptical orbit of the earth around the sun this 'constant' will vary, as given by

$$I'_{sc} = I_{sc}\{1 + 0.033 \cos[(360 - Y)/370]\} \qquad (5.61)$$

where I'_{sc} is the corrected solar constant (W m^{-2}), I_{sc} is the solar constant

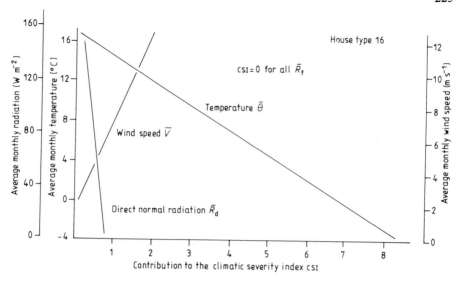

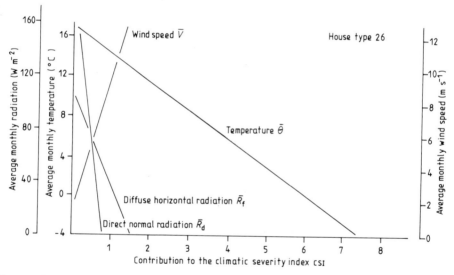

Figure 5.31 Contribution of the climatic parameters to the CSI for house types 16 and 26.

evaluated at the mean sun–earth distance, and *Y* is the year day number (January 1 = 1).

The value assigned to the mean solar constant, as used by NASA, is $1.353 \times 10^3 \ \text{W m}^{-2}$ (Thekaekara 1973) and its spectral composition, that is the wavelength-dependent irradiance, is as shown in figure 5.34. Whilst the entire spectrum spans the range from x-rays ($< 0.01 \ \mu$m) to radio waves (> 100 m), some 99.9% of the total energy is contained within the range 0.217–$10.94 \ \mu$m.

For terrestrial applications (such as those described in §5.3.2) the

Table 5.19 Each equation gives the same total CSI.

Climate condition	Contribution to the CSI for house type ...		
	16	26	30
1 $\bar{\theta}$ = 15.3	0.75	0.67	0.67
$\bar{R}_d$ = 155.2	0.16	0.08	0.20
$\bar{R}_f$ = 94.4	0.02	0.18	0.05
$\bar{V}$ = 2.7	0.12	0.11	0.12
Total	1.05	1.04	1.04
2 $\bar{\theta}$ = −1.9	7.63	6.61	6.78
$\bar{R}_d$ = 0.6	0.77	0.80	1.60
$\bar{R}_f$ = 3.3	0.02	1.49	0.41
$\bar{V}$ = 10.6	1.60	1.11	1.24
Total	10.02	10.01	10.03

extraterrestrial intensity will require modification on the basis of sun position, sky conditions and site features. As the radiation penetrates the earth's atmosphere, scattering (and subsequent back reflection) and absorption will occur, so that some portion of the radiation flux will not reach the earth's surface. Unsworth (1975) has produced data for the irradiation of the earth's surface under clear sky

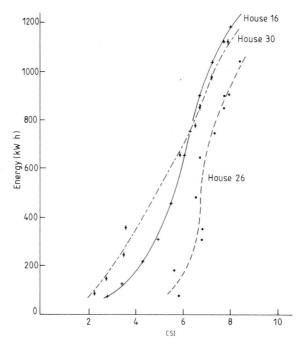

Figure 5.32 CSI/energy relationship for three house types.

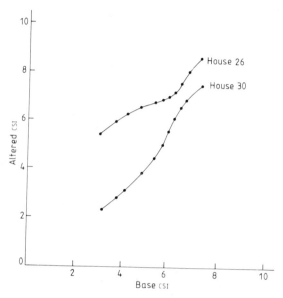

Figure 5.33 Energy equity curves (relative to house type 16).

conditions and Rodgers *et al* (1978) have fitted the following expression:

$$I_{dn} = I'_{sc} \exp\left[\sum_{i=0}^{3} \left(\sum_{j=0}^{2} b_{ij} W^j\right) m^i\right] \exp(-\tau m) \qquad (5.62)$$

where I_{dn} is the direct normal solar intensity under cloudless sky conditions (W m^{-2}), b_{ij} is a constant as given in table 5.20, W is the level of atmospheric precipitable water content (mm), m the air mass level or atmospheric path length, and τ the atmospheric turbidity coefficient. The precipitable water content is defined as the water level which would result from the condensation of all water vapour contained in a vertical column extending from the earth's surface to the outer limits of the atmosphere. For the UK, Rodgers *et al* have produced, from

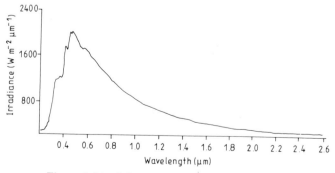

Figure 5.34 Solar spectral irradiance curve.

Table 5.20 Constants b–g for use with equations (5.62)–(5.65).

i	0	1	2	3	4	5	6	7
	e 3.67985	c −6.468	c 1.056	c −0.128				
		d −3.492	d 2.049	d 0.579				
		e −24.4465	e 154.017	e −742.181	e 2263.36	e −3804.89	e 2261.05	
		f 47.382	f 29.671	f −15.8621	f 4.3463	f −0.57764	f 0.03472	f −0.0007362
		g 1.8313	g −3.7082	g 4.1233	g −0.6409	g 0.02855		
j	g 297.0							
0	b −0.129641	b −0.0642111	b −0.0046883	b 0.000844097				
1	b 0.00412828	b −0.00801046	b −0.00220414	b −0.000191442				
2	b −0.0112096	b 0.0153069	b −0.00429818	b 0.000374176				

Meteorological Office data, the following expression:

$$W = 10.44 + \sum_{i=1}^{3} [c_i \cos(\omega i) + d_i \cos(\omega i)] \qquad (5.63)$$

where c and d are constants as given in table 5.20, and $\omega = 2\pi y/366$. The air mass is related to the solar altitude and site elevation such that

$$m = \exp[h(-0.0017h - 0.1174)]/\sin a$$

where h is the site elevation (km) and a the solar altitude. For low altitude angles an alternative expression is suggested based on a least squares fit applied to air mass data extracted from the Smithsonian Meteorological Tables (List 1951):

$$m = \exp[h(-0.0017h - 0.1174)]\exp[e_0 + \sum_{i=1}^{6} e_i(\sin a)^i] \qquad \text{for } a < 10° \qquad (5.64)$$

where e is a constant as given in table 5.20.

The turbidity coefficient is a measure of the atmospheric aerosol effect (Unsworth 1975) and is mainly influenced by the prevailing airstream. Typical values are given in table 5.21 and elsewhere (Souster *et al* 1979) for the UK.

Table 5.21 Typical values for the turbidity coefficient.

Weather condition	τ
Clear winter	0.09
Clear summer, country	0.19
Clear summer, town	0.28
Hazy	0.37
Heavy warm hazy	0.45

With regard to diffuse horizontal intensities under cloudless sky conditions, Rodgers *et al* have produced the following expression, based on earlier work by Parmelee (1954) which established that, for a fixed solar altitude, a linear relationship exists between diffuse and direct horizontal irradiance:

$$I_{\text{fh}} = 2 + \left(\sum_{i=1}^{7} f_i(a/10)^i\right)\{1 + 0.033 \cos[(360 - Y)/370]\}$$

$$- \left(10^{-3} \sum_{i=0}^{5} g_i(a/10)^i\right) I_{\text{dn}} \sin a \qquad (5.65)$$

where f and g are constants as given in table 5.20.

For cloudy sky conditions a number of mathematical models have been derived. Whereas the foregoing model is the basis of the recommended EEC clear sky model (Page 1979), the model of Krochmann (1979) is being developed as the standard cloudy sky technique.

In the US, the ASHRAE Task Group on Energy Requirements for Heating and Cooling (ASHRAE 1975) have suggested a procedure based on the work of Kimura and Stephenson (1969). The direct normal intensity under cloudy skies, I_{dnc}, can be determined from

$$I_{dnc} = I_{dn}F\left(1 + \frac{I_{fh}}{I_{dn}\sin a}\right).$$

The F factor is a function of the cloud type and amount and is given by

$$F = (1 - \text{CC}/10)\left(\frac{\sin a}{p + \sin a} + \frac{(q - 1)}{[0.691 - 0.137\sin a + 0.394(\sin a)^2]}\right) \qquad (5.66)$$

where p and q are constants as given in table 5.22, and CC is the cloud cover.

Assuming that information on cloud type and amount is available for four reference vertical layers—as specified in the US standard—then the cloud cover can be obtained from

$$\text{CC} = T_{ca} - 0.5\left[\left(\sum_{j=1}^{4}C_j\right)_{cirrus} + \left(\sum_{j=1}^{4}C_j\right)_{cirrostratus} + \left(\sum_{j=1}^{4}C_j\right)_{cirrocumulus}\right] -$$

where T_{ca} is the total cloud amount and C_j is the estimated cloud amount (on a scale 0–10) for the jth layer of the stated cloud type.

Table 5.22 Constants p, q, r and s for use with equations (5.66) and (5.67).

Season	p	q	r	s
Spring		1.06	0.012	− 0.0084
Mar 21	0.071			
Apr 21	0.097			
May 21	0.121			
Summer		0.96	0.033	− 0.0106
June 21	0.134			
Jul 21	0.136			
Aug 21	0.122			
Autumn		0.95	0.030	− 0.0108
Sep 21	0.092			
Oct 21	0.073			
Nov 21	0.063			
Winter		1.14	0.003	− 0.0082
Dec 21	0.057			
Jan 21	0.058			
Feb 21	0.060			

The cloudy sky diffuse horizontal intensity, I_{fhc}, is then obtained from

$$I_{\text{fhc}} = (I_{\text{dn}} \sin a + I_{\text{fh}})(\text{CCF} - F) \qquad (5.67)$$

where CCF is a cloud cover factor $= q + r\,\text{CC} + s\,\text{CC}^2$, and q, r and s are constants as given in table 5.22. The quantities I_{dnc} and I_{fhc} (or I_{dn} and I_{fh}) which emerge can then be used as the starting point for the evaluation of building solar loading by the techniques of §5.3.2.

References

ABACUS 1983 Specification and Prototypical Implementation of a Computer-Based Multichromatic Lighting Model *Research Grant Case for Support* (Glasgow: ABACUS, University of Strathclyde)

AJ 1954 The Split Flux Method *Architects J.* **120** 335

ASHRAE 1975 Procedure for Determining Heating and Cooling Loads for Computerizing Energy Calculations *ASHRAE Task Group Report* (New York)

—— 1981 *Fundamentals Volume*

Alamdari F and Hammond G P 1983 Improved Data Correlation for Buoyancy-Driven Convection in Rooms *Rep. SME/J/83/01* (Cranfield: Cranfield Institute of Technology, Applied Energy Group)

Anderson B 1974 Meteorological Data for Design of Building and Installation *Rep. 89, Danish Build. Res. Inst.*

Born M and Wolf E 1965 *Principles of Optics* (3rd edn) (London: Pergamon)

Brown 1962a Natural Convection Through Rectangular Openings in Partitions—1: Vertical Partitions *J. Heat Mass Transf.* **5**

—— 1962b Natural Convection Through Rectangular Openings—2: Horizontal Partitions *J. Heat Mass Transf.* **5** 869–81

Chung B T F and Samitra P S 1972 Radiation Shape Factors from Plane Point Sources *J. Heat Transf.* **94** 328–30

Clarke J A 1977 Environmental Systems Performance *PhD Thesis* University of Strathclyde

—— 1982 *ESP Documentation Set* (Glasgow: ABACUS, University of Strathclyde)

Cockcroft J P 1979 Heat Transfer and Air Flow in Buildings *PhD Thesis* University of Glasgow

Cole R J 1976 The Longwave Radiation Incident Upon the External Surface of Buildings *Building Services Engineer* **44** 195–206

Cooper I and Crisp V H C 1983 Barriers to the Exploitation of Daylighting in Building Design *Proc. Int. Daylighting Conf.*

Dick J B 1950 The Fundamentals of Natural Ventilation of Houses *J. IHVE* **18** 123–34

Dresler A 1954 *Trans. Illum. Eng. Soc.* **19** 50

Duffie J A and Beckman W A 1980 *Solar Engineering of Thermal Processes* (New York: Wiley)

Etheridge D W 1977 Crack Flow Equations and Scale Effect *Building and Environment* **12** 181–9

Foley J D and Van Dam A 1982 *Fundamentals of Interactive Computer Graphics* (Reading, Mass: Addison-Wesley)

Fujii T and Imura H 1970 Natural Convection Heat Transfer from a Plate with Arbitrary Inclination *J. Heat Mass Transf.* **15** 755–67

Hanson T, Smith F, Summers D and Wilson C B 1982 Computer Simulation of Wind Flow Around Buildings *Computer-Aided Design* **14**(1) 27–31

Harris-Bass J, Lawrence P and Kavarana B 1974 Adventitious Ventilation of Houses *Proc. Br. Gas. Symp. on Ventilation of Housing*

Hay J E 1979 Study of Shortwave Radiation on Non-Horizontal Surfaces *Rep. 79–12* (Downsview, Ontario: Atmospheric Environmental Service)

Heavens O S 1955 *Optical Properties of Thin Solid Films* (London: Butterworth)

Holmes M J and Hitchin E R 1978 An Example Year for the Calculation of Energy Demand in Buildings *J. Build. Serv. Eng.* (January)

Hopkinson R G, Rogers D F and Adams J A 1963 *Mathematical Elements for Computer Graphics* (London: McGraw-Hill)

Hottel H C 1930 Radiant Heat Transmission *J. Mech. Eng.* **52** 699–704

Howson R P, Ridge M I, Bishop C A and Cottman G P 1984 The Production of Selective Optical Coatings on Plastic Sheet *Proc. EEC Seminar on Energy Saving in Buildings* (Dordrecht: Reidel) pp 421–33

IHVE (now CIBS) 1973 Some Fundamental Data Used by Building Services Engineers *Rep. Working Group 6, Building Services Subcommittee of CACCI* (London)

Ito N, Kimura K and Oka J 1972 A Field Experiment Study on the Convective Heat Transfer Coefficient on Exterior Surface of a Building *ASHRAE Trans.* (2225) 184–91

Jacob M 1949 *Heat Transfer* (New York: Wiley)

Kimura K and Stephenson D G 1969 Solar Radiation on Cloudy Days *ASHRAE Trans. Pt I* 227–33

Klucher T M 1979 Evaluation of Models to Predict Insolation on Tilted Surfaces *J. Solar Energy* **23**(2) 111–14

Kreith F 1973 *Principles of Heat Transfer* (3rd edn) (New York: Harper and Row)

Kreyszig E 1979 *Advanced Engineering Mathematics* (4th edn) (New York: Wiley)

Krochmann 1979 Calculation Method and its Comparison with Other Results *Final Rep. Contract no 270-77-ESP* (Berlin: Institut für Lichttechnik, Technische Universität)

Launder B E and Spalding D B 1972 *Mathematical Models of Turbulence* (New York: Academic)

List R J (ed) 1951 *Smithsonian Meteorological Tables* (6th edn) (Washington DC: Smithsonian Inst.)

Lund H 1976 Requirements and Recommendations for the Test Reference Year *Proc. CIB Symp.*

Ma C C Y and Iqbal M 1983 Statistical Comparison of Models for Estimating Solar Radiation on Inclined Surfaces *J. Solar Energy* **31**(2) 313–17

McAdams W H 1954 *Heat Transmission* (New York: McGraw-Hill)

McLean R 1982 *GLIM: General Light Inter-Reflection Model; User Manual* (Cambridge: Applied Research of Cambridge)

Malinowski H K 1971 Wind Effect on the Air Movement Inside Buildings *Proc. 3rd Conf. Wind Effects on Buildings and Structures* 125–34

Markus T A, Clarke J A and Morris E N 1984 Climatic Severity *Technical Rep., Dept Architecture and Build. Sci.* (Glasgow: University of Strathcylde)

Min T C, Schutrum L F, Parmelee G V and Vouris J D 1956 Natural Convection and Radiation in a Panel-Heated Room *ASHRAE Trans.* **62** 337–58

Moore G and Numan M Y 1982 *Form Factors: The Problem of Partial Obstruction* (Cambridge: Cambridge University, Martin Centre Report)

—— 1983 *Specular Reflection of Radiation* (Cambridge: Cambridge University, Martin Centre Working Paper)

Moult A 1980 CAFE—A Computer Program to Calculate the Flow Environment *Proc. CAD 80*

Nielsen P V 1982 Mathematical Models for Room Air Distribution *Proc. Univ. Liège Conf. Systems Simulation in Buildings* (Liège)

Optical Society of America 1978 *Handbook of Optics* (New York: McGraw-Hill)

Page J K 1979 Predetermination of Irradiation on Inclined Surfaces for Different European Centres *Final Rep. Contr. no 297-77-UK* (Sheffield: University of Sheffield)

Page J K, Thompson J L and Smith W A 1983 A Meteorological Database System for Architectural and Building Engineering Designers *Users' Handbook* vol 1, 2nd edn (SERC Publication, September 1983)

Parmelee G V 1954 Irradiation of Horizontal and Vertical Surfaces by Diffuse Solar Radiation from Cloudless Skies *Trans. ASHVE* **60** 341–56

Patankar S V 1980 *Numerical Heat Transfer and Fluid Flow* (New York: McGraw-Hill)

Patankar S V and Spalding D B 1972 A Calculation Procedure for Heat, Mass and Momentum Transfer in Three-Dimensional Parabolic Flows *J. Heat Mass Transf.* **15** 1787–1806

Pilkington Glass Environmental Advisory Service 1973 *Thermal Transmission of Windows*

Pleijel G 1952 *J. Sci. Instrum.* **29** 137

Purdie C 1983 Computer Aided Visual Impact Analysis *PhD Thesis* University of Strathclyde

Rodgers G G, Souster C G and Page J K 1978 Development of an Interactive Computer Program SUN1 *Rep. IRN BS28* (Sheffield: University of Sheffield)

Rogers D F and Adams D A 1976 *Mathematical Elements for Computer Graphics* (New York: McGraw-Hill)

Rousseau and Mathieu 1973 *Problems in Optics* (London: Pergamon)

Saito H and Matsuo Y 1974 Standard Weather Data for SHASE Computer Program *Proc. 2nd Symp. Use of Computers for Environ. Eng. related to Build.*

Souster C G, Page J K and Colquhoun I D 1979 Climatological values of the Turbidity Coefficient in the UK for Different Classes of Radiation Day *Proc. UK ISES Conf. C18* pp 1–16

Stamper E 1977 Weather Data *ASHRAE J.* p 47 (February)

Stearn D D 1982 VISTA: Visual Impact Simulation Technical Aid *Proc. Eurographics 82* (Amsterdam: North-Holland) pp 333–7

Stockmar A W 1983 Developments in Computer Aided Lighting Design *Proc. PARC83*

Sutherland I E, Sproull R F and Schumacher R A 1976 A Characterization of Ten Hidden-Surface Algorithms *Computing Surveys* **6**(1) 1–55

Swinbank W C 1983 Longwave Radiation from Clear Skies *J. R. Meteorol. Soc.* **89** 339–48

Tamura G T and Wilson A G 1967 Pressure Difference Caused by Chimney Effect in Three High Rise Buildings *ASHRAE Trans.* **73**

Temps R C and Coulson K L 1977 Solar Radiation Incident on Slopes of Different Orientations *J. Solar Energy* **19**(2) 179–84

Thekaekara M P 1973 Solar Energy Outside the Earth's Atmosphere *J. Solar Energy* **14** 109–27

Tragenza P R 1983 Predicting Daylighting from Skies of Random Cloudiness *Proc. Int. Daylighting Conf.*

Unsworth M H 1975 Variations in the Shortwave Radiation Climate in the UK *Proc. Conf. UK Meteorol. Data and Solar Energy* (International Solar Energy Society (UK Section))

Vasicek A 1960 *Optics of Thin Films* (Amsterdam: North-Holland)

Walton G N 1982 Airflow and Multiroom Thermal Analysis *ASHRAE Trans.* **88** 2

Welty J R 1974 *Engineering Heat Transfer* (London: Wiley)

Wong H Y 1977 *Heat Transfer for Engineers* (London: Longman)

6

Plant Simulation

The ASHRAE Task Group on Energy Requirements for Heating and Cooling of Buildings (ASHRAE 1976) have defined plant (or system) simulation as

> ... predicting the operating quantities within a system (pressures, temperatures, energy- and fluid-flow rates) at the condition where all energy and material balances, all equations of state of working substances, and all performance characteristics of individual components are satisfied.

They also state that whilst

> In such a simulation, the thermal capacities of equipment are crucial ... the dynamic response of most systems is much more rapid than that of the building.

And so, for this reason,

> ... a steady-state simulation of the system is adequate for most energy calculations.

An increasing proportion of the modelling community no longer accepts this conclusion and is concerned to develop dynamic models of plant components when combined to form a system. This chapter is included to demonstrate the formulation of such a dynamic model by the application of the theory of chapter 3 to some example plant systems. The plant models to result are expressed in a form suitable for full integration with the previously derived building model. It should be noted, however, that dynamic plant modelling is a recent development field and much research is currently under way to establish efficacious nodal schemes and appropriate techniques for the prediction of complex heat transfer coefficients.

The modelling approach, as presented in chapters 3 and 4 in relation to building energy flowpaths, is essentially a three-stage process: system subdivision (discretisation) into a number of elemental finite regions is followed by the establishment of an equivalent equation set, one equation for each node (to represent region flow and storage characteristics), which can be solved simultaneously and repeatedly. That is, building operational, geometrical and constructional information, zone-by-zone, is transformed (by software) into an equivalent resistance–capacitance network by locating region-characteristic simulation equations in some overall system matrix at the dictates of some in-built indexing scheme which respects the spatial integrity of the entire system.

For a ten-zone building problem, a model of this kind will be attempting, at each computational time-step t $(0 < t < 1;$ t variable), the simultaneous processing of a system matrix of up to 2000×2000 matrix elements. Efficient matrix processing can only be achieved by the procedure of matrix partitioning in which the overall matrix is partitioned into a number of component matrices matching exactly the physical scheme but excluding unfilled matrix elements (indicating the non-existence of an inter-region flow potential). Extracted component matrices are then processed independently—but only if their contents have changed since the previous time-step—to produce the characteristic equations (CEs) of chapter 4. Combined CEs are then further processed, in the context of any imposed system control, to complete the simultaneous solution. By this means only some 0.1–10% (depending on the problem size) of the total number of matrix elements are ever actually processed (operated on) to achieve simultaneous solution. This matrix technique is the kernel of the combined processing of building and plant systems.

From a purely modelling viewpoint there is little difference between the simulation of building and plant energy flows. In the former case, modelling involves the mathematical representation of the time- and space-dependent energy and mass flows between different building regions. If these flows are considered analogous to the energy and mass exchange between different plant components or, depending on the required simulation complexity, between different internal regions of the same component, then a dynamic simulation model can be derived to represent intra- and inter-component energy flow as before.

This chapter demonstrates application of the chapter 3 technique to three system types: a simple air conditioning system incorporating the usual heating, cooling and dehumidifying components; an active solar system incorporating collector, remote storage and auxiliary heat exchanger; and a domestic wet central heating system incorporating boiler, radiators and a hot water cylinder. The combination of the plant matrices to result with the previously derived building matrix is then discussed.

6.1 Sequential versus simultaneous approaches

There are two approaches to plant simulation: sequential and simultaneous.

In a sequential model, each component will be replaced by some equivalent input/output relationship so that, when connected to comprise a system, the output from one component, in the calculation stream, becomes the input to the next. Thus, at some point in time, a simulation can commence at a known boundary—perhaps ambient conditions or the building-side load as generated previously by a building model—with calculation progressing in a prescribed manner and serially throughout the entire system relying, perhaps, on successive substitution or other iterative methods for overall network solution. Individual components are therefore replaced by 'black box' algorithms which can be simple (the assessment of fan heat pick-up from manufacturer's data for example) or complex (the iterative solution of a differential equation set representing storage tank behaviour or the selection and application of some non-linear, empirical relationship representing a humidification process). This technique has distinct advantages: different modelling methods, simple or complex, can be used for different plant components, allowing piecemeal system development as technical knowledge improves; and the underlying modularity allows component models to be appended and updated whilst minimising subsequent integration difficulties. Problems will arise, however, when control dynamics are to be incorporated or where component evaluation relies on information as yet uncomputed. It is also usual to experience difficulty when recirculating loops are present and various component linking protocols are being pursued by sequential modellers. Most plant models currently operating will fall into the sequential category and the specific techniques employed to overcome these problems are described elsewhere (Quick 1982, Hanby 1984).

Simultaneous plant modelling is a new development field involving, at its most complex, the representation of plant components by discrete nodal schemes and the derivation of energy and mass flow equation sets which represent whole-system, inter-node exchanges over time and space dimensions. The plant matrix to emerge is then integrated with the multi-zone building matrix (see figure 4.5 of §4.1.3), subjected to any control function based on nodal condition statements, and the combined matrix solved simultaneously and repeatedly as the model steps through its finite time increments. An approach of moderate complexity is to formulate the system matrix relative to a single node component treatment and then to rely on independent 'black box' algorithms, operating in parallel, to represent internal component performance. Component thermal capacities are still included and the technique remains simultaneous but the modelling task is less complex. And each component model can utilise any available theory. The plant system matrix to result in either case *is* the system linking protocol and so the problems associated with the sequential approach are overcome. The integration of building and plant matrices (representing building and plant node capacities and inter-node exchanges), for simultaneous processing, is greatly simplified if both employ the same technical 'engine', and for this reason the implicit finite difference approach of chapter 3 is the basis of the following plant model derivations.

Returning to figure 4.5, and ignoring for the present the matrix portion shown dashed, this multi-zone building matrix is composed of a number of zone matrices with plant interaction terms as detailed in figure 4.4. It is important to note that the q_p terms are the building-side requirements to maintain any specified time-dependent environmental condition after full allowance is made for the location of the zone plant interaction point and thermostat (that is the nodal temperature(s) being controlled during the simulation). At this stage in the formulation only the most rudimentary account is taken of plant response times, plant control and plant-side inefficiencies. It is these q_p terms that are replaced by the plant matrix systems to allow the detailed tracing of each and every energy 'packet' as it navigates the plant network.

The simultaneous modelling approach is demonstrated in relation to three plant systems:

The single node component treatment is applied to a simple air conditioning arrangement and the required component models formulated. A more detailed multi-node equation set is then established for one component for introduction to the system matrix to replace the previously assumed component model.

An active solar system is then considered at the full multi-node modelling level.

And, lastly, a multi-node model is developed for a domestic wet central heating system.

In each case the simulation equations are derived for each characteristic nodal type not already considered in chapter 3, and the overall system matrix formulated and presented in a manner suitable for integration with the building matrix.

6.2 Air conditioning systems

The dynamic simulation of air conditioning systems is complicated by the fact that the working fluid (air) is not homogeneous but comprised of a dry air and vapour mix. Also, a large number of components may be present with complex inter-component connections.

Some existing energy packages have air conditioning system modelling capabilities, although often these are based on steady state design considerations. At the present time, a number of models are emerging or planned which have the ability to predict the transient performance of air conditioning systems. In the UK, the Science and Engineering Research Council are funding several groups to develop such models (SERC 1984) and, in the US, Honeywell have developed the GEMS system (Benton *et al* 1982) and NBS the HVACSIM+ system (Kelly *et al* 1984).

This section is concerned to demonstrate the application of the implicit finite volume technique of chapter 3 to air conditioning system simulation. A linking protocol is described and a simulation model is derived for a packaged air handling unit by single node component representation coupled with component models which represent internal component behaviour. One component, the cooling coil, is then extracted and a more detailed multi-node model derived for reintroduction to the overall system matrix equation. The demonstration of component model derivation for the numerous component types found within air conditioning systems (terminal units, chillers, cooling towers, VAV boxes, etc) is outside the scope of this book. Such component models have been established for a range of typical components (ABACUS 1982) and detailed energy and mass balance equation sets have been established for components such as compressors, condensers, evaporators and heat exchangers as present within refrigeration plant (James and Marshall 1973).

Consider the packaged air handling unit of figure 6.1. Outside air at temperature θ_o, humidity ratio g_o and enthalpy h_o is mixed with zone return air at temperature θ_r, humidity ratio g_r and enthalpy h_r and passed to a chilled water cooler, a humidifier and a re-heater to achieve the required zone supply conditions to offset the zone sensible and latent loads. In the usual way it is possible to establish, for each component, an energy balance for any arbitrary time ξ:

For component 1

$$m_o h_o + m_r h_r - m_1 h_1 + q_{e1} = \left. \frac{d(\bar{\varrho}_1 V_1 h_1)}{dt} \right|_{t=\xi} \tag{6.1}$$

For component 2

$$m_1 h_1 - m_2 h_2 - m_c h_c + q_{e2} - q_{x2} = \left. \frac{d(\bar{\varrho}_2 V_2 h_2)}{dt} \right|_{t=\xi} \tag{6.2}$$

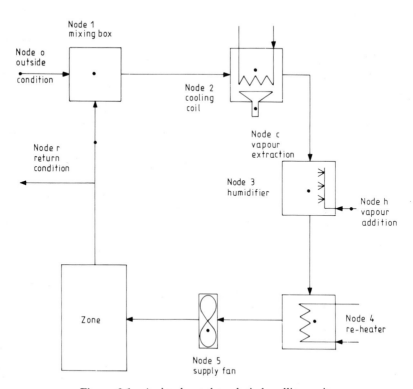

Figure 6.1 A simple packaged air handling unit.

For component 3

$$m_2h_2 + m_hh_h - m_3h_3 + q_{e3} = \frac{d(\bar{\varrho}_3 V_3 h_3)}{dt}\bigg|_{t=\xi} \qquad (6.3)$$

For component 4

$$m_3h_3 - m_4h_4 + q_{e4} + q_{x4} = \frac{d(\bar{\varrho}_4 V_4 h_4)}{dt}\bigg|_{t=\xi} \qquad (6.4)$$

For component 5

$$m_4h_4 - m_5h_5 + q_{e5} = \frac{d(\bar{\varrho}_5 V_5 h_5)}{dt}\bigg|_{t=\xi} \qquad (6.5)$$

where m is the mass flowrate of air/vapour mixture (kg s^{-1}), h the mixture specific enthalpy (J kg^{-1}), q_{ei} the component i heat exchange with surroundings (W), q_{x2} the cooling coil total heat transfer (W) and q_{x4} the re-heater coil total heat transfer (W); $\bar{\varrho}_i$ is the volume weighted mean density of component i (kg m^{-1}), V_i is the total volume of component i (m^3), o,r relate to ambient and zone air states respectively, c relates to the cooler moisture extract, and h relates to the humidifier moisture addition.

Since, at the present time, each component is represented by a single node, the component thermal capacity must be expressed as a function of the average component state as represented by the nodal enthalpy. A more refined approach results from the introduction of a multi-node component representation since, then, the capacity of each intra-component region (coil water, air/vapour mix, casing) can be treated separately. Such a refinement is demonstrated later in this section in the case of the cooler. For the current one-node model the mean density $\bar{\varrho}_i$ is given by a volume weighted average of the densities of the N component regions such that:

$$\varrho_i = \sum_{j=1}^{N}(\varrho_j V_j)\bigg/\sum_{j=1}^{N}(V_j).$$

A mass balance, component-by-component, for the dry air and vapour separately, will yield at any time ξ:

For component 1

$$m_o^d + m_r^d - m_1^d = 0\,|_{t=\xi} \qquad (6.6)$$

$$m_o^d g_o + m_r^d g_r - m_1^d g_1 = 0\,|_{t=\xi} \qquad (6.7)$$

For component 2

$$m_1^d - m_2^d = 0\,|_{t=\xi} \qquad (6.8)$$

$$m_1^d g_1 - m_2^d g_2 - m_c = \frac{d(\varrho_L V_c)}{dt}\bigg|_{t=\xi} \qquad (6.9)$$

For component 3

$$m_2^d - m_3^d = 0 \,|_{t=\xi} \tag{6.10}$$

$$m_2^d g_2 - m_3^d g_3 + m_h = \frac{d(\varrho_L V_h)}{dt}\bigg|_{t=\xi} \tag{6.11}$$

For component 4

$$m_3^d - m_4^d = 0 \,|_{t=\xi} \tag{6.12}$$

$$m_3^d g_3 - m_4^d g_4 = 0 \tag{6.13}$$

For component 5

$$m_4^d - m_5^d = 0 \tag{6.14}$$

$$m_4^d g_4 - m_5^d g_5 = 0 \tag{6.15}$$

where m^d is the mass flowrate of dry air (kg s^{-1}), g the humidity ratio (kg kg^{-1}), ϱ_L the density of liquid remaining in cooler (kg m^{-3}); V_c is the cooler residual liquid volume (m^3) and V_h the humidifier residual liquid volume (m^3); m_c is the cooler vapour extraction rate (kg s^{-1}) and m_h the humidifier vapour addition rate (kg s^{-1}).

The energy simulation equation for each node (component) is now obtained by an equal weighting of the explicit and implicit finite difference forms of equations (6.1)–(6.5):

For component 1

$$[2\bar{\varrho}_1(t + \delta t)V_1 + m_1(t + \delta t)\delta t]h_1(t + \delta t) - \delta t q_{e1}(t + \delta t)$$
$$= [2\bar{\varrho}_1(t)V_1 - m_1(t)\delta t]h_1(t)$$
$$+ m_o(t + \delta t)\delta t h_o(t + \delta t) + m_r(t + \delta t)\delta t h_r(t + \delta t) + m_o(t)\delta t h_o(t)$$
$$+ m_r(t)\delta t h_r(t) + \delta t q_{e1}(t).$$

Now, since a building equation set is not being considered at present, it is not possible to expand the q_{ei} terms by introducing an exchange resistance, based on fundamental considerations, between component and surrounding building nodes. The assumption is therefore made that the $q_{ei}(t + \delta t)$ exchange can be evaluated by independent means—perhaps based on previous time-step component conditions—and so can be removed to the equation right-hand (or known) side. In the same way the $h_o(t + \delta t)$ and $h_r(t + \delta t)$ terms are treated as known quantities since they relate, in the current treatment (an open system), to system boundary conditions. The return air state will of course be represented by the building matrix in combined processing schemes.

With reference to figure 6.2, which shows the overall system energy balance matrix equation, the equation for component 1 becomes:

$$a_{11}h_1(t + \delta t) = b_{11}h_1(t) + c_1 \tag{6.16}$$

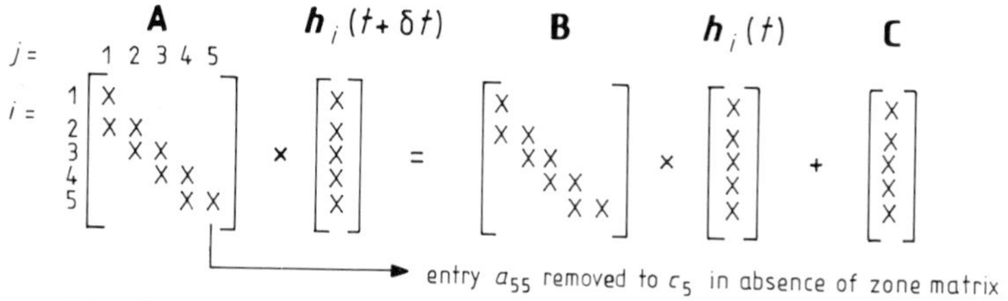

Figure 6.2 System energy balance matrix equation $\mathbf{A}h(t + \delta t) = \mathbf{B}h(t) + \mathbf{C}$.

where

$$a_{11} = 2\bar{\varrho}_1(t + \delta t)V_1 + m_1(t + \delta t)\delta t$$
$$b_{11} = 2\varrho_1(t)V_1 - m_1(t)\delta t$$
$$c_1 = m_o(t + \delta t)\delta th_o(t + \delta t) + m_r(t + \delta t)\delta th_r(t + \delta t) + m_o(t)\delta th_o(t)$$
$$+ m_r(t)\delta th_r(t) + \delta t[q_{e1}(t + \delta t) + q_{e1}(t)].$$

For component 2

$$[2\bar{\varrho}_2(t + \delta t)V_2 + m_2(t + \delta t)\delta t]h_2(t + \delta t) - m_1(t + \delta t)\delta th_1(t + \delta t)$$
$$+ m_c(t + \delta t)\delta th_c(t + \delta t) - \delta tq_{e2}(t + \delta t) + \delta tq_{x2}(t + \delta t)$$
$$= [2\bar{\varrho}_2(t)V_2 - m_2(t)\delta t]h_2(t) + m_1(t)\delta th_1(t) - m_c(t)\delta th_c(t)$$
$$+ \delta tq_{e2}(t) - \delta tq_{x2}(t).$$

As with component 1, the future time-row energy exchange with the surroundings is removed to the equation right-hand side. Also, in the absence of a more detailed component model, the coil total heat transfer and condensate exit condition must be independently assessed from the free-standing component algorithm.

The component 2 equation becomes:

$$a_{21}h_1(t + \delta t) + a_{22}h_2(t + \delta t) = b_{21}h_1(t) + b_{22}h_2(t) + c_2 \qquad (6.17)$$

where

$$a_{21} = -m_1(t + \delta t)\delta t$$
$$a_{22} = 2\bar{\varrho}_2(t + \delta t)V_2 + m_2(t + \delta t)\delta t$$
$$b_{21} = m_1(t)\delta t$$
$$b_{22} = 2\bar{\varrho}_2(t)V_2 - m_2(t)\delta t$$
$$c_2 = -m_c(t + \delta t)\delta th_c(t + \delta t) - m_c(t)\delta th_c(t)$$
$$+ \delta t[q_{e2}(t + \delta t) + q_{e2}(t) - q_{x2}(t + \delta t) - q_{x2}(t)].$$

For component 3

$$[2\bar{\varrho}_3(t + \delta t)V_3 + m_3(t + \delta t)\delta t]h_3(t + \delta t) - m_2(t + \delta t)\delta th_2(t + \delta t)$$
$$- m_h(t + \delta t)\delta th_h(t + \delta t) - \delta tq_{e3}(t + \delta t)$$
$$= [2\bar{\varrho}_3(t)V_3 - m_3(t)\delta t]h_3(t) + m_2(t)\delta th_2(t) + m_h(t)\delta th_h(t) + \delta tq_{e3}(t).$$

Again in the absence of a multi-node humidifier model, the vapour supply term $m_h h_h(t + \delta t)$ must be independently assessed, and so this equation becomes:

$$a_{32}h_2(t + \delta t) + a_{33}h_3(t + \delta t) = b_{32}h_2(t) + b_{33}h_3(t) + c_3 \qquad (6.18)$$

where

$$a_{32} = -m_2(t + \delta t)\,\delta t$$
$$a_{33} = 2\bar{\varrho}_3(t + \delta t)V_3 + m_3(t + \delta t)\,\delta t$$
$$b_{32} = m_2(t)\,\delta t$$
$$b_{32} = 2\bar{\varrho}_3(t)V_3 - m_3(t)\,\delta t$$
$$c_3 = m_h(t + \delta t)\,\delta t h_h(t + \delta t) + m_h(t)\,\delta t h_h(t) + \delta t[q_{e3}(t + \delta t) + q_{e3}(t)].$$

For component 4

$$[2\bar{\varrho}_4(t + \delta t)V_4 + m_4(t + \delta t)\,\delta t]h_4(t + \delta t) - m_3(t + \delta t)\,\delta t h_3(t + \delta t)$$
$$- \delta t q_{e4}(t + \delta t) - \delta t q_{x4}(t + \delta t)$$
$$= [2\bar{\varrho}_4(t)V_4 - m_4(t)\,\delta t]h_4(t) + m_3(t)\,\delta t h_3(t) + \delta t q_{e4}(t) + \delta t q_{x4}(t)$$

which gives:

$$a_{43}h_3(t + \delta t) + a_{44}h_4(t + \delta t) = b_{43}h_3(t) + b_{44}h_4(t) + c_4 \qquad (6.19)$$

where

$$a_{43} = -m_3(t + \delta t)\,\delta t$$
$$a_{44} = 2\bar{\varrho}_4(t + \delta t)V_4 + m_4(t + \delta t)\,\delta t$$
$$b_{43} = m_3(t)\,\delta t$$
$$b_{44} = 2\bar{\varrho}_4(t)V_4 - m_4(t)\,\delta t$$
$$c_4 = \delta t[q_{e4}(t + \delta t) + q_{e4}(t) + q_{x4}(t + \delta t) + q_{x4}(t)].$$

For component 5

$$[2\bar{\varrho}_5(t + \delta t)V_5 + m_5(t + \delta t)\,\delta t]h_5(t + \delta t) - m_4(t + \delta t)\,\delta t h_4(t + \delta t) - \delta t q_{e5}(t + \delta t)$$
$$= [2\bar{\varrho}_5(t)V_5 - m_5(t)\delta t]h_5(t) + m_4(t)\,\delta t h_4(t) + \delta t q_{e5}(t)$$

and this becomes

$$a_{54}h_4(t + \delta t) + a_{55}h_5(t + \delta t) = b_{54}h_4(t) + b_{55}h_5(t) + c_5 \qquad (6.20)$$

where

$$a_{54} = -m_4(t + \delta t)\,\delta t$$
$$a_{55} = 2\bar{\varrho}_5(t + \delta t)V_5 + m_5(t + \delta t)\,\delta t$$
$$b_{54} = m_4(t)\,\delta t$$
$$b_{55} = 2\bar{\varrho}_5(t)V_5 - m_5(t)\,\delta t$$
$$c_5 = \delta t[q_{e5}(t + \delta t) + q_{e5}(t)].$$

The matrix equation of figure 6.2 will normally be combined with its single-zone (in this example) matrix counterpart (as given by figure 4.1). For the present purpose, the two matrix systems are decoupled so that the time-series of required supply conditions is assumed a known boundary condition. Therefore the component 5 equation is written:

$$a_{54}h_4(t + \delta t) = b_{54}h_4(t) + b_{55}h_5(t) + c_5'$$

where

$$c_5' = c_5 - [2\bar{\varrho}_5(t + \delta t)V_5 + m_5(t + \delta t)\delta t]h_5(t + \delta t).$$

The form of the figure 6.2 matrix system is simple for the linear problem considered here. For more complex arrangements, involving recirculating loops, more complex matrix structures will result, but for any system two possibilities exist for matrix solution:

The q_{x2} and q_{x4} terms (plus the internal component heat transfer quantities for all other components in more complex systems) of the **C** matrix can be assessed by independent component models based on anticipated component performance requirements as assessed from the solution of the system matrix equation (see later).

Alternatively, the q_{x2} and q_{x4} terms can be removed to the future time-row side of the matrix equation—**A**h$(t + \delta t)$—and replaced by an expanded multi-node equation set. This substitute for the uni-node treatment then represents internal nodal regions by more fundamental thermodynamic considerations.

The remainder of this section outlines a solution method for the former possibility and then demonstrates the formulation of a detailed component representation to allow the latter.

By the same reasoning that was applied to the formulation of the energy balance matrix, a mass balance equation can be formulated by taking an equal weighting of the explicit and implicit forms of equations (6.6)–(6.15):

For component 1

$$m_1^d(t + \delta t) = m_o^d(t + \delta t) + m_r^d(t + \delta t) + m_o^d(t) + m_r^d(t) - m_1^d(t)$$

and, therefore, with reference to figure 6.3, this becomes:

$$d_{11}m_1^d(t + \delta t) = e_{11}m_1^d(t) + f_1$$

and

$$m_1^d(t + \delta t)g_1(t + \delta t) = m_o^d(t + \delta t)g_o(t + \delta t) + m_r^d(t + \delta t)g_r(t + \delta t)$$
$$+ m_o^d(t)g_o(t) + m_r^d(t)g_r(t) - m_1^d(t)g_1(t)$$

and, therefore,

$$d_{22}[m_1^d(t + \delta t)g_1(t + \delta t)] = e_{22}[m_1^d(t)g_1(t)] + f_2.$$

For component 2

$$m_1^d(t + \delta t) - m_2^d(t + \delta t) = -m_1^d(t) + m_2^d(t)$$

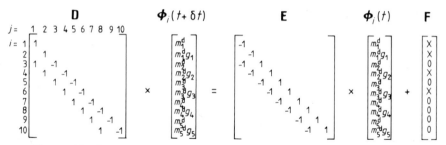

Figure 6.3 System mass balance matrix equation $\mathbf{D}\boldsymbol{\phi}(t + \delta t) = \mathbf{E}\boldsymbol{\phi}(t) + \mathbf{F}$.

and, therefore,

$$d_{31}m_1^d(t + \delta t) + d_{33}m_2^d(t + \delta t) = e_{31}m_1^d(t) + e_{33}m_2^d(t)$$

and

$$[m_1^d(t + \delta t)g_1(t + \delta t)] - [m_2^d(t + \delta t)g_2(t + \delta t)] - m_c(t + \delta t) - 2\varrho_L(t + \delta t)V_c/\delta t$$
$$= - [m_1^d(t)g_1(t)] + [m_2^d(t)g_2(t)] + m_c(t) - 2\varrho_L(t)V_c/\delta t.$$

Again, in the absence of a detailed component model, some independent assessment must be made of the cooler vapour extraction rate based on known cooler conditions, as detailed later. The component 2 vapour equation becomes:

$$d_{42}[m_1^d(t + \delta t)g_1(t + \delta t)] + d_{44}[m_2^d(t + \delta t)g_2(t + \delta t)]$$
$$= e_{42}[m_1^d(t)g_1(t)] + e_{44}[m_2^d(t)g_2(t)] + f_4$$

where

$$f_4 = m_c(t) + m_c(t + \delta t) + 2V_c[\varrho_L(t + \delta t) - \varrho_L(t)]/\delta t.$$

For component 3

$$m_2^d(t + \delta t) - m_3^d(t + \delta t) = - m_2^d(t) + m_3^d(t)$$

and, therefore,

$$d_{53}m_2^d(t + \delta t) + d_{55}m_3^d(t + \delta t) = e_{53}m_2^d(t) + e_{55}m_3^d(t)$$

and

$$[m_2^d(t + \delta t)g_2(t + \delta t)] - [m_3^d(t + \delta t)g_3(t + \delta t)] + m_h(t + \delta t) - 2\varrho_L(t + \delta t)V_h/\delta t$$
$$= - [m_2^d(t)g_2(t)] + [m_3^d(t)g_3(t)] - m_h(t) - 2\varrho_L(t)V_h/\delta t$$

and, therefore,

$$d_{64}[m_2^d(t + \delta t)g_2(t + \delta t)] + d_{66}[m_3^d(t + \delta t)g_3(t + \delta t)]$$
$$= e_{64}[m_2^d(t)g_2(t)] + e_{66}[m_3^d(t)g_3(t)] + f_6$$

where

$$f_6 = -m_h(t) - m_h(t + \delta t) + 2V_h[\varrho_L(t + \delta t) - \varrho_L(t)]/\delta t.$$

For component 4

$$m_3^d(t + \delta t) - m_4^d(t + \delta t) = -m_3^d(t) + m_4^d(t)$$

and, therefore,

$$d_{75}m_3^d(t + \delta t) + d_{77}m_4^d(t + \delta t) = e_{75}m_3^d(t) + e_{77}m_4^d(t)$$

and

$$m_3^d(t + \delta t)g_3(t + \delta t) - m_4^d(t + \delta t)g_4(t + \delta t) = -m_3^d(t)g_3(t) + m_4^d(t)g_4(t)$$

and, therefore,

$$d_{86}[m_3^d(t + \delta t)g_3(t + \delta t)] + d_{88}[m_4^d(t + \delta t)g_4(t + \delta t)]$$
$$= e_{86}[m_3^d(t)g_3(t)] + e_{88}[m_4^d(t)g_4(t)].$$

For component 5

$$m_4^d(t + \delta t) - m_5^d(t + \delta t) = -m_4^d(t) + m_5^d(t)$$

and since m_5^d is the known (in this example) zone mass flowrate:

$$d_{97}m_4^d(t + \delta t) + d_{99}m_5^d(t + \delta t) = e_{97}m_4^d(t) + e_{99}m_5^d(t)$$

and

$$m_4^d(t + \delta t)g_4(t + \delta t) - m_5^d(t + \delta t)g_5(t + \delta t) = -m_4^d(t)g_4(t) + m_5^d(t)g_5(t)$$

and, therefore,

$$d_{108}[m_4^d(t + \delta t)g_4(t + \delta t)] + d_{1010}[m_5^d(t + \delta t)g_5(t + \delta t)]$$
$$= e_{108}[m_4^d(t)g_4(t)] + e_{1010}[m_5^d(t)g_5(t)].$$

It is important to note that the mass balance matrix is a statement of nodal mass balance and that in complex circuits involving recirculating loops nodes will be required to represent the branching points. Even then, matrix solution will not provide a unique result in the absence of statements concerning the diversion ratios prevailing at each branching point. Two approaches are possible. A flow simulation model can be established in which individual flowstreams and fittings are expressed in terms of characteristic flow equations. This flow model can then be solved simultaneously and in parallel with the mass and energy balance matrix solution stream. The derivation of such an equation set, involving many empirical relationships for pipe, duct and fitting flow behaviour, is beyond the scope of this book. The second approach is to introduce simple statements on the actual or desired diversion ratio at each branching point. Thus, in the current example, the component 1 recirculating statement would be: $m_o/m_r = 0.15$ (say). In more complex circuits a statement of this kind will be required for each branching point.

If nodes located downstream from a control valve do not experience a positive flowrate until some time after valve operation due to network inertia, this can be modelled by delaying the introduction of a non-zero mass flowrate to the matrix coefficient entry until some later matrix formulation depending on the node location, fluid velocity and simulation time-step.

The matrix equations of figures 6.2 and 6.3 are now solved for any time-step in the context of component algorithms which establish the q and m terms as present within the **C** and **F** matrices and on the basis of user specified control objectives. One method, of many possible methods, is to proceed as follows:

Step 1

At each time-step establish

$$\mathbf{A}h(t + \delta t) = \mathbf{B}h(t) + \mathbf{C}$$
$$\mathbf{D}\phi(t + \delta t) = \mathbf{E}\phi(t) + \mathbf{F}$$

and initialise $q_{ei}(t + \delta t)$, $q_{x2}(t + \delta t)$, $q_{x4}(t + \delta t)$, $m_c(t + \delta t)$ and $m_h(t + \delta t)$ to zero.

Step 2

Assume no humidification or dehumidification and determine circuit humidity ratios, $g_i'(t + \delta t)$, from:

$$\phi(t + \delta t) = \mathbf{D}^{-1}[\mathbf{E}\phi(t) + \mathbf{F}]. \tag{6.21}$$

Set $\Delta g = (g_5' - g_5)$ where g_5 is the desired humidity ratio (kg kg^{-1}).

$$\begin{array}{lll} \text{If} & \Delta g < 0 & \text{go to step 3} \\ & > 0 & \text{go to step 4} \\ & = 0 & \text{go to step 5.} \end{array}$$

Step 3

Humidification required; determine $m_h(t + \delta t)$ to give required $g_5(t + \delta t)$ from iterative application of equation (6.21). Then go to step 5.

Step 4

Dehumidification required; determine $m_c(t + \delta t)$ to give required $g_5(t + \delta t)$ from iterative application of equation (6.21). From cooler model determine minimum $q_{x2}(t + \delta t)$ to give required $g_2(t + \delta t)$.

Step 5

Estimate $q_{ei}(t + \delta t)$ from known component conditions (unless building matrix is present so that component nodes are coupled to surroundings at both time-rows).

Step 6

With $q_{x4}(t + \delta t)$ remaining at zero and $q_{x2}(t + \delta t)$ set at value determined at step 4 (or zero if step 4 bypassed) determine circuit enthalpies, $h_i(t + \delta t)$, from

$$h(t + \delta t) = \mathbf{A}^{-1}[\mathbf{B}h(t) + \mathbf{C}]. \tag{6.22}$$

Set $\Delta h = (h'_5 - h_5)$, where h_5 is the desired supply enthalpy (J kg^{-1}).

$$\text{If} \qquad \Delta h < 0 \qquad \text{go to step 7}$$
$$> 0 \qquad \text{go to step 8}$$
$$= 0 \qquad \text{go to step 9.}$$

Step 7

Re-heat required; determined $q_{x4}(t + \delta t)$ from iterative application of equation (6.22). Then go to step 9.

Step 8

Further cooling, then humidification required; determine $q_{x2}(t + \delta t)$ from iterative application of equation (6.22). From cooler model assess new $m_c(t + \delta t)$ and determine new $m_h(t + \delta t)$ to give required $g_5(t + \delta t)$ from iterative application of equation (6.21).

Step 9

The desired supply conditions are now achieved, corresponding to:

$$\text{minimum [absolute value of } q_{x2}(t + \delta t) + q_{x4}(t + \delta t)]$$

that is for minimum cooler and re-heater energy.

Since the duty and entering and leaving air states, corresponding to this statement, are now known, each component model can be used to assess any internal operating conditions (such as outlet water temperatures for any given inlet condition, the effect of coil flowrate modification, or the impact of changing coil transfer areas or other design parameters). Should the component be unable to perform as required then it can be set to its limit capability (that is, in this example system, one or more of $q_{x2}(t + \delta t)$, $q_{x4}(t + \delta t)$, $m_c(t + \delta t)$ or $m_h(t + \delta t)$ can be fixed) and the effect on supply conditions established from equations (6.21) and (6.22). In the presence of a building matrix, any deviation from the required supply conditions will be manifest in an environmental penalty for some time after. If component constraints are imposed prior to simulation, then the limit condition may be indicated when the cooler model is invoked at steps 4 or 8. In this case, the supply conditions will not be met and the simulation will proceed with a correspondingly greater demand at subsequent time-steps until the building load diminishes. Note that the procedure of steps 1 through 9 is independent of component location since the matrix representations define the linking protocol. Section 6.5 discusses the establishment of the matrix template for any system configuration.

The remaining part of this section describes the formulation of component models; firstly in a form suitable for use as described in the foregoing solution procedure, and secondly in a form which allows the removal of the $q_{xi}(t + \delta t)$ terms to the future time-row for 'explicit' treatment in the implicit equation scheme to achieve first principle component simulation. The component selected for this latter treatment is the cooler of the system of figure 6.1.

Free-Standing Component Models
It is possible to derive an entirely free-standing component algorithm to represent the performance capabilities of any component. This algorithm then operates in tandem with the matrix reduction previously described to limit component capabilities if the system demands are too great. That is, the algorithm represents component operation whilst the matrix scheme represents component inertia and inter-component connections. Many algorithmic formulations are possible and one is given here to illustrate the technique.

Consider the annotated counterflow cooling coil schematic of figure 6.4. The following procedure—based on the sensible heat ratio method—can be used to calculate chilled water coil performance from inlet conditions:

1   At any time-step, the following quantities are known: inlet water temperature and mass flow rate, θ_{wi} and m_w; inlet air dry bulb temperature, mass flowrate, humidity ratio and enthalpy, θ_{ai}, m_a, g_{ai} and h_{ai}; air, water and metal thermal resistances, R_a, R_w and R_m; coil surface area, A; atmospheric pressure, P_a; and air and water specific heats, C_{pa} and C_{pw}.

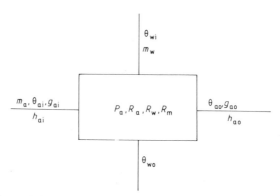

Figure 6.4   Quantities defining the state of a cooling coil.

2   Calculate the coil bypass factor β from

$$\beta = \exp[-A/(C_{pa}m_aR_a)].$$

3   Guess coil effectiveness E, perhaps previous time-step value.
4   Guess sensible heat ratio, SHR.
5   Calculate the coil U-value, U, the number of heat transfer units, NTU, and capacity-rate ratio, CRR, from:

$$\phi_1 = (m_aC_{pa})/\text{SHR} \qquad U = 1/[(R_a\text{SHR}) + R_m + R_w]$$
$$\phi_2 = m_wC_{pw} \qquad \text{NTU} = AU/C_{min}$$
$$C_{min} = \min(\phi_1, \phi_2) \qquad \text{CRR} = C_{min}/C_{max}.$$
$$C_{max} = \max(\phi_1, \phi_2)$$

6 Establish if guessed E and SHR match by applying:

$$E = \{1 - \exp[-\text{NTU}(1 - \text{CRR})]\}/\{1 - \text{CRR} \exp[-\text{NTU}(1 - \text{CRR})]\}.$$

If $\text{CRR} \rightarrow 1$, $E = \text{NTU}/(1 + \text{NTU})$.

7 If not matched, return to step 4 and iterate until SHR is established corresponding to guessed E.

8 Evaluate coil heat transfer from:

$$Q = C_{\min}E(\theta_{ai} - \theta_{wi}).$$

9 From Q calculate outlet air enthalpy, h_{ao}.

10 Calculate the saturation enthalpy, h_s, at coil surface temperature from:

$$h_s = (h_{ao} - \beta h_{ai})/(1 - \beta).$$

11 Determine the coil surface temperature, θ_s, and saturation humidity ratio, g_s, from the saturation enthalpy and atmospheric pressure.

12 Calculate outlet air temperature, θ_{ao}, and humidity ratio, g_{ao}, from:

$$\theta_{ao} = \beta(\theta_{ai} - \theta_s) + \theta_s \qquad g_{ao} = \beta(g_{ai} - g_s) + g_s.$$

13 Calculate the corresponding sensible heat ratio, SHR$'$, from:

$$\text{SHR}' = (\theta_{ai} - \theta_{ao})C_{pa}/(h_{ai} - h_{ao}).$$

14 Compare SHR$'$ with SHR and, if different, return to step 3 and iterate until agreement obtained.

15 Eventually (perhaps after changing coil parameters to achieve desired coil performance) terminate algorithm, and insert Q in system matrix equation to give final circuit enthalpies and humidity ratios.

Component matrix representation

As an alternative to operating with a free-standing component algorithm, it is possible to introduce a number of equations to the system matrix equation to represent, directly, internal component processes. For example, Holmes (1982) has suggested the following simple coil model, consisting of two first-order ordinary differential equations:

$$C_w \frac{d\theta_1}{dt} = \frac{\theta_0 - \theta_1}{R_1} - \frac{\theta_1 - \theta}{R_{mw}} \tag{6.23}$$

$$C_m \frac{d\theta}{dt} = \frac{\theta_1 - \theta}{R_{mw}} - \frac{\theta}{R_a + R_4} \tag{6.24}$$

$$\theta = \frac{\theta_2(R_a + R_4)}{R_4}$$

where $\theta_0 = \theta_{wi} - \theta_{ai};$ $\theta_1 = \theta_{wo} - \theta_{ai};$ $\theta_2 = \theta_{ao} - \theta_{ai};$ $R_1 = 1/(m_w C_{pw});$ R_{mw} is the metal + water film thermal resistance; $R_4 = 1/(m_a C_{pa});$ C_w is the water thermal capacity (J K^{-1}) and C_m the metal thermal capacity.

Equations (6.23) and (6.24) define a 2-node component model. A finite difference approximation applied to these equations, by the technique of chapter

3, will give nodal expressions in a form suitable for incorporation within the system matrix. Figure 6.5 shows the modified form of the figure 6.2 matrix equation.

The following section demonstrates the formulation of detailed component models by applying the heat balance method to the equipment types found in active solar and wet central heating systems.

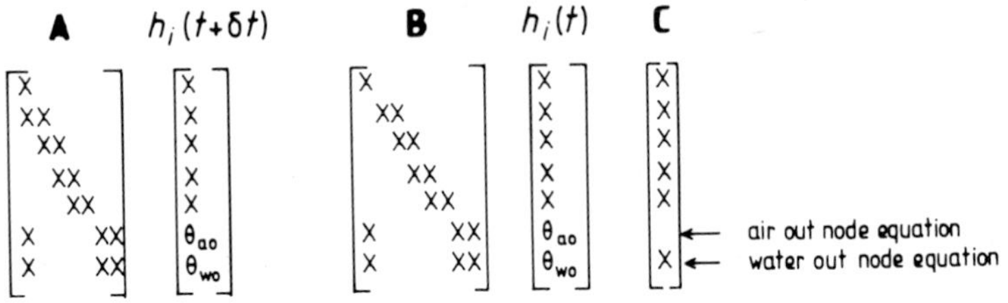

Air equation

$$-\left(\frac{2C_m(R_a + R_4)}{\delta t R_4} + \frac{(R_a + R_4)}{R_{mw}R_4}\right)\theta_{ai}(t + \delta t)$$

$$+\left(\frac{2C_m(R_a + R_4)}{\delta t R_4} + \frac{(R_a + R_4)}{R_{mw}R_4} + \frac{1}{R_{mw}}\right)\theta_{ao}(t + \delta t) - \frac{1}{R_{mw}}\theta_{wo}(t + \delta t)$$

$$= -\left(\frac{2C_m(R_a + R_4)}{\delta t R_4} - \frac{(R_a + R_4)}{R_{mw}R_4}\right)\theta_{ai}(t)$$

$$+\left(\frac{2C_m(R_a + R_4)}{\delta t R_4} - \frac{(R_a + R_4)}{R_{mw}R_4} - \frac{1}{R_{mw}}\right)\theta_{ao}(t) + \frac{1}{R_{mw}}\theta_{wo}(t)$$

Water equation

$$\left(\frac{2C_w}{\delta t} - \frac{(R_a + R_4)}{R_{mw}^2} + \frac{1}{R_{mw}}\right)\theta_{ai}(t + \delta t) - \frac{(R_a + R_4)}{R_{mw}^2}\theta_{ao}(t + \delta t)$$

$$+\left(\frac{2C_w}{\delta t} + \frac{1}{R_1} + \frac{1}{R_{mw}}\right)\theta_{wo}(t + \delta t)$$

$$= -\left(\frac{2C_w}{\delta t} - \frac{1}{R_{mw}} + \frac{(R_a + R_4)}{R_{mw}^2}\right)\theta_{ai}(t) - \frac{(R_a + R_4)}{R_{mw}^2}\theta_{ao}(t)$$

$$+\left(\frac{2C_w}{\delta t} - \frac{1}{R_1} - \frac{1}{R_{mw}}\right)\theta_{wo}(t) + \frac{1}{R_1}[\theta_{wi}(t) + \theta_{wi}(t + \delta t)]$$

Figure 6.5 The matrix equation of figure 6.2 with two differential equations added to represent air- and water-side regions.

6.3 Active solar systems

Figure 6.6 summarises the main elements of an active solar system. The flat plate collector can supply some remote space and/or water heating load directly or, in times of excess, can communicate with some remote thermal store. A number of

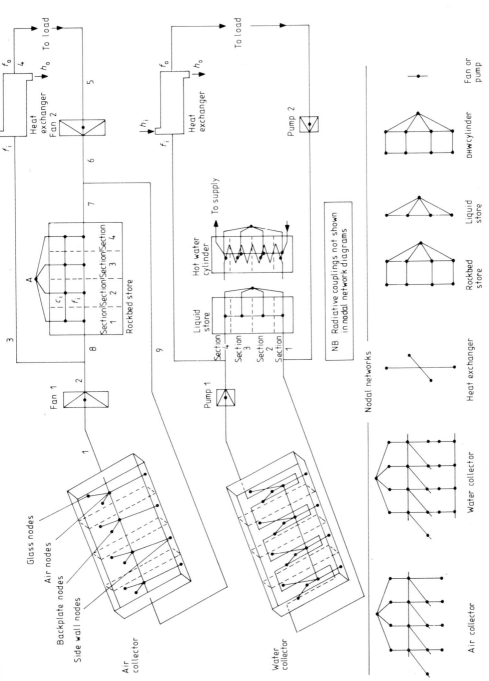

Figure 6.6 Active solar system components and nodal scheme.

commonly encountered difficulties necessitate a detailed modelling approach. These include: the temperature dependency of fluid thermophysical properties; the variation of absorber plate temperature with surface position; the inclusion of multi-substrate collectors allowing multiple passes of the collector fluid; the presence of selective coatings applied to transparent and/or opaque surfaces; storage medium complexities involving stratification or phase change phenomena; and the control dynamics inherent in collector tracking and energy diversion between load demand and remote storage.

It is possible to address these problems by the technique of chapters 3 and 4. Many of the implicit simulation equations to result will be identical to equations derived in chapter 3. For example, collector back plate nodes will adhere to the formulation of equation (3.10) whilst glazing nodes are given by equation (3.5) and contained fluid nodes by equation (3.12). Of course, in this application, the convection coefficient terms of surface and fluid node equations will be evaluated by the forced convection theory of §5.5.2.

McLean (1982) has formulated such a simulation model for the system of figure 6.6. Considering those nodal types not previously addressed in chapter 3:

Heat exchanger

A four-node model is used to represent sensible heat exchange between 'cold' and 'hot' fluid regions within the exchanger. Theoretically, the maximum rate of heat transfer between the two fluids is given by:

$$q = C_{min}(\theta_{hi} - \theta_{ci})$$

where $C_{min} = \min(m_h C_{ph}, m_c C_{pc})$, m is the mass flow rate (kg s^{-1}), C_p the specific heat capacity (J kg^{-1} °C^{-1}), θ_{hi} the hot fluid inlet temperature (°C), θ_{ci} the cold fluid inlet temperature (°C), and h, c the hot and cold fluid suffix respectively. In reality, the heat transfer rate is less than this theoretical maximum as expressed by the exchanger effectiveness given by:

$$E = \text{(actual heat transfer rate)/(theoretical maximum)}.$$

Table 6.1 gives some common expressions for exchanger effectiveness as a function of flow geometry.

A representative equation can now be derived for each fluid stream by substituting the nodal flux quantities in the primitive finite volume heat balance relationship of equation (3.1), first at the present time-row of an arbitrary time-step, and then at the future time-row. Concatenation of the two expressions to emerge gives the nodal simulation equation as with the formulations of chapter 3.

For any fluid stream, f, equation (3.1) becomes, at the future time-row:

$$\frac{\varrho_f(t + \delta t)C_{pf}(t + \delta t)\delta V_f}{\delta t}\theta_f(t + \delta t) - \frac{\varrho_f(t)C_{pf}(t)\delta V_f}{\delta t}\theta_f(t)$$

$$= m_f(t + \delta t)C_{pf}(t + \delta t)\theta_{fi}(t + \delta t) - m_f(t + \delta t)C_{pf}(t + \delta t)\theta_{fo}(t + \delta t)$$

$$\pm EC_{min}(t + \delta t)[\theta_{hi}(t + \delta t) - \theta_{ci}(t + \delta t)] + q_L(t + \delta t) \qquad (6.25)$$

Table 6.1 Heat exchanger effectiveness expressions (from Kays and London 1964).

Flow geometry	Effectiveness

Parallel $\Longrightarrow$ $\dfrac{1 - \exp[-N(1 + C)]}{(1 + C)}$

Counterflow $\rightleftharpoons$ $\dfrac{1 - \exp[-N(1 - C)]}{1 - C \exp[-N(1 - C)]}$

Crossflow

C_{max}, C_{min} unmixed $\quad 1 - \exp\dfrac{C}{N^{-0.22}}[\exp(-N^{0.78}C) - 1]$

C_{max}, C_{min} mixed $\quad \left(\dfrac{1}{1 - \exp(-N)} + \dfrac{C}{1 - \exp(-NC)} - \dfrac{1}{N}\right)^{-1}$

C_{max} mixed, C_{min} unmixed $\quad (1/C)\{1 - \exp[C(1 - \exp(-N))]\}$

C_{max} unmixed, C_{min} mixed $\quad 1 - \exp\{[1 - \exp(-NC)]/C\}$

Shell and tube
one shell pass
2, 4, 6 tube passes

$2\left[1 + C + (1 + C^2)^{0.5}\left(\dfrac{1 + \exp[-N(1 + C^2)^{0.5}]}{1 - \exp[-N(1 + C^2)^{0.5}]}\right)\right]^{-1}$

N number of transfer units (NTU)
$\quad UA/C_{min}$
C $\quad C_{min}/C_{max}$ (CCR)

where ϱ_f is the bulk fluid density (kg m^{-3}), C_{pf} is the bulk fluid specific heat (J kg^{-1} °C^{-1}), δV_f is the fluid volume (m^3), δt the time-step (s), m_f the fluid mass flow rate (kg s^{-1}), θ_f the bulk fluid temperature (°C), θ_{fi} the inlet fluid temperature (°C), θ_{fo} the outlet fluid temperature (°C) and q_L represents miscellaneous exchanger losses to the surroundings (W); and, at the present time-row,

$$\frac{\varrho_f(t + \delta t)C_{pf}(t + \delta t)\delta V_f}{\delta t}\theta_f(t + \delta t) - \frac{\varrho_f(t)C_{pf}(t)\delta V_f}{\delta t}\theta_f(t) = m_f(t)C_{pf}(t)\theta_{fi}(t)$$

$$- m_f(t)C_{pf}(t)\theta_{fo}(t) \pm EC_{min}(t)[\theta_{hi}(t) - \theta_c(t)] + q_L(t). \qquad (6.26)$$

Assuming that the bulk fluid temperature can be expressed as a weighted average of inlet and outlet conditions such that

$$\theta_f = \alpha\theta_{fi} + (1 - \alpha)\theta_{fo} \qquad (6.27)$$

then combining equations (6.25) and (6.26) gives, after rearrangement, for both hot (f = h) and cold (f = c) fluids:

$$[2\alpha\varrho_f(t + \delta t)C_{pf}(t + \delta t)\delta V_f - m_f(t + \delta t)C_{pf}(t + \delta t)\delta t$$
$$+ EC_{min}(t + \delta t)\delta t]\theta_{fi}(t + \delta t) + [2(1 - \alpha)\varrho_f(t + \delta t)C_{pf}(t + \delta t)\delta V_f$$
$$+ m_f(t + \delta t)C_{pf}(t + \delta t)\delta t]\theta_{fo}(t + \delta t) - EC_{min}(t + \delta t)\delta t\theta_{xi}(t + \delta t) - \delta tq_L(t + \delta t)$$
$$= [2\alpha\varrho_f(t)C_{pf}(t)\delta V_f + m_f(t)C_{pf}(t)\delta t - EC_{min}(t)\delta t]\theta_{fi}(t)$$
$$+ [2(1 - \alpha)\varrho_f(t)C_{pf}(t)\delta V_f - m_f(t)C_{pf}(t)\delta t]\theta_{fo}(t) + EC_{min}(t)\delta t\theta_{xi}(t)$$
$$+ q_L(t)\delta t$$

where x = c for the hot fluid equation and x = h for the cold fluid equation.

If accuracy considerations render equation (6.27) unacceptable, then further fluid volume subdivision will be necessary with the foregoing procedure repeated for each hot/cold fluid volume pairing.

Sensible storage units: rockbeds
Consider figure 6.6 which shows a rockbed store segmented into a number of isothermal volumes. Each volume has a two-node representation; a bulk fluid node and a bulk capacity node.

Equation (3.12) defines the nodal simulation equation for the fluid node so that for any rockbed segment i receiving fluid from segment $i - 1$:

$$[2\varrho_{fi}(t + \delta t)C_{fi}(t + \delta t)\delta V_{fi} + h_{ci}(t + \delta t)\delta A_{ci}\delta t$$
$$+ v_{i-1,i}(t + \delta t)\varrho'_{i-1,i}(t + \delta t)C'_{i-1,i}(t + \delta t)\delta t]\theta_{fi}(t + \delta t)$$
$$- h_{ci}(t + \delta t)\delta A_{ci}\delta t\theta_{ci}(t + \delta t)$$
$$- v_{i-1,i}(t + \delta t)\varrho'_{i-1,i}(t + \delta t)C'_{i-1,i}(t + \delta t)\delta t\theta_{fi-1}(t + \delta t) - \delta tq_{fi}(t + \delta t)$$
$$= [2\varrho_{fi}(t)C_{fi}(t)\delta V_{fi} - h_{ci}(t)\delta A_{ci}\delta t - v_{i-1,i}(t)\varrho'_{i-1,i}(t)C'_{i-1,i}(t)]\theta_{fi}(t)$$
$$+ h_{ci}(t)\delta A_{ci}\delta t\theta_{ci}(t) + v_{i-1,i}(t)\varrho'_{i-1,i}(t)C'_{i-1,i}(t)\delta t\theta_{fi-1}(t) + \delta tq_{fi}(t) \qquad (6.28)$$

where h_{ci} is the convective heat transfer coefficient connecting segment fluid to segment capacity (W m^{-2} °C^{-1}), δA_{ci} the exposed surface area of segment capacity (m^2), δV_{fi} the segment fluid volume (m^3), $v_{i-1,i}$ the fluid volume flowrate from segment $i - 1$ to segment i (m^3 s^{-1}), θ_{fi} the temperature of fluid node in segment i (°C), θ_{ci} the temperature of capacity node in segment i, θ_{fi-1} the temperature of fluid node in segment $i - 1$ and q_{fi} any additional heat extraction/addition to the fluid of segment i (W); ϱ_{fi}, C_{fi} are the density and specific heat of segment i fluid (kg m^{-3}, J kg^{-1} °C^{-1}), $\varrho'_{i-1,i}$, $C'_{i-1,i}$ the density and specific heat of fluid evaluated at the mean temperature of the two segments.

Occasionally a preheat liquid is obtained by passing conduits through the rockbed. The q_{fi} terms facilitate the modelling of such a device by allowing the removal of heat as a function of any thermostatic or time-based schedule.

In many modelling applications it is difficult to assess the h_{ci}, δA_{ci} and δV_{fi} terms directly, and so it is desirable to operate with a volumetric convective heat transfer coefficient defined as:

$$h_{vi} = h_{ci}\delta A_{ci}/\delta V_{fi}.$$

Empirical relationships have been derived (Lof and Hawley 1948) which give the volumetric coefficient as a function of rockbed geometry and flow data such that:

$$h_{vi} = 650(\varrho v/A_b d)^{0.7}$$

where A_b is the rockbed cross-sectional area (m^2), d is the equivalent spherical diameter of rockbed particles (m), and the equivalent diameter is found from

$$d = (6V_p/\pi N)^{1/3}$$

where V_p is the net volume of particles (m^3) and N is the number of particles. The rockbed bulk capacity node is represented by a modified form of equation (3.11):

$$[2\varrho_{ci}(t+\delta t)C_{ci}(t+\delta t) + h_{vi}(t+\delta t)\delta t]\theta_{ci}(t+\delta t) - h_{vi}(t+\delta t)\delta t\theta_{fi}(t+\delta t)$$
$$- \delta t q_{ci}(t+\delta t)/\delta V_{ci}$$
$$= [2\varrho_{ci}(t)C_{ci}(t) - h_{vi}(t)\delta t]\theta_{ci}(t) + h_{vi}(t)\delta t\theta_{fi}(t) + \delta t q_{ci}(t)/\delta V_{ci} \qquad (6.29)$$

where ϱ_{ci}, C_{ci} are the density and specific heat of segment i capacity and δV_{ci} is the volume of segment material (m^3).

Sensible storage units: liquids
Although thermal stratification will occur to a lesser extent with liquid storage than with rockbed storage, it is still an important phenomenon. As with the rockbed case, the liquid store will require subdivision, although in a vertical direction as shown in figure 6.6. Any elemental volume node is then represented by equation (3.12). With reference to figure 6.6 and expanding the $\sum^M$ and q_1 terms of equation (3.12), the representative segment i simulation equation is given by

$$[2\varrho_{fi}(t+\delta t)C_{fi}(t+\delta t)\delta V_i + U_{ci}(t+\delta t)\delta A_{si}\delta t + v_s(t+\delta t)\varrho'_{s,i}(t+\delta t)C'_{s,i}(t+\delta t)\delta t$$
$$+ v_{i-1,i}(t+\delta t)\varrho'_{i-1,i}(t+\delta t)C'_{i-1,i}(t+\delta t)\delta t]\theta_i(t+\delta t)$$
$$- U_{ci}(t+\delta t)\delta A_{si}\delta t\theta_c(t+\delta t)$$
$$- v_s(t+\delta t)\varrho'_{s,i}(t+\delta t)C'_{s,i}(t+\delta t)\delta t\theta_s(t+\delta t)$$
$$- v_{i-1,i}(t+\delta t)\varrho'_{i-1,i}(t+\delta t)C'_{i-1,i}(t+\delta t)\delta t\theta_{i-1}(t+\delta t)$$
$$- \delta t[q_e(t+\delta t) + q_k(t+\delta t)]$$
$$= [2\varrho_{fi}(t)C_{fi}(t)\delta V_i - U_{ci}(t)\delta A_{si}\delta t - v_s(t)\varrho'_{s,i}(t)C'_{s,i}(t)\delta t$$
$$- v_{i-1,i}(t)\varrho'_{i-1,i}(t)C'_{i-1,i}(t)\delta t]\theta_i(t) + U_{ci}(t)\delta A_{si}\delta t\theta_c(t)$$
$$+ v_s(t)\varrho'_{s,i}(t)C'_{s,i}(t)\delta t\theta_s(t) + v_{i-1,i}(t)\varrho'_{i-1,i}(t)C'_{i-1,i}(t)\delta t\theta_{i-1}(t)$$
$$- \delta t[q_e(t) + q_k(t)]$$

where U_{ci} is the global thermal transmittance value of tank walls (including insulation, if any) (W m^{-2} $^\circ$C^{-1}), δA_{si} the total tank surface area associated with segment i (m^2), v_s the volume flowrate (direct cylinder case) of supply to segment from collector (m^3 s^{-1}), $v_{i-1,i}$ the volume flowrate coupling between segments $i-1$ and i (m^3 s^{-1}), q_e the possible electrical resistance heat input (W), q_k the possible heat injection from collector fluid flowing through conduit (indirect cylinder case) $(W) = A_k U_k (\theta_s - \theta_i)$; A_k is the conduit surface area associated with segment i (m^2) and U_k is the conduit overall thermal transmittance (W m^{-2} °C^{-1}).

For the case of an indirect cylinder, an additional node per segment will be required to represent the change in collector fluid condition as it passes through the conduit associated with the segment volume. Equation (6.28) is the representative simulation equation for this nodal type (with $v_{i-1,i}$ representing the conduit flowrate, h_{ci} replaced by the conduit overall thermal transmittance, ci representing the segment fluid and fi representing the conduit fluid).

Latent storage units
It is usual to neglect thermal stratification in a latent storage unit and so the entire unit can be modelled by a two-node representation: a phase change material node and a conduit/fluid node. Unit subdivision will, however, increase modelling accuracy by allowing the conduit fluid temperature to vary with store position and reducing discretisation errors. As before, equation (6.28)) is the appropriate conduit fluid equation (with the terms changed as for the sensible liquid store and ci representing the phase change material).

The phase change material is represented by equation (6.29) but with the heat generation term expanded to account for the latent energy absorbed or released during the phase change, so that, for any material node i:

$$[2\varrho_{ci}(t+\delta t)C_{ci}(t+\delta t) + h_{vi}(t+\delta t)\delta t]\theta_{ci}(t+\delta t) - h_{vi}(t+\delta t)\delta t\theta_{fi}(t+\delta t)$$
$$- \delta t[q_{ci}(t+\delta t) + q_x(t+\delta t)]$$
$$= [2\varrho_{ci}(t)C_{ci}(t) - h_{vi}(t)\delta t]\theta_{ci}(t) + h_{vi}(t)\delta t\theta_{fi}(t) + \delta t[q_{ci}(t) + q_x(t)] \qquad (6.30)$$

where q_{ci} refers to miscellaneous store losses to surroundings, and q_x is the latent flux stored or released (both W).

The q_x term is used to account for the latent energy currently in store. During the sensible cooling or heating phase $q_x(\xi) = 0$. When the phase change temperature is reached $q_x(t + \delta t)$ is used to ensure that isothermal conditions prevail so that $\theta_{ci}(t + \delta t) = \theta_{ci}(t)$. In this way, the total latent energy in store is known at any time from the summation history given by $\sum^n q_x(t + \delta t)\delta t$ (q_x positive or negative at any time-row n). If the summated energy at any time exceeds the latent heat of fusion (mh_{fg}) or the summation is reduced to zero, then q_x is set to zero in equation (6.30) and sensible cooling/heating can recommence.

The active solar system of figure 6.6 can now be made discrete by distributing nodes in a manner which reflects the importance of the energy exchanges to be modelled. One possible scheme is shown superimposed on the system of figure 6.6. Table 6.2 gives the complete difference equation set for all active solar system nodal types and figure 6.7 demonstrates the matrix system to result for the example system considered here. As with the air conditioning formulation a nodal mass balance matrix can now be formed which, in conjunction with branching point statements, can be used to determine the mass flowrates, from time-series boundary values, as the simulation proceeds.

Table 6.2 Characteristic simulation equations and coefficient formula for active solar system nodal types.

Node location	Characteristic simulation equation
Collector walls (homogeneous element node and node between two homogeneous elements)	$-A_1(t+\delta t)\theta_{I-1}(t+\delta t)+A_2(t+\delta t)\theta_I(t+\delta t)-A_3(t+\delta t)\theta_{I+1}(t+\delta t)-A_4(t+\delta t)q_I(t+\delta t)$ $=A_1(t)\theta_{I-1}(t)+A_5(t)\theta_I(t)+A_3(t)\theta_{I+1}(t)+A_4(t)q_I(t)$
Internal surface nodes, rockbed bulk node and latent storage node	$-B_1(t+\delta t)\theta_{I-1}(t+\delta t)+B_2(t+\delta t)\theta_I(t+\delta t)-B_3(t+\delta t)\theta_{I+1}(t+\delta t)-\sum_{j=1}^{N}B_{4,j}(t+\delta t)\theta_j(t+\delta t)$ $-B_5(t+\delta t)q_I(t+\delta t)=B_1(t)\theta_{I-1}(t)+B_6(t)\theta_I(t)+B_3(t)\theta_{I+1}(t)+\sum_{j=1}^{N}B_{4,j}(t)\theta_j(t)+B_5(t)q_I(t+\delta t)$
Collector fluid node, heat exchanger node, rockbed fluid node and liquid store fluid node	$-\sum_{f=1}^{N}C_f(t+\delta t)\theta_f(t+\delta t)+C_1(t+\delta t)\theta_b(t+\delta t)-\sum_{o=1}^{M}C_o(t+\delta t)\theta_o(t+\delta t)-C_2(t+\delta t)q_I(t+\delta t)$ $=\sum_{f=1}^{N}C_f(t)\theta_f(t)+C_3(t)\theta_b(t)+\sum_{o=1}^{M}C_o(t)\theta_o(t)+C_2(t)q_I(t)$

$\theta_f=$ bulk fluid temperature
$\theta_b=$ fluid flowstream temperature
$\theta_o=$ boundary or outer surface temperature

(continued)

Table 6.2 (*continued*)

Coefficient formula

$A_1(\xi) = (k_A(\xi)R_c(\xi) + 2\delta x_{I-1,I})\delta t/\delta x_{I-1,I}R_c(\xi)\delta_{I-1,I+1}$
$A_2(\xi) = 2\varrho_I(\xi)C_I(\xi) + A_1(\xi) + A_3(\xi)$
$A_3(\xi) = (k_B(\xi)R_c(\xi) + 2\delta x_{I+1,I})\delta t/\delta x_{I+1,I}R_c(\xi)\delta_{I-1,I+1}$
$A_4(\xi) = \delta t/(\delta_{I-1,I+1}\,\delta_{J-1,J+1}\,\delta_{K-1,K+1})$
$A_5(\xi) = 2\varrho_I(\xi)C_I(\xi) - A_1(\xi) - A_3(\xi)$
$q_I(\xi) = q_P(\xi)$

$\qquad A_1(\xi) = k(\xi)\,\delta t/\delta x_I^2$
$\qquad A_2(\xi) = 2\varrho_I(\xi)C_I(\xi) + A_1(\xi) + A_2(\xi)$
$\qquad A_3(\xi) = k(\xi)\,\delta t/\delta x_I^2$
$\qquad A_4(\xi) = \delta t/\delta x_I\,\delta x_J\,\delta x_K$
$\qquad A_5(\xi) = 2\varrho_I(\xi)C_I(\xi) - A_1(\xi) - A_3(\xi)$
$\qquad q_I(\xi) = q_P(\xi)$

for boundary node between homogeneous elements   for homogeneous element centre node

$B_1(\xi) = k_A(\xi)\,\delta t/\delta x_{I-1,I}\delta_{I-1,I+1}$
$B_2(\xi) = 2\varrho_I(\xi)C_I(\xi) + B_1(\xi) + B_3(\xi) + \displaystyle\sum_{j=1}^{N} B_{4,j}(\xi)$
$B_3(\xi) = h_c(\xi)\,\delta t/\delta_{I,I-1}$
$B_{4,j}(\xi) = h_{cj,I}(\xi)\,\delta t/\delta_{I,I-1}$
$B_5(\xi) = \delta t/(\delta_{I,I-1}\,\delta_{J-1,J+1}\,\delta_{K-1,K+1})$
$B_6(\xi) = 2\varrho_I(\xi)C_I(\xi) - B_1(\xi) - B_3(\xi) - \displaystyle\sum_{j=1}^{N} B_{4,j}(\xi)$

$\qquad B_1(\xi) = 0$
$\qquad B_2(\xi) = 2\varrho_I(\xi)C_I(\xi) + B_3(\xi)$
$\qquad B_3(\xi) = h_v(\xi)\,\delta t$
$\qquad B_{4,j}(\xi) = 0$
$\qquad B_5(\xi) = \delta t/\delta V$
$\qquad B_6(\xi) = 2\varrho_I(\xi)C_I(\xi) - B_3(\xi)$
$\qquad q_I(\xi) = q_P(\xi) + q_x(\xi)$
$\qquad q_x(\xi)$ used only in latent storage applications for rockbed and latent storage nodes

$q_I(t+\delta t) = q_P(t+\delta t)$
$q_I(t) = q_P(t) + q_{SI}(t) + q_{RI}(t) + q_{SI}(t+\delta t) + q_{RI}(t+\delta t)$
for internal surface node

$$C_f(\xi) = v_f(\xi)\varrho_f(\xi)C_j(\xi)\delta t/\delta V_1$$

$$C_1(\xi) = 2\varrho_f(\xi)C_I(\xi) + \sum_{f=1}^{N} C_f(\xi) + \sum_{o=1}^{M} C_o(\xi)$$

$$C_o(\xi) = h_{c,o}(\xi)A_o\delta t/\delta V_1$$

$$C_2(\xi) = \delta t/\delta V_1$$

$$C_3(\xi) = 2\varrho_f(\xi)C_I(\xi) - \sum_{f=1}^{N} C_f(\xi) - \sum_{o=1}^{M} C_o(\xi)$$

$$q_I(\xi) = q_p(\xi)$$

for collector fluid node, rockbed fluid node, and liquid store node (for which $q_I = q_c + q_k$; $h_{c,o} = U_o$ and $M = 2$, $o = 1$ for inlet fluid, $o = 2$ for stratified store contents).

$$C_{f=1}(\xi = t + \delta t) = \frac{2\alpha\varrho_I(\xi)C_I(\xi)\delta V}{\delta t} - m(\xi)C(\xi) + C_{\min}(\xi)$$

$$C_{f=2}(\xi = t + \delta t) = \frac{2(1-\alpha)\varrho_I(\xi)C_I(\xi)\delta V}{\delta t} + m(\xi)C(\xi)$$

$$C_1(\xi) = C_{\min}(\xi)$$

$$C_2(\xi) = 1$$

$$C_{f=1}(\xi = t) = \frac{2\alpha\varrho_I(\xi)C_I(\xi)\delta V}{\delta t} + m(\xi)C(\xi)$$

$$C_{f=2}(\xi = t) = \frac{2(1-\alpha)\varrho_I(\xi)C_I(\xi)\delta V}{\delta t} - m(\xi)C(\xi)$$

$$C_o(\xi) = 0$$

$$q_I(\xi) = q_e(\xi)$$

for heat exchanger

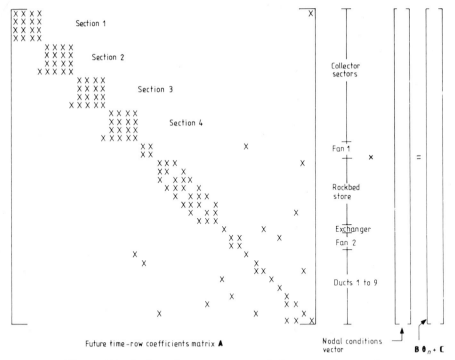

Future time-row coefficients matrix **A**   Nodal conditions vector

B θ$_n$ **+ C**

Figure 6.7   Active solar system energy balance matrix equation.

6.4   Wet central heating systems

Figure 6.8 shows the main components of a domestic wet central heating system: a boiler, radiators and hot water cylinder linked by a connecting tube subsystem and subjected to some control action. As with the active solar system a number of technical complexities can be identified which will often invalidate a non-dynamic modelling approach: complex flow regimes can be established within radiators leading to surface temperature variation; thermostatic valve operation requires dynamic modelling techniques to predict accurately system regulation; thermal stratification will occur within the hot water cylinder; hydraulic circuit connection is often complex, especially in multi-zone installations; the various levels of air/fuel mixes will affect operating efficiency; and the losses associated with flue gases can be utilised as a source of heat recovery to improve operating efficiency.

Tang (1985) has established such a dynamic model—based on the implicit finite volume heat balance technique—capable of simulating any boiler/radiator/cylinder configuration at any time-step. Many of the required nodal equations are identical to the formulations of chapter 3. For example, connecting tube fluid nodes are given by equation (3.15), radiator external surface nodes are given by an equation of form similar to equation (3.11), and some boiler and radiator

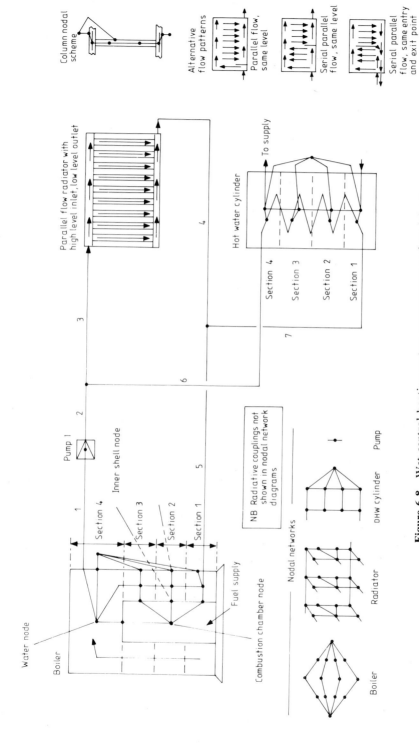

Figure 6.8 Wet central heating system components and nodal scheme.

water nodes conform to the formulations of equation (3.12) or (3.16). The following derivations therefore consider only those nodal types for which simulation equations have not yet been derived.

Boilers

The simulation equation set, relating to the multi-node approximation of a boiler, will depend on the type of boiler being modelled. For the purposes of the derivation given here, only one boiler type is considered: a tubeless, steel shell, non-steam producing, domestic boiler for which a number of assumptions can be made:

It is assumed that the combustion products are thoroughly mixed so that the combustion chamber can be represented by a single node.

The heat source, at the average combustion chamber temperature, is considered as a radiating plane parallel to the heat transfer surface separating the combustion products from the boiler water.

Ground heat transfer is negligible in comparison to the combustion chamber to shell exchange.

The nodal scheme applied to the boiler is shown in figure 6.8. Four vertical subdivisions are considered with node points placed to represent the outer shell, the inner shell and the water volumes associated with each subdivision. An external air node and combustion chamber node are separately established.

The simulation equation pertaining to the boiler room air node (A) and to the boiler water nodes (w1 to w4) are already established—as given by equations (3.12) or (3.15) (with modified surface heat transfer coefficients, as described later in this section, in the case of water nodes)—and are not considered further. For the combustion chamber node (G) the right-hand side of equaton (3.1) can be re-expressed as:

$$\sum_{j=1}^{N} h_{\mathrm{T}ij}(\theta_{ij} - \theta_{\mathrm{G}})\,|_{t=\xi} + \varrho_{\mathrm{G}}C_{\mathrm{p}\mathrm{G}}v_{\mathrm{G}}(\theta_{\mathrm{A}} - \theta_{\mathrm{G}})\,|_{t=\xi} + \overline{W}q_{\mathrm{G}}\,|_{t=\xi} \qquad (6.31)$$

where $h_{\mathrm{T}ij}$ is the total heat transfer coefficient for inner side surface of inner shell node ij (W m^{-2}°C^{-1}), θ_{ij} the inner shell node temperature (°C), θ_{G} the temperature of the combustion chamber, ϱ_{G} the density of combustion products (kg m^{-3}), $C_{\mathrm{p}\mathrm{G}}$ the specific heat of combustion products (J kg^{-1}°C^{-1}), v_{G} the volume flowrate of supply air (m^3 s^{-1}), θ_{A} the supply air temperature, $\overline{W}$ the fuel supply rate (kg s^{-1}), q_{G} the fuel heat content (J kg^{-1}) and N the number of water-side subdivisions (4 in this case).

Equating equation (6.31) to the capacity term of equation (3.1) ($\varrho_{\mathrm{G}}C_{\mathrm{p}\mathrm{G}}\,\delta V_{\mathrm{G}}\,\partial\theta_{\mathrm{G}}/\partial t$) and combination of the resulting formulations for some arbitrary present ($\xi = t$) and future ($\xi = t + \delta t$) time-rows gives, after rearrangement, the usual simulation equation form:

$$\left[2\varrho_{\mathrm{G}}(t + \delta t)C_{\mathrm{p}\mathrm{G}}(t + \delta t)\delta V_{\mathrm{G}} + \delta t \sum_{j=1}^{N} h_{\mathrm{T}ij}(t + \delta t) \right.$$

$$\left. + v_{\mathrm{G}}(t + \delta t)\varrho_{\mathrm{G}}(t + \delta t)C_{\mathrm{p}\mathrm{G}}(t + \delta t)\delta t \right]\theta_{\mathrm{G}}(t + \delta t)$$

$$-\delta t \sum_{j=1}^{N} h_{\mathrm{T}ij}(t+\delta t)\theta_{ij}(t+\delta t)$$

$$-v_{\mathrm{G}}(t+\delta t)\delta t\varrho_{\mathrm{G}}(t+\delta t)C_{\mathrm{pG}}(t+\delta t)\theta_{\mathrm{A}}(t+\delta t) - \overline{W}(t+\delta t)q_{\mathrm{G}}(t+\delta t)\delta t$$

$$= \left[2\varrho_{\mathrm{G}}(t)C_{\mathrm{pG}}(t)\delta V_{\mathrm{G}} - \delta t \sum_{j=1}^{N} h_{\mathrm{T}ij}(t) - \delta t v_{\mathrm{G}}(t)\varrho_{\mathrm{G}}(t)C_{\mathrm{pG}}(t)\right]\theta_{\mathrm{G}}(t) + \delta t \sum_{j=1}^{N} h_{\mathrm{T}ij}(t)\theta_{ij}(t)$$

$$+ \delta t\varrho_{\mathrm{G}}(t)C_{\mathrm{pG}}(t)v_{\mathrm{G}}(t)\theta_{\mathrm{A}}(t) - \overline{W}(t)q_{\mathrm{G}}(t) \qquad (6.32)$$

where δV_{G} is the volume of the combustion chamber (m^3).

For the inner shell nodes, adjacent to the vertical water segments (i1 to i3), heat transfer (at the water-side surface) will occur by local forced convection nucleate pool boiling since the node temperature will be greater than the water saturation temperature. Here the assumption is made that the bulk water temperature is sub-cooled. The water-side heat transfer is given by

$$(q/A)_{\mathrm{total}} = (q/A)_{\mathrm{boiling}} + (q/A)_{\mathrm{forced\ convection}}$$

where q is the heat flux (W) and A the transfer area (m^2).

Many heat transfer texts give empirical relationships for both subcooled nucleate pool boiling and forced convection heat transfer coefficients so that:

$$(q/A)_{\mathrm{total}} = (h_{\mathrm{c}} + h_{\mathrm{b}})(\theta_{\mathrm{w}j} - \theta_{ij})$$

where h_{c} is the forced convection coefficient (W m$^{-2}\,^{\circ}$C^{-1}) and h_{b} the subcooled nucleate boiling coefficient.

Holman (1981) suggests that h_{c} can be obtained from the following empirical relation:

$$Nu = 0.019Re^{0.8}Pr^{0.4}$$

where Nu, Re and Pr are as defined in §5.5.

For nucleate pool boiling Rohsenow (1952) correlated experimental data to obtain an expression from which h_{b} can be determined:

$$\frac{C(\theta_{\mathrm{w}j} - \theta_{\mathrm{SAT}})}{h_{\mathrm{fg}}Pr} = C_{\mathrm{sf}}\left[\frac{(q/A)_{\mathrm{boiling}}}{\mu h_{\mathrm{fg}}}\left(\frac{g\sigma}{g(\varrho_{\mathrm{w}} - \varrho_{\mathrm{v}})}\right)^{1/2}\right]^{0.33}$$

where C is the specific heat of saturated liquid (J kg$^{-1}\,^{\circ}$C^{-1}), θ_{SAT} the saturation temperature ($^{\circ}$C), $\theta_{\mathrm{w}j}$ the bulk water temperature, h_{fg} the enthalpy of vaporisation (J kg^{-1}), Pr the Prandtl number of saturated liquid, C_{sf} is a constant 0.013 for water to copper (from Holman 1981). Further, μ is the liquid viscosity (kg m^{-1}s^{-1}), g the gravitational constant (m s^{-2}) and σ the surface tension at liquid/vapour interface (N m^{-1}); ϱ_{w} is the density of saturated liquid (kg m^{-3}) and ϱ_{v} the density of saturated vapour.

Table 6.3 gives some simplified empirical relationships for h_{b} in relation to boiling at submerged surfaces at atmospheric pressure. For non-atmospheric pressures an empirical modification is required given by

$$h_{\mathrm{b}}' = h_{\mathrm{b}}(p/p_{\mathrm{AT}})^{0.4}$$

where h_{b}' is the boiling coefficient at pressure p and h_{b} is the boiling coefficient at atmospheric pressure p_{AT}.

Equation (3.1), for any node ij, becomes

$$h_{Tij}A_{ij}(\theta_G - \theta_{ij})\big|_{t=\xi} + (h_{cij} + h_{bij})A_{ij}(\theta_{wj} - \theta_{ij})\big|_{t=\xi} \qquad (6.33)$$

where A_{ij} is the transfer surface area associated with node ij (m^2).

Substitution of equation (6.33) in equation (3.1) at some arbitrary present and future time-row gives the explicit and implicit heat balance formulations respectively which, when combined, yield:

$$\{2\varrho_{ij}(t + \delta t)C_{pij}(t + \delta t)\,\delta V_{ij} + \delta t h_{Tij}(t + \delta t)A_{ij}$$
$$+ [h_{cij}(t + \delta t) + h_{bij}(t + \delta t)]A_{ij}\delta t\}\theta_{ij}(t + \delta t) - \delta t h_{Tij}(t + \delta t)A_{ij}\theta_G(t + \delta t)$$
$$- [h_{cij}(t + \delta t) + h_{bij}(t + \delta t)]A_{ij}\delta t\theta_{wj}(t + \delta t)$$
$$= \{2\varrho_{ij}(t)C_{pij}(t)\,\delta V_{ij} - \delta t h_{Tij}(t)A_{ij} - [h_{cij}(t) + h_{bij}(t)]A_{ij}\delta t\}\theta_{ij}(t)$$
$$+ \delta t h_{Tij}(t)A_{ij}\theta_G(t) + [h_{cij}(t) + h_{bij}(t)]A_{ij}\delta t\theta_{wj}(t). \qquad (6.34)$$

Table 6.3 Simplified relationships for h_b for water at atmospheric pressure (from Holman 1981).

Surface	h_b	
Horizontal	$1042(\theta_w - \theta_{SAT})^{1/3}$	$(q/A)_b < 16\,kW\,m^{-2}$
	$5.56(\theta_w - \theta_{SAT})^3$	$16 < (q/A)_b < 240\,kW\,m^{-2}$
Vertical	$537(\theta_w - \theta_{SAT})^{1/7}$	$(q/A)_b < 3\,kW\,m^{-2}$
	$7.96(\theta_w - \theta_{SAT})^3$	$3 < (q/A)_b < 63\,kW\,m^{-2}$

With the top section inner shell node (i4), the water-side surface velocity can be assumed zero so that only subcooled boiling occurs. Thus in equation (6.34) $h_{cij}(\xi) = 0$. With the outer shell nodes (o1 to o4) only forced convection takes place at the water-side surface so that equation (6.34) still defines the characteristic nodal equation but with subscript ij replaced with oj ($j = 1, 2, 3, 4$). Setting $h_{boj}(\xi) = 0$ and $\theta_G(\xi) = \theta_A(\xi)$ gives for any node ij:

$$[2\varrho_{oj}(t + \delta t)C_{poj}(t + \delta t)\,\delta V_{oj} + \delta t h_{Toj}(t + \delta t)A_{oj} + h_{coj}(t + \delta t)A_{oj}\delta t]\theta_{oj}(t + \delta t)$$
$$- \delta t h_{Toj}(t + \delta t)A_{oj}\theta_A(t + \delta t) - h_{coj}(t + \delta t)A_{oj}\delta t\theta_{wj}(t + \delta t)$$
$$= [2\varrho_{oj}(t)C_{poj}(t)\,\delta V_{oj} - \delta t h_{Toj}(t)A_{oj} - h_{coj}(t)A_{oj}\delta t]\theta_{oj}(t) + \delta t h_{Toj}(t)A_{oj}\theta_A(t)$$
$$+ h_{coj}(t)A_{oj}\delta t\theta_{wj}(t). \qquad (6.35)$$

In this case the total heat transfer coefficient $h_{Toj}(\xi)$ can be separated into convective and radiant parts as was done in §3.2.2 for building surface layers. This will facilitate the linking of boiler outer shell nodes (representing the boiler casing) with surrounding room surface nodes to increase the accuracy with which longwave radiation exchanges are modelled.

Radiators

Figure 6.8 also shows a typical parallel flow, flat plate radiator with high level inlet and low level outlet connections. The top horizontal tube is connected to an identical bottom tube by a number of finned columns. For any column, connected to its immediate top and bottom tube segment, a 7-node scheme is established as shown in figure 6.8. The inter-connection of each column node-set gives the complete radiator model. Alternatively, the radiator columns can be grouped so that centrally located columns, and the groupings to either side, will be treated as single regions to give a complete radiator model consisting of 21 nodes as shown. In either case, only three node types are present: water nodes conform to equation (3.15); horizontal (unfinned) tube wall nodes conform to equation (6.35) derived for the boiler outer shell node case; and column tube wall nodes requiring a modification to equation (6.35) to account for the presence of fins. For this last nodal type the total heat transfer coefficient of equation (6.35) is replaced by an enhanced coefficient, $h_{\mathrm{T}oj}^{*}(\xi)$, given by

$$h_{\mathrm{T}oj}^{*}(\xi) = \frac{1}{A_{\mathrm{T}}} \left[h_{\mathrm{T}oj}(\xi)A_{oj} + 2(h_{\mathrm{T}oj}(\xi)pkA_{\mathrm{f}})^{1/2} \tanh ml \right]$$

$$+ \frac{h_{\mathrm{r}oj}(\xi) \left[\theta_{\mathrm{A}}(\xi) - 0.5 \left(\sum_{i=1}^{M} f_{oj \to i} \theta_{si}^{4} \right)^{0.25} \right] \left[h_{\mathrm{T}oj}(\xi)A_{oj} + 2(h_{\mathrm{T}oj}(\xi)pkA_{\mathrm{f}})^{1/2} \tanh ml + 2kA_{\mathrm{f}}/\cosh ml \right]}{A_{\mathrm{T}} h_{\mathrm{T}oj}(\xi) [\theta_{oj}(\xi) - \theta_{\mathrm{A}}(\xi)]}$$

where A_{oj} is the surface area of finned tube base, that is the transfer area associated with node oj (m^2), p is the fin perimeter (m), k the conductivity of fin material (W m^{-1} °C^{-1}), A_{f} the fin cross-sectional area, $m = (h_{\mathrm{T}oj}(\xi)p/kA_{\mathrm{f}})^{0.5}$, l is the fin length (m), $h_{\mathrm{r}oj}$ the radiative heat transfer coefficient for radiator surface (W m^{-2} °C^{-1}) $= h_{\mathrm{T}oj} - h_{\mathrm{c}oj}$, $f_{oj \to i}$ is the view factor between node oj and room surface node i, M is the number of room surface nodes in visual communication with node oj, θ_{si} is the temperature of surface oi (°C) and $A_{\mathrm{T}} = A_{oj} + lp/2$. Figure 6.9 summarises these quantities for a single finned column segment.

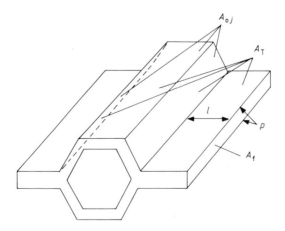

Figure 6.9 A single finned column segment.

Table 6.4 Characteristic simulation equations and coefficient formula for wet central heating system nodal types.

Node location	Characteristic simulation equation
Radiator shell, pump casing and pipe nodes	$-A_1(t+\delta t)\theta_b(t+\delta t) + A_2(t+\delta t)\theta_m(t+\delta t) - A_3(t+\delta t)\theta_o(t+\delta t) - A_4(t+\delta t)q_I(t+\delta t)$ $= A_1(t)\theta_b(t) + A_5(t)\theta_m(t) + A_3(t)\theta_o(t) + A_4(t)q_I(t+\delta t)$
Boiler combustion chamber, boiler water, pipe water and pump water nodes	$-\sum_{f=1}^{N} B_f(t+\delta t)\theta_f(t+\delta t) + B_1(t+\delta t)\theta_b(t+\delta t) - \sum_{o=1}^{M} B_o(t+\delta t)\theta_o(t+\delta t)$ $-B_2(t+\delta t)q_I(t+\delta t) = \sum_{f=1}^{N} B_f(t)\theta_f(t) + B_3(t)\theta_b(t) + \sum_{o=1}^{M} B_o(t)\theta_o(t) + B_2(t)q_I(t)$
Water cylinder fluid and surface nodes	As table 6.2

θ_b = bulk fluid temperature
θ_f = fluid flowstream temperature
θ_o = boundary or outer surface temperature
θ_m = membrane or shell temperature

Coefficient formula

$A_1(\xi) = h_{T,I}A_I\delta t/\delta V$
$A_2(\xi) = 2\varrho_I(\xi)C_I(\xi) + A_1(\xi) + A_3(\xi)$
$A_3(\xi) = h_{T,o}A_o\delta t/\delta V$
$A_4(\xi) = \delta t/\delta V$
$A_5(\xi) = 2\varrho_I(\xi)C_I(\xi) - A_1(\xi) - A_3(\xi)$

$B_o(\xi) = h_{T,o}(\xi)A_I\delta t/\delta V$

$B_1(\xi) = 2\varrho(\xi)C_f(\xi) + \sum_{f=1}^{N} B_f(\xi) + \sum_{o=1}^{M} B_o(\xi)$

$B_f(\xi) = m_f(\xi)C_f(\xi)\delta t/\delta V$

$B_2(\xi) = \delta t/\delta V$

$B_3(\xi) = 2\varrho_I(\xi)C_I(\xi) - \sum_{f=1}^{N} B_f(\xi) - \sum_{o=1}^{M} B_o(\xi)$

Hot water cylinder

This is treated in the same manner as the liquid sensible storage unit of the active solar system formulation by considering natural circulation of the internal water.

Pumps

The heat loss characteristics of pumps can be included by considering a pump as an equivalent length of connecting tube with an internal energy source derived from electrical power conversion into heat.

The overall equation set can now be formed. Table 6.4 gathers together the simulation equations for each nodal type and figure 6.10 demonstrates the matrix system to result (assuming the simplified radiator formulation). As with the previous systems it is possible to formulate nodal mass balance equations to determine, as a parallel operation to energy matrix inversion, the nodal mass flow rates to allow energy matrix coefficient assessment at each time-step rather than relying on preassigned values.

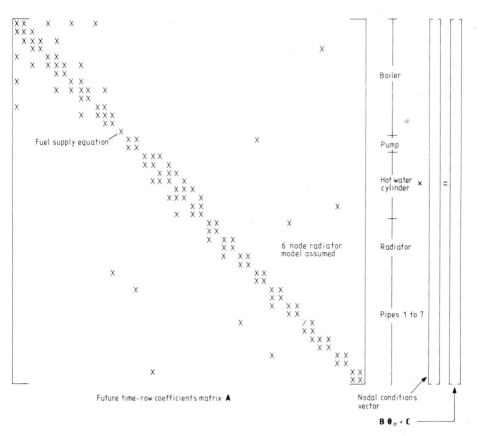

Figure 6.10 Wet central heating system energy balance matrix equation.

6.5 Simultaneous building/plant modelling

Any matrix equation, representing a connected plant system, can now be interlocked with any multi-zone building matrix as shown previously in figure 4.5. This is done by uniquely locating the plant matrix before introducing the cross-coupling coefficients which link plant and building nodes. As a general rule, the plant matrix equation will be substantially smaller than the building matrix since building capacity will demand a greater number of nodal regions and hence matrix equation entries. For example, the total number of nodes in the central heating system of figure 6.8 is approximately 150. An average sized house, modelled by the technique of chapters 3 and 4, will require approximately 1000 nodal equations. It is, therefore, appropriate to treat the plant system as one matrix rather than attempting partitioning as was demonstrated for building-side components.

In general, the combined matrix equation will not yet define a closed system, that is the matrix will contain more unknowns than equations. This is overcome by including control information either in the form of user-supplied statements of control objectives or by inserting additional control equations in the matrix to represent control loops linking node variables to heat injection, fuel supply or flow conditions. This of course implies that nodes have been located to represent sensor and actuator points of action and that equations exist to represent controller behaviour. This is discussed further in the following section.

Simultaneous solution of the combined matrix equation gives the overall system time-series solution, defining building and plant component status throughout the simulated period. One solution method is to proceed in the manner of chapter 4:

1 Each building zone partitioned matrix is processed (as described in chapter 4) to generate the zone characteristic equation (CE).
2 All zone CEs are gathered together to give the whole building CE matrix equation as demonstrated in figure 4.15 for a three-zone problem.

At this stage two possibilities exist for onward processing, depending on the form of the control information as described above.

3a Solution of the CE matrix under user-specified control objectives (time of day control node desired temperatures for example) will produce the heating or cooling loads required to be met by the plant system.
4a The plant matrix equation can now be solved, iteratively, until these loads are met or component limiting operation is encountered.
5a Plant nodal conditions then define component status to achieve the desired, or as near as possible desired, conditions.

Alternatively, if control equations are incorporated in the system matrix equation:

3b As 3a.

4b The plant matrix is solved in terms of the additional control equations to give the plant future time-row nodal states.

5b Insertion of the now known plant interaction terms in the building CE matrix equation gives the future time-row building state.

6b If this differs from that assumed to allow CE matrix solution at step 3b, then steps 3b through 5b can be re-applied until agreement is achieved.

The first procedure is useful in determining component sizes to meet stated performance objectives, whilst the second procedure allows the prediction of operational performance under specified control regimes.

If the partitioning technique is used, it is important to note that the combined coefficients matrix should not be established and held as a two-dimensional array since, in most applications, it will be very large. Instead, it is usual to create a pointer vector, the elements of which hold the address of each individual coefficient entry. This vector is the matrix template and can be used in conjunction with coefficient generators to establish the matrix equation at each computational time-step.

6.6 Control system simulation

In any plant system a number of control loops will exist, each one acting to control some region property (mass flowrate, fuel supply rate, etc) on the basis of detected signals which define some deviation from the controlled variable set point. Traditionally the approach to this problem is to establish a simplified model of the process—a simple first-order lag model with gain and dead time is not uncommon—and then to combine this with a controller model to give some overall transfer function for the system (by the Laplace transform technique of chapter 2, for example).

Underlying this section is the assumption that the process model is already established in matrix form as described in previous sections. The objective then is to solve this matrix system in terms of control system characteristics or, alternatively, to incorporate a model of the control loop(s) within the matrix equation. For the present purpose, a control loop is considered to be comprised of: a sensor detecting some nodal property; a comparator/controller causing some control action in response to the detected signal; and an actuator attempting to regulate some nodal condition on the basis of the controller instructions.

The basic operational principles underlying these elements, and their application to control the operation of building plant, is described in a number of texts (Letherman 1981, Wolsey 1975) and current research activity is concerned to develop *adaptive* control systems which rely on an in-built model (usually contained on a single micro chip) of the process they control (Dexter 1983 and 1984). For this reason, no attempt is made here to define control terminology (see BS 1967) or to discuss the philosophy of control application. Instead this section describes how, *if the relevant control details are available*, control theory is used to dictate the solution of the combined building/plant matrix system previously derived in this chapter and chapters 3 and 4.

There are two techniques which can be used for this purpose:

1   At each time-step as a simulation proceeds, the nodal property detected by the sensor is fed to some independent algorithm representing controller response. Note that this property must be one of the variables held in the matrix equation future time-row state vector.

   The controller algorithm then acts to fix or limit some other nodal property, via the actuator node, prior to matrix reformulation for the current time-step. In this way, simulation control can be achieved on the basis of a prevailing control point deviation (proportional control), control point deviation rate of change (derivative control) or the control point deviation past history (integral control) of any individual or multiple node test—that is, the sensor could be a room stat sensing air temperature or, perhaps, sensing some weighting of air and surface temperatures. Controller types can be combined (P + I etc) and the effects of dead times and sensor/actuator response rates (if known) included by incorporating a *software* memory facility so that any control action can be delayed until some later matrix inversion with current action arising from previous sensor detection.

   For a proportional controller, the control algorithm will take the form

$$\theta_p = K_p \theta_1 \tag{6.36}$$

where θ_1 is the deviation of the input signal from the set point, θ_p the change in output signal and K_p the proportional gain factor.

   The proportional gain is ideally a constant and is dependent on the adjustment of the controller. To improve controller response or to remove offset, derivative and/or integral control is often added to the basic proportional control action. Derivative action is defined by

$$\theta_d = K_p T_d (d\theta_1/dt)$$

where T_d is the derivative action time; and, for integral action,

$$\theta_i = \frac{K_p}{T_i} \int \theta_1 \, dt$$

where T_i is the integral action time. Any mixed scheme can now be established:

$$\theta_{p+d+i} = K_p [\theta_1 + T_d(d\theta_1/dt) + (1/T_i)\int \theta_1 \, dt]$$

and the controller algorithm follows from a knowledge of the controller gain and derivative and integral action times. Note, however, that this information may be subject to extreme uncertainty and is often difficult to obtain from the manufacturers of control equipment.

2   An alternative procedure is to locate an equation set—representing sensor, controller and actuator operation—within the building/plant matrix

equation. Consider a thermostatic radiator valve as required by the wet central heating system of §6.4. Here the sensor is a wax-filled capsule which can expand against a spring to cause the valve to throttle water flow. The valve stroke is, therefore, a continuous function of the sensed temperature deviation and so control action is effectively proportional. Valve manufacturers will normally have data available which describe the relationship between sensed temperature and valve position and between valve position and flow rate. These data can then be re-expressed in the form of figure 6.11 so that, accepting linearity for the present purpose, the proportional gain can be determined against any operating pressure.

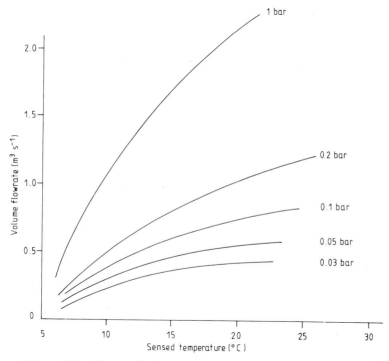

Figure 6.11 Characteristic curves for a thermostatic radiator valve.

Applying a simple energy balance to the wax capsule gives the following first-order ordinary differential equation:

$$\varrho C V \frac{d\theta}{dt} = \sum^{i} h_i A_s(\theta_i - \theta_1) + q$$

where ϱ is the wax density, C the wax specific heat, V the capsule volume, h_i the heat transfer coefficient associated with heat exchange i, θ_1 the sensor temperature, θ_i the surroundings temperature (air or surface), and q the additional heat gain (from water flow, for example).

Replacing the derivative by a finite difference approximation, and by the usual process of present and future time-row equation concatenation, the final sensor simulation equation emerges as

$$[2\varrho(t + \delta t)C(t + \delta t)V + h_c(t + \delta t)A_s\,\delta t + h_r(t + \delta t)A_s\,\delta t]\theta_1(t + \delta t)$$
$$- h_c(t + \delta t)A_s\,\delta t\theta_A(t + \delta t) - h_r(t + \delta t)A_s\,\delta t\theta_s(t + \delta t) - q(t + \delta t)\,\delta t$$
$$= [2\varrho(t)C(t)V - h_c(t)A_s\,\delta t - h_r(t)A_s\,\delta t]\theta_1(t) + h_c(t)A_s\,\delta t\theta_A(t)$$
$$+ h_r(t)A_s\,\delta t\theta_s(t) + q(t)\,\delta t. \tag{6.37}$$

This equation defines sensor response and can be incorporated within the matrix equation giving, at each time-step, the sensor temperature to allow mass flowrate assessment (from equation (6.36)) for the following time-step. Note that equation (6.36) cannot be included in the matrix equation since the control variable, mass flowrate, is not held in the matrix state vector but is an element of the coefficient entries. In other control scenarios such an insertion would be possible. For example, in the matrix equation of figure 6.10, the boiler fuel supply rate is held in the state vector. Equation (6.37) can be added to this matrix to represent (say) a room stat with a boiler on/off or proportional control equation acting to link the stat node to the boiler fuel supply rate term of the combustion node equation (equation 1 of figure 6.10). In the case of proportional action, since control is a continuous function of the sensor temperature, the control equation can be entirely relocated within the future time-row coefficients matrix to influence simultaneous solution. Of course, if the sensor temperature moves outside the proportional band then the matrix equation must be re-solved with the limiting valve position assumed.

Any number of control loops can be imposed on the simulation allowing control system evaluation and realistic energy consumption estimates. This may not be a trivial exercise.

6.7 Future developments in systems simulation

Dynamic systems simulation is a new development field and, as yet, no modelling system exists which can provide a user with the range of components and the operational flexibility demanded in practice. A number of developments are however under way (Kelly *et al* 1984, Clarke 1984 and LBL 1984) which should lead to the next generation capable of simulating, dynamically, *any* network of connected plant and building components. In view of the underlying complexity of this task, it is likely that in the long term any successful model will be forced to rely on an expert system interface to direct network construction and the formulation of simulation objectives and constraints. Future systems will very likely possess the following features:

A central database containing a range of component descriptions and allowing additional components to be entered at different levels of detail.

An agreed procedure for the concatenation of components to define the real or hypothesised plant network.

And a generalised solution technique which allows transient simulation under full dynamic control.

At the present time an international exercise is under way to construct such a components database (IEA 1983) and collaborative ventures are being pursued which may lead to standards for simulation modelling of this kind.

References

ABACUS 1982 Component Specifications of Air Conditioning Equipment *ABACUS Publication* (Glasgow: University of Strathclyde)

ASHRAE 1976 *Procedures for Simulating the Performance of Components and Systems for Energy Calculations* (ASHRAE Task Group on Energy Requirements for Heating and Cooling of Buildings)

Benton R *et al* 1982 Generalised Modelling and Simulation Software Tools for Building Systems *Proc. CIB W67 3rd Int. Symp. on Energy Conservation in the Built Environment* (Dublin)

BS 1967 *Glossary of Terms Used in Automatic Controlling and Regulating Systems* (London: British Standards Institution)

Clarke J A 1984 Development of a Prototypical Component-Based Energy Modelling System *SERC Grant Application Case for Support, ABACUS Publication* (Glasgow: University of Strathclyde)

Dexter A L 1983 Self-Tuning Control Algorithm for Single-Chip Microcomputer Implementation *Proc. IEE* **130** part D no 5

Dexter A L and Graham W J 1984 A Simple Self-Tuning Controller for Heating Plant Applications *Proc. Int. Symp. on the Performance of HVAC Systems and Controls in Buildings* (Building Research Establishment)

Hanby V 1984 Various Internal Reports (University of Loughborough: Department of Civil Engineering)

Holman J P 1981 *Heat Transfer* (New York: McGraw-Hill)

Holmes M J 1982 The Simulation of Heating and Cooling Coils for Performance Analysis *Proc. Int. Conf. on Systems Simulation in Buildings, University of Liège, December 1982*

IEA 1983 International Energy Agency *Annex 10 Newsletter*, Laboratoire de Physique du Batîment (Liège: University of Liège)

James R W and Marshall S A 1973 Dynamic Analysis of a Refrigeration System *J. Inst. Refrigeration*

Kays W M and London A L 1964 *Compact Heat Exchangers* (New York: McGraw-Hill)

Kelly G, Park C, Clark D and May W 1984 HVACSIM+, a Dynamic Building/HVAC/Control System Simulations Model *Proc. Workshop on HVAC Controls Modelling and Simulation, Georgia Institute of Technology, Atlanta*

LBL 1984 *The Energy 1 Project* (Berkeley, Calif.: Lawrence Berkeley Laboratory)

Letherman K M 1981 *Automatic Controls for Heating and Air Conditioning* (Oxford: Pergamon)

Lof G O G and Hawley R W 1948 Unsteady State Heat Transfer Between Air and Loose Solids *Ind. Eng. Chem.* **40** 1061–66

McLean D J 1982 The Simulation of Solar Energy Systems *PhD Thesis* University of Strathclyde

Quick J P 1982 Simultaneous Solution of Room Response and Plant Performance *Proc. Int. Conf. on Systems Simulation in Buildings, University of Liège, December 1982*

Rohsenow W M 1952 A Method of Correlating Heat Transfer Data for Surface Boiling Liquids *Trans. ASME* **74** 969

SERC 1984 *Energy in Buildings Specially Promoted Programme: Report for the Period 1979–84* (SERC Environment Committee)

Tang D C 1985 The Simulation of Wet Central Heating Systems *PhD Thesis* University of Strathclyde

Wolsey W H 1975 *Basic Principles of Automatic Control* (London: Hutchinson Educational)

7

Software development

The objective of this chapter is to consider the general structuring of the theory of the preceding chapters in a form suitable for software organisation into a computer-based energy modelling system. A generalised logic structure is presented, which attempts to maximise modularity and strike a balance between primary and secondary memory requirements, and separate sections are included on discrete aspects of the proposed system such as input, output, permanent and temporary databases, simulation control, programming language, operating system, hardware considerations and indicative target performance statistics.

To gain general acceptance by the building design profession—a profession as yet with little experience of the use of CAD techniques—a modelling system must be technically comprehensive to accommodate the range of conventional and non-conventional design problems requiring study. It must also be flexible in use to permit application at any stage in the design process, from sketch through detailed design, and exceptionally user-friendly to alleviate the problems inherent in the use by a community who demand intensive but infrequent model access.

To accommodate these aspects, the finished system will be computationally demanding, with the potential for many combinatorial execution routes when in use. It is important, therefore, to select an appropriate driving mechanism and to realise from the outset that a high percentage (perhaps about 50%) of the system software will be concerned with user interface matters and the related background organisation.

Recalling the software development process introduced in the chapter 1 introduction (Maver and Ellis 1982):

1 Research into model needs, methods, algorithms and organisation. This leads to a research prototype embodying the fundamental laws governing heat flow.
2 Development of a pilot applications program based on the research findings and which offers a reasonable user interface.
3 Validation of the model to test the physical assumptions and the selected numerical scheme.

4   Implementation trials to test the robustness, relevance and efficacy of the software in the real-world, real-time context of design practice.
5   Improvement of the software and documentation with respect to commercial standards by the incorporation of the lessons learned through the validation and trial implementation studies.
6   Commercial exploitation.

This development process will require a resource of certainly no less than some 10 man years.

Previous chapters can be viewed as the outcome of the research stage. This chapter addresses the development and improvement stages and, finally, chapter 8 addresses validation and implementation in practice.

7.1 Structuring the overall system

Any energy modelling system intended for use as a front-line design decision-making aid will have a number of distinct functional modules, perhaps held as separate program modules or combined to form multi-function programs:

Input data management
This module will allow the establishment of a building description in a form suitable for subsequent simulation processing. By some means (see §7.2.1) the module will request and accept geometrical, constructional (thermophysical properties), plant and zone operational data. It is usual to expend much software effort on exhaustive input data checking, with a number of user-oriented checking aids (visual and numerical) made available throughout the process. Module output must include a disk file-set storing all input data (and perhaps data derived therefrom) in a form suitable for subsequent recall, display and editing via inbuilt facilities.

Permanent system databases
A number of recurring data types can be located in standardised, fast access system databases and corresponding software developed to facilitate a range of management operations such as appending, listing and editing.

A constructions database will be required to hold the basic thermophysical properties of a range of construction primitives (for example see appendix B) likely to constitute the building fabric.

A profiles database will be required to hold the range of typical or actual profiles which describe, in condensed or time-series form, time-dependent occupancy, lighting and plant controller variations for different spatial or plant types.

A climatological database will be required to hold the time-series climatic collections for selection, according to some severity criteria, as the boundary condition for any simulation.

Lastly, a plant components database will be required to hold the data which describe the range of components on offer by the modelling system.

Simulation

This module will accept the filed data structure produced by the input management module and apply its own exhaustive checks similar to those performed at input time. This is necessary to overcome the possible introduction of data errors if users bypass system input procedures to operate directly on the input data via software external to the energy modelling system (for example system editors or other applications software).

Simulation coordination and control possibilities should be maximised and an extensive information system established to keep the user informed of simulation progress.

Results output to secondary memory should be flexible to allow various levels of user-dictated save sets.

Results recovery

The recovery of the saved data should be possible in a number of ways to accommodate the user's experience level and perceived objectives.

Interrogation facilities allow rapid result scanning to locate or compute specific data items such as maximum or minimum values of a single parameter, occurrence times, simple summations or integrations, period mean values and so on. Such synoptic data help to answer such preliminary design questions as, 'does the building overheat?', 'if so, where?' or 'how much cooling is required for comfort?'

A statistical facility allows the removal of the time dimension and so will promote a better 'understanding' of a result set of formidable size or complexity. Frequency analysis, principal component evaluation, linear regression and curve fitting are all useful tools in this respect and give answers to such questions as, 'how often is the plant unable to cope at its present rating?'; or 'what design element has the most impact on energy requirements'?

Monochromatic graphical display, allowing the construction of multiple parameter profiles against time or the display of one parameter against another, will serve to explain prevailing causal relationships and will lead to more informed design decisions. The introduction of colour display capabilities allows results to be displayed in a more meaningful form, for example, pie charts and Sankey diagrams. In addition, colour allows a departure into more 'experiential' display forms such as pseudo thermograph perspective images.

Tabular output—perhaps the last resort—is nevertheless essential for all validation work or projects in which numerical perception is deemed important.

System utilities

Many of the theoretical subsystems cannot be rigorously analysed at the early design stages because of the paucity of appropriate input data. For this reason one attractive option—with the virtue of memory saving and computational efficiency—is to remove these subsystems from the simulation module for treatment as free-standing system utilities. This allows, in the absence of specific data, the simulation module to operate either on the basis of default information or on the basis of some simplified calculation procedure applied to the flowpath concerned. As the design hypothesis evolves, and the appropriate data become available, the free-standing utility is invoked to create a related time-series data structure for access and use throughout the simulation. This treatment can be applied to a number of subsystems.

External surface shading prediction is often not possible at an early design stage due to the absence of facade and surrounding obstructions data.

Internal surface insolation prediction and solar patch movement require ray tracing which has little meaning in the case of an approximated or incomplete geometrical specification. Inter-surface view factor evaluation, likewise, is meaningless if final geometry, including furnishings and other inter-surface obstructions, has yet to be finalised. Air flow simulation is only possible when a detailed knowledge of leakage distribution has been acquired. Finally, a spectral analysis of multi-layered glazing systems (to predict angle-dependent shortwave properties) is only possible when detailed elemental properties can be specified.

Each of these modules will, when invoked, generate a temporary database (temporary in the sense that it relates only to the currently defined problem) for subsquent access by the simulation module at run time.

Temporary system databases

In the current scenario these would include a shading and insolation database, a view factor database, an air flows database, a window properties database and, of course, the input management database describing the problem for simulation.

The shading/insolation database should be flexible to allow profiles for different facades, windows and dates to be predicted and saved as required. This allows the transmission of detailed information to the simulation module for some critical facade and date, with all other facade/date combinations treated on the basis of a default assumption (zero shading and defined plane insolation say), accepted until more detailed data are available.

In a similar way the air flows database (containing infiltration and zone-to-zone air coupling time-series) must be able to expand to accommodate new zones, zone subdivision or the addition of new leakage paths. Note that the simulation module must also possess air flow simulation capabilities if exact temperature driven predictions are required.

Figure 7.1 shows one possible system layout which includes the foregoing modules. Many other layouts are possible—for example, combination of the input and output modules, the simulation and output modules or the simulation and utility modules—and some of the modules are often crudified or omitted completely. But in view of the processor capabilities of the emerging low cost 16 and 32 bit virtual memory systems, in the author's opinion the system of figure

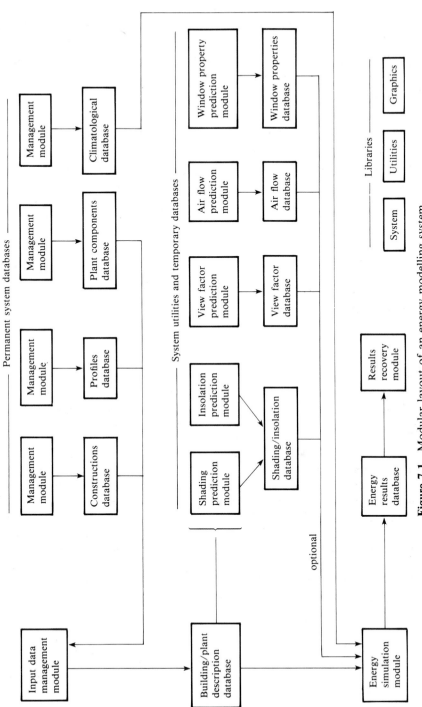

Figure 7.1 Modular layout of an energy modelling system.

7.1 should be regarded as the minimum modular subdivision. Indeed small machine operating systems such as UNIX will permit (and encourage) substantial additional modularisation. The simulation module can, for example, be further divided into modules addressing input, simulation control, each technical subsystem (solar, convection etc), results transfer to database and so on.

Further, since many of the modules will share common computational elements, it is desirable to locate such elements in a system library. Table 7.1 identifies an example set of such elements and indicates the modules requiring each. Two additional libraries will be required if the intention is to design a system which can be controlled interactively with a high degree of graphical display:

A utilities library containing the software modules to control interaction with the user. These will include file handling utilities, device-dependent text control and screen erasing with graphics terminals.

A graphics library containing the software modules to allow the production of graphical displays.

Table 7.1 Examples of common module requirements.

Element	Required by
Sun position	Simulation, shading and insolation
Error handling	All
Input data checking	All
File handling	All
Geometrical operations	Input management, simulation and utilities

Table 7.2 summarises the general requirements of such libraries, which are available as proprietary packages from a number of sources: for example, UTIL and TEKLIB (ABACUS 1981), suitable for a limited range of machine and terminal types, and GINO (CADC 1976) offering extended graphical capabilities and less terminal dependency but greater memory overhead. Care should be taken when selecting any graphics package in view of the current emergence of the GKS graphics standard (Hopgood *et al* 1983, GKS 1982).

It is also essential to derive a commands protocol to 'drive' the system both interactively and in background (or batch) mode.

In the latter case it is necessary to arrange that the entire system can be disconnected from a user terminal and driven instead from a filestore of commands established prior to system use. This allows the system to be applied in a parametric study to process many design combinations in the search for some optimum configuration. In this mode the simulation module will build a results database with any output request directed to text or plot files for later display by interactive means. The batch facility is also invaluable in the case of long term simulation, especially on machines which offer batch discount rates and so penalise interactive users.

For interactive operation a number of control techniques exist, ranging from simple 'question and answer' sequences to structured syntax systems. An approach of intermediate complexity involves the use of a hierarchical menu protocol. At any time the user is given access to a menu of commands, each active at that time. Selection of a command invokes the appropriate computational sequence before control is returned to the user via the same menu or another menu lower or higher in the tree structure. Figure 7.18 shows the menu structure and available commands for the output module of the ESP simulation system (Clarke 1982).

The following section expands each of the modules of figure 7.1 to reveal module function, organisation and runtime operation.

Table 7.2 Example requirements of system utility and graphics libraries.

Utility library:	polygon areas
	sun position
	coordinate transformations
	error management
	menu manipulation
	terminal dialogue and organisation
	file handling
	data/filename decoding
Graphics library:	device drivers
	cursor position
	hardcopy
	graph scaling and display

7.2 Software requirements

Since the development of any computer model is subject to personal logic preference, coding style, adopted language and a number of subjective criteria relating to eventual system use, it is impossible to prescribe one system format which would be acceptable to all potential developers. Instead, the example material which follows relates to one particular system—the ESP package developed at the ABACUS unit within the Department of Architecture and Building Science at the University of Strathclyde between 1977 and 1984 (Clarke 1982)—and is included to demonstrate the possible elements of the modules outlined in the previous section. The hope is that by study of a fully working system an astute reader can accept those portions of the system judged to be efficacious whilst attempting improvement of the remainder. Appendix E lists the ESP system documentation set and describes one strategy for system use.

7.2.1 Input data management

Figure 7.2 shows one example logic structure for the input module. The main controlling routine initialises the permanent system databases as required by the input module and displays the main menu of commands to cover the column 2 options. Selection of any command will lead to a further menu containing the command options relating to the items of column 3. Here we consider each of the column 2 options.

Zone description

The basic template of the building description is a zone, defined as any volume contained by a number of planar bounding surfaces, of any inclination, and containing internal objects which represent capacity or obstruction items. For each zone in some multi-zone configuration (comprising a building in whole or part), three data types are essential and, once entered and checked, are transferred to a zone's database for later access by the simulation module, by a system utility module (such as shading prediction) or by the input module itself to allow zone modification via the inbuilt editing facilities. All geometric data are held in the standard vertex form of §5.1, although a number of alternative techniques can be employed to ease the burden of input, especially in the case of complex geometries. A standardised form also allows all subsequent computational operations to be tailored to the same data structure to reinforce further module commonality. In essence there are three geometrical input techniques:

Digitising methods are attractive in cases where detailed drawings exist or where the proposed geometry is largely rectilinear, so that plan digitising is the predominant operation.

At an early design stage, where geometrical form is often under study, it is useful to operate with volume primitives and techniques exist to allow operations such as extrusion, rotation, scaling, concatenation and translation of primitives created by graphical means and held in some associated database. It is possible by this technique to establish rapidly a realistic approximation to some complex layout.

Although not obvious until tried, it is in some cases easier to use manual input techniques. In this case a number of shape types can be defined which dictate the level of input required to fully specify body geometry. For example, a simple rectangular shape is fully specified by 7 data items (x, y and z site location coordinates, length, width and height, and orientation) whereas a more complex shape with inclined surfaces will require topographical and topological information as described in §5.1.

Further information on these techniques is given elsewhere (Foley and Van Dam 1982, Bernard 1983, Parkins 1978 and Stearn 1983).

It is convenient to define window and door geometry as 'tiles' associated with the zone surfaces since this allows zone relocation without the need for window/door coordinate modification. In such a scheme these tiles are defined relative to a local xz coordinate system.

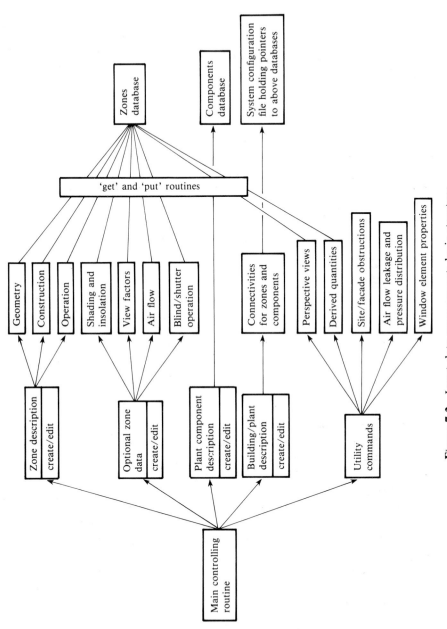

Figure 7.2 Input data management: logic structure.

Regardless of the input technique, routines will be required to check that the geometry data are within acceptable range. Table 7.3 lists these essential checks.

Construction data will include the number of elements in each multi-layered construction, the basic thermophysical properties of each element (conductivity, density and specific heat), the thickness of each element, the absorptivity and emissivity of all surface layers, the angle-dependent shortwave properties (transmission, absorption and reflection) of all transparent elements and data to describe the nature of special regions such as ventilated cavities and phase change materials. It is sometimes advantageous to combine (by the theory of §5.3.3) the shortwave properties of all transparent elements comprising a window system to allow treatment of the window system as one region to reduce computing demands. This is the function of the spectral analysis utility of figures 7.1 and 7.11.

Table 7.3 Essential geometrical checks.

Range checks on input data
  —number of vertices
  —number of surfaces
  —number of windows
  —number of doors
  —dimensions

Range checks on data derived from input data
  —zone azimuths and elevations
  —areas of opaque and transparent portions
  —volumes
  —surface intersections

Multi-zone connectivities
  —adjacent surface connections
  —window and door positions

Operational data will include the profiles which define zone occupancy, lighting, miscellaneous casual sources, plant control and air movement. Each profile will define the time variation of the related process. In the case of heat gain profiles a convective/radiant split is required to determine the portion which acts convectively (at an air point) and the portion which is applied as a radiant input (to surface nodes) to be transmitted through time by the building's inertia. Plant control profiles must be accompanied by information on the nature of the controller and the location and characteristics of the heat input or extract. This is described in §7.2.4. Air flow profiles are default profiles to be used when building leakage characteristics are unknown or unspecified and therefore prediction at simulation time, or by the pre-simulation air flow utility, is not possible. Data defining the coupled zones and the thermostatic and time tests to be applied to any air movement are also required.

Table 7.4 Essential constructional checks.

Range of checks on input data
—number of constructions/elements
—thermophysical properties
—surface finishes

Range checks on data derived from input data
—Fourier numbers
—U-values

Check constructional details exist for all components defined at geometry input time.

Profiles are selected from a permanent profiles database and can exist in two forms:

Typical weekly profiles deemed to hold for any specified period (month, season, etc). These can be regarded as design profiles.

Long term profiles to allow the specification of known data, perhaps acquired from a monitoring scheme.

Tables 7.4 and 7.5 list the essential checks associated with the foregoing constructional and operational data respectively. Appendix F gives the annotated contents of the zones database for a one-zone problem as produced by ESP.

Table 7.5 Essential operational checks.

Range checks on input data
—air flow
—casual gains
—radiant/convective splits
—environmental control

Multi-zone checks on leakage distribution and zone/plant locations.

Optional zone data

Four data sets can, optionally, be associated with any zone to allow an increase in computational rigour. These include shading/insolation, view factor, air flow and blind/shutter operation. All four data sets can be created or edited by simple question and answer sequences. In addition, the first three can be generated by prediction via the system utility modules. For the periods covered, these data will supersede any corresponding default data defined under the 'zone description' menu pick.

The blind/shutter data allow the imposition of time-dependent blind/shutter operation on the windows in any zone. This consists of replacement thermophysical properties relating to one or more zone windows.

The other data sets are described in §7.2.3.

Plant component description
The required plant components, pertaining to the current simulation request, are extracted from the component's database and relocated in a temporary database associated with the current problem. Due to the multiplicity of component types, a description of the data requirements is outside the scope of this book. Appendix F, however, gives some example component data.

Building description
At this stage the user specifies the zones and plant components which together will participate in some simulation. This allows the specification of a number of configurations—to focus on single zones, groups of zones or entire buildings—selected from the zone/component set previously defined. This facility is essential to allow a phased simulation study in which the findings from specific zone or zone cluster analyses are incorporated in the overall building data set.

The connectivity between zones, between components and between zones and components must now be defined. Zone connectivity can be software determined if at creation time the zone geometries have been given relative to the same coordinate system. All other connectivities must be entered numerically.

Lastly site location and exposure information is specified along with character strings used later for output identification purposes.

All building description data are held in a 'systems configuration file' in the form of pointers to the relevant zone and component data sets.

Utility commands
A number of utility commands exist to aid data checking and to establish the input data for other system utility modules.

The generation of perspective views of the building network is essential for checking of zone location by visual means. The output from such a menu pick is a disk file containing the zone vertex information in a form suitable for use by a free-standing perspective drawing package known as BIBLE (Parkins 1979).

The ability to display quantities derived from the zone input data is a further aid to data checking. Required software, which is common to other system modules and so should be located in the system library, will include: surface areas; contained volumes; planar angles; construction reference U-values; thermal diffusivities; and ventilation conductances.

Additional utilities include: the ability to establish the geometrical data describing facade and surrounding obstruction features, relative to the same coordinate system as the target building, for use by the shading prediction module; the definition of building leakage and pressure distribution for use by the free-standing air flow module or by the energy simulation module; and the definition of the properties of a series of glazing and thin-film elements which together constitute a multi-layered window system for processing, via the spectral analysis module, to determine overall properties.

These data sets are discussed in greater detail in §7.2.3 where the utility modules themselves are considered.

7.2.2 Permanent databases and related software

Four permanent databases are considered and their related management modules described.

Constructions database

Figure 7.3 shows the logic structure for the constructions database management module. Two separate databases are identified: a primitives database containing the thermophysical properties of a number of homogeneous elements which can be combined to form composite multi-layered systems representing walls, floors, roofs and windows; and a composites database containing a number of user-specified multi-layered constructions for inclusion in the simulation exercise.

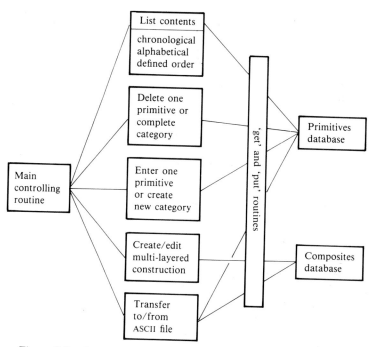

Figure 7.3 Constructions database management: logic structure.

The former database is precious whereas the latter can be updated or edited to suit the project or range of projects in hand. Both databases should be binary random access to facilitate good interaction during access by the management program and by the input module during the building description process. Appendices B and F describe database contents.

Listing, deletion and creation facilities are offered for both databases and a number of checking features are incorporated to ensure that the archived data cannot be accidently destroyed. A facility also exists to allow the transfer of database contents to disk ASCII file (and back) for transmission to other installations.

Profiles database

Figure 7.4 shows the logic structure for the profiles database management module. This allows the inspection, manipulation and organisation of miscellaneous profiles representing the time-related percentage variation of some heat source or desired control function. At input time, any heat source magnitude or temperature set point can then be associated with the 100% condition. This means that an occupancy profile is independent of the number of occupants and so is a valid profile for all building zones with the same occupancy pattern.

The profiles database should also be binary random access. Appendix F describes database organisation.

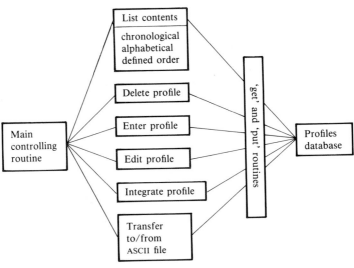

Figure 7.4 Profiles database management: logic structure.

Plant components database

Figure 7.5 shows the logic structure for the management module of the plant components database. Again a binary random access file is preferred with several management facilities. Unlike the other permanent databases, each entry (plant component) will have a different data set depending on the component type and the complexity with which it is modelled. Appendix F describes database organisation.

Climatological database

Figure 7.6 shows the logic structure for the climatic database management module. This allows the creation, organisation and analysis of a number of

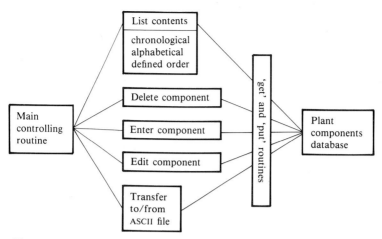

Figure 7.5 Plant components database management: logic structure.

climatic collections which can be declared a boundary condition at simulation time.

This binary random access database will contain a number of annual collections relating to different geographical locations. For each site several collections will

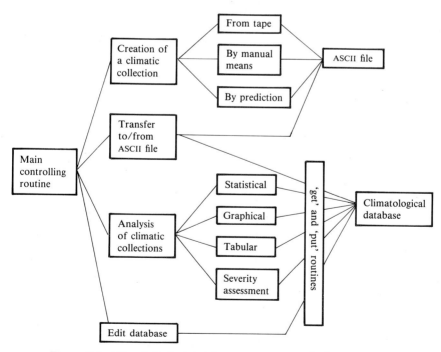

Figure 7.6 Climatological database management: logic structure.

(typically) exist to represent different levels of severity and each collection will (typically) correspond to a one hour frequency time-series since this is consistent with the recording frequency adopted by most meteorological stations. It is often convenient to hold these data in 'integerised' form and for this reason variables such as dry and wet bulb temperature and wind speed can be retained as tenth units. To minimise disk storage requirements, it is also convenient to hold a number of databases, each containing data valid within a national boundary, a region or a site.

Considering each of the column 2 options of figure 7.6:

Creation of a collection Normally a number of annual hourly collections will be available from the meteorological collection agencies in computer-readable form. Facilities are required to process these data and perform the range of consistency checks and conversions often necessary. Two additional facilities will be required for use in cases where such data are not available or where synthetic boundary condition tests are required.

The 'manual' create facility allows any sequence of weather data to be entered directly by the user from the terminal. Alternatively, the fitting of curves to specified synoptic data (such as maximum and diurnal range values) and the ability to generate constant values and repeating profiles will greatly reduce the labour in generating synthetic weather influences.

Creation by prediction is required to generate realistic profiles based on a scant knowledge of site conditions. This is possible in the case of design solar data relating to clear sky conditions or to skies of known cloud cover (see §5.8.3).

The output from the 'creation' routines is directed to a system ASCII file for later transfer (as a positive action) to a climate database of the user's choice.

Transfer to/from an ASCII file In addition to the transfer of created data to a climate database, transfer in the opposite direction to a printable disk file allows the transfer of data to other installations. The transfer facility should be sophisticated to allow the mixing of real, manually entered and predicted data.

Analysis of climatic collections A number of analysis options exist. Statistical analysis allows the identification of maximum and minimum values of any identified parameter within any specified period and from any defined climate database. It is also possible to identify sequences of days which exceed or fall below some specified criterion. This is useful for the selection of design or typical conditions. The production of average data, either for all hours over the period considered or separately for each similar hour, allows the extraction of typical low, high or average days of real data as assessed on the basis of some scoring relative to the period average. A range of other functions, such as percentile classification or profile integration (degree days in the case of temperature), will also contribute to a better understanding of the climatic influence and so assist in the selection of meaningful data for simulation purposes.

Graphical display techniques are necessary to promote an understanding of the temporal aspects of climate so as to stress the building in the required manner (minimum outside temperatures occurring prior to plant start-up for example).

Tabular output allows a perusal of the individual data items and, lastly, severity assessment techniques (such as those of §5.8.2) are the prerequisite of the selection of long term collections for representative energy consumption prediction.

Editing of database values In many cases a database collection will require slight modification. For example, most variables are acceptable, but the wind direction requires slight modification to make allowance for microclimatic effects. To protect the source data, held in the database, it is essential that such an editing operation is applied to the ASCII file with subsequent transfer to a climate database for simulation use.

Appendix F shows the organisation of the ESP climatological database.

7.2.3 System utilities and temporary databases

Five system utilities are considered and the databases they produce described. In each case the utility can be used merely to investigate the related subsystem as an aid to the specification of a realistic default scheme at the building description stage. In this case a temporary database need not be produced, since the energy simulation module of figure 7.1 will operate on the basis of the default alone.

External surface shading prediction
Figure 7.7 gives the logic structure for the shading prediction module. The hourly shading profiles, for external opaque and transparent surfaces, as caused by surrounding and facade obstructions, are predicted by the theory of §5.2. One or more of these profiles can then be transferred directly to the shading/insolation database for subsequent recall and use during energy simulation. Alternatively, any specified profile can be simplified and input to the shading/insolation database via the input management module.

In the case of an annual simulation of a large building the hourly shading data will be prohibitive (approximately 2 Mb for a 10-zone problem with each zone bounded by 15 surfaces, each having 10 windows). Either the shading prediction software must be incorporated within the energy simulation module, to permit shading determination separately at each computational time-step, or the data requirements restricted to manageable limits. Now, since the shading patterns on any one day will differ little from those of days immediately preceding and following, it is perhaps acceptable to assume that the shading patterns for such a day are valid for a sequence of days up to (say) a month. The database need then only contain 12 data sets, each comprised of the user-selected opaque and transparent surface shading patterns determined for the day which falls closest to the average monthly solar declination. The shading prediction module has the following facilities in operation:

Site plan and perspective display showing the relationship between the target building and the facade/surrounding obstruction features.

The display of solar azimuth versus altitude plots to aid decisions on deliberate obstruction location prior to shading prediction.

Hour-by-hour shading computation for the whole site case and for a single window considered in isolation to facilitate study of window solar control devices.

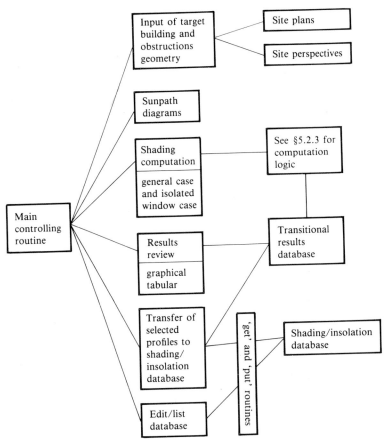

Figure 7.7 External surface shading prediction: logic structure.

Graphical and tabular review capabilities with the potential for communication with advanced colour graphics devices to allow the introduction of realism in the display of shaded facades.

A transfer and editing mechanism to allow management of the shading/insolation database.

Note that all predicted values are stored, in the first instance, in a transitional results database. It is from this database that shading patterns are selected and transferred to the main database for later use. This allows the archiving of all predictions and the extraction and use of any subset to satisfy the simulation objectives in hand. Appendix F gives the format of the obstructions geometry database.

Internal surface insolation prediction

Figure 7.8 gives the insolation prediction module logic. The hourly insolation profiles for each zone window dictate which internal surface(s), opaque or transparent, will receive direct radiation transmitted through the window at any time and after modification by any shading data previously predicted by the shading module. As before, selected profiles can be transferred to the shading/insolation database or, alternatively, input to this database in some simplified form (to help in the interpretation of the energy simulation results) via the input management module.

Since the shading and insolation prediction modules are similar in function, it may be attractive to combine the two modules into a single program. Note,

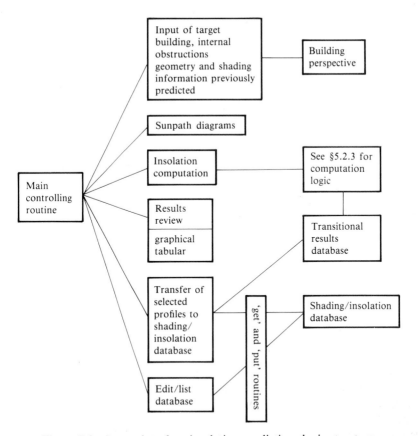

Figure 7.8 Internal surface insolation prediction: logic structure.

however, that it should still be possible to utilise either subsystem separately to accommodate any mismatch in available information as the design evolves.

Since both prediction modules can communicate, piecemeal over time, with the shading/insolation database, the internal organisation and related software will be correspondingly complicated. Appendix F shows the format of this database.

View factor prediction

Figure 7.9 gives the logic structure of the view factor module. For each surface pair within any zone, the black body view factor is evaluated by the theory of §5.4. Results review consists of a matrix of inter-surface view factors and error data relating to the deviation of the results from compliance with the reciprocity rule due to numerical accuracy associated with the adopted discretisation scheme. Error magnitude is dependent on geometrical complexity and for complex zone shapes 10–15% deviation from reciprocity is not unreasonable. Appendix F gives the view factor database format.

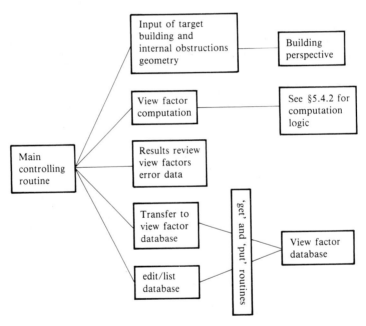

Figure 7.9   View factor prediction: logic structure.

Air flow prediction

Figure 7.10 shows the logic structure for the air flow module. The information on building leakage is established via the input management module and consists of the coefficients and exponents relating to the characteristic flow equations defining the problem (see §5.6.1). The time-series data on wind speed and direction are extracted from the climatological database for the user-specified period and location. The pressure distribution is specified in terms of non-dimensional pressure coefficients, one set for each external surface exposure. A coefficient set consists of 16 values for each of the 22.5° compass segments and is defined relative to the surface normal vector so that the set is independent of absolute surface orientation. Database facilities exist which contain a number of design coefficient sets. Additional coefficient sets, established by wind tunnel testing or other means, can be appended.

Again a transitional database is used to minimise core memory demands whilst ensuring result archiving and flexibility in final database development. Appendix F gives the air flows database format and describes the leakage and pressure distribution databases.

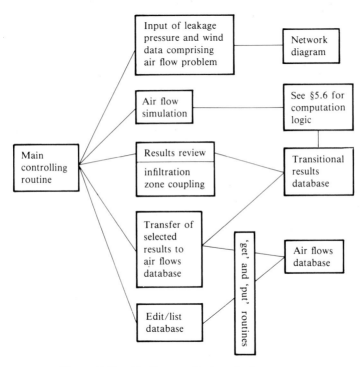

Figure 7.10 Air flow prediction: logic structure.

Window properties prediction

Figure 7.11 shows the logic structure for the window properties module. This module is used to analyse any window configuration where the elements are plane and parallel, allowing an undistorted view. The overall transmission, reflection and absorption of the window system is predicted by the theory of §5.3.3 as a function of the angle of incidence of the incoming shortwave radiation and in terms of the prevailing solar spectrum.

Appendix F describes the substrate and thin film properties required for prediction purposes. Once specified a spectral analysis is possible against some default solar spectrum. Most glazing manufacturers quote performance data relative to a solar spectrum for air mass 2. That is for a rural site under clear sky conditions with a solar zenith angle of 60 degrees. Alternative spectra can be established on the basis of user-specified site, sky and zenith angle data.

Result review exists at two levels. Full output allows the display (both graphically and in tabular form) of the predicted transmission, reflection and absorption for any angle of incidence and spectral wavelength. Abbreviated output gives the same incidence-dependent information averaged over the entire solar spectrum ($\sim 0.3-5$ μm). Appendix F shows the format of the window properties database.

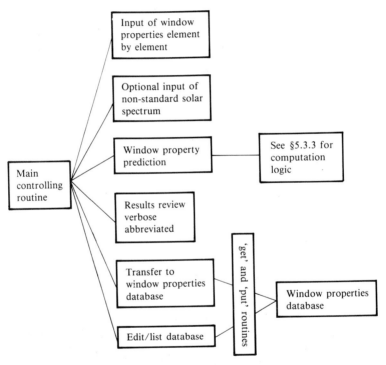

Figure 7.11 Window properties prediction: logic structure.

7.2.4 *Energy simulation*

Figure 7.12 shows the structure of the energy simulation module. Three distinct sections are considered: the pre-simulation input utility; the simulation itself; and a trace and batch facility to aid program development, validation and technique study, and to ease the burden of parametric simulation work.

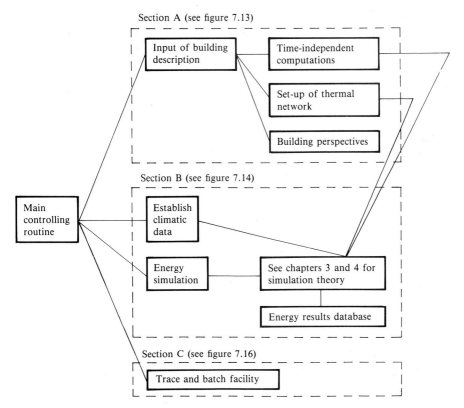

Figure 7.12 Energy simulation: general logic structure.

Pre-simulation input utility

Figure 7.13 gives the essential elements for this section of the simulation module. Considering each element in turn:

A1: The system configuration file, established by the input module, is accessed to establish the building and plant network for simulation. An extensive check list is applied to the data at this point and all detected discrepancies classified as fatal or non-fatal. The former type cause run abort whereas in the latter case computation can continue after user intervention.

A2: The zones and components databases are now accessed and the descriptive data read and checked as in A1. To minimise memory requirements, zones and components are read and the time-independent computations of A4 performed. In this way, only the information required at simulation time is held in core for all participating zones and components.

A3: Any optional databases, as indicated by the pointers held in the system configuration file, are now made ready for subsequent access throughout the simulation.

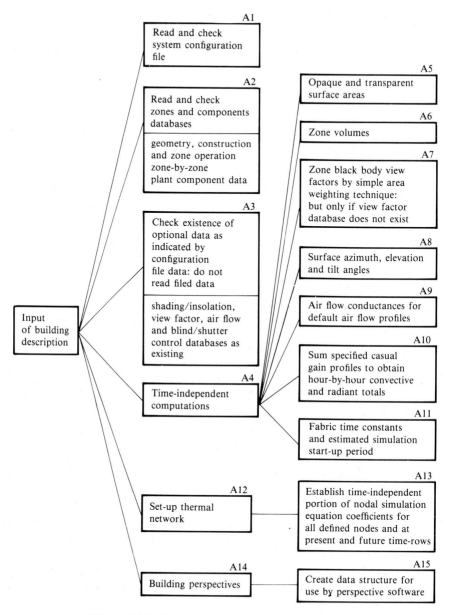

Figure 7.13 Expansion of section A of figure 7.12.

A4: All time-independent computations can now be applied to the established data.

A5: The area of each zone bounding surface is found from the theory of §5.1. Component heat transfer areas are also evaluated at this point.

A6: Zone volumes are found from the formulation of §5.1 with appropriate reduction to account for zone contents.

A7: If a view factor database is not available for use with the current simulation then approximate view factors are obtained from the simplified area weighting technique of §5.4.2.

A8: Surface azimuth, elevation and tilt angles (angles α, β and γ respectively in figure 5.5) are determined from the formulations of §5.1.

A9: The hourly ventilation conductance is computed for all default infiltration and zone-coupled air movement. This conductance is given by:

$$C_v = \varrho C n V / 3600$$

where ϱ is the air density (kg m^{-3}), C the air specific heat (J kg^{-1} °C^{-1}), n the default air change rate (hr^{-1}) and V the zone volume (m^3).

A10: All casual gain profiles are combined after modification by the appropriate convective/radiant split. It is at this point that decisions are taken on the application point(s) of the radiant portion.

A11: The time constant of each multi-layered construction is evaluated from a method by Mackey and Wright (1946):

$$\tau = \frac{(k\varrho C)_o \left(R_o - 0.1R_i - 0.1\sum_{j=1}^{M} R_j\right) + 1.1\left((\varrho Cx)_i + \sum_{j=1}^{M}(\varrho Cx)_j\right)}{U}$$

where k is the conductivity (W m^{-1}K^{-1}), ϱ the density (kg m^{-3}), C the specific heat capacity (J kg^{-1}K^{-1}), R the thermal resistance (W m^{-2}K^{-1}), x the element thickness (m); and o, i, j refer to the outermost, innermost and M central layers, respectively. The simulation start-up period—required to eliminate the effects of the arbitrary assigned initial nodal conditions—is empirically related to the maximum time constant.

A12: The thermal network is now established in terms of a matrix indexing which ensures that all nodal equations are properly located in relation to each other.

A13: The time-independent portion of all nodal equation coefficients is evaluated in terms of those system properties which can be considered constant throughout the simulation.

A14 (and A15) Facilities exist to create display files for access by perspective viewing software so that zone location can be checked prior to simulation.

Simulation methodology

Figure 7.14 details the simulation methodology, which essentially is concerned to evaluate each subsystem in order to complete the set-up of the time-dependent portion of the equation coefficients commenced in A13. These coefficients are then assigned to a partitioned matrix cluster and the matrix solution implemented for the time-step. Considering each element in turn:

B1: The simulation period start and finish day is input and converted to a start

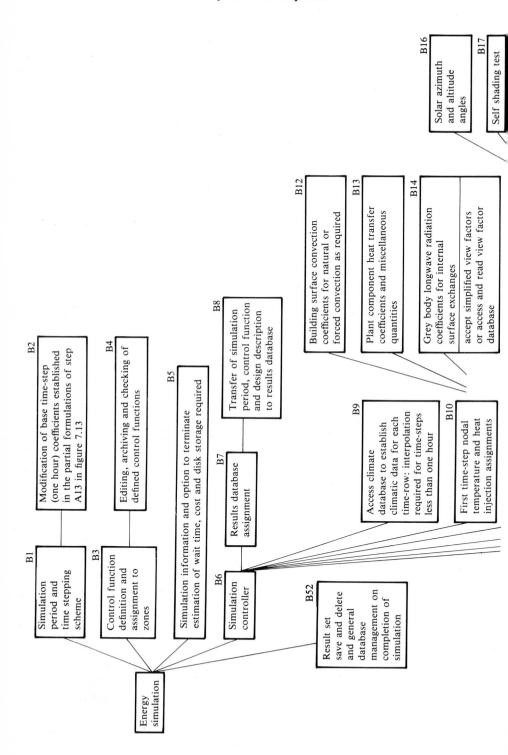

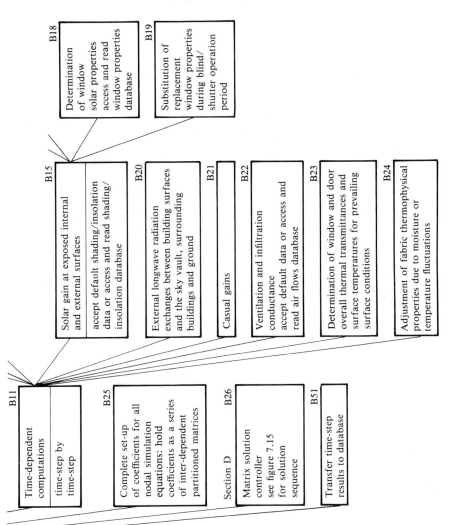

Figure 7.14 Expansion of section B of figure 7.12.

and finish year day number which takes account of the required start-up period. The computational time-step is input. Solution stability is guaranteed for any time-step but small values (~10 minutes) will be required to allow frequent control evaluation to avoid over- or under-shoot within a time-step.

B2:   In the partial evaluations of step A13 a one hour default time-step is assumed. All coefficients are now modified by the ratio 'actual:default' time-step in the case of a sub-hour time-stepping scheme.

B3:   (and B4) A number of editing, archiving and checking facilities exist to allow the interactive definition of the control functions, if not already specified during the building description process. Any control function is then associated with one or more building zones to define the time-dependent environmental conditions and capacity restrictions for that zone as the simulation proceeds.

The building plant control strategy is therefore comprised of one or more control functions. A control function is a data set describing the time-dependent operation of zone-side plant components as well as the location of control and plant interaction nodes.

If no control functions are defined then the strategy is 'free-float' and the entire system will have no control imposed at any time. Alternatively, some zones may have no associated control function, whilst other zones obey some imposed function. Once created (interactively) a control function can be transferred to a control function file for later recall. Listing and editing facilities exist to allow the inspection and modification of a control function before it is associated with a selected zone. A control function is comprised of the following information:

(a) The location of the control point. A zone can be controlled according to the condition of any node point within its physical boundary. Typical locations include air nodes, surface nodes, intra-capacity nodes (if the zone represents an electrical storage unit for example) or some fictitious node sensing a user-specified weighting of some or all system nodes. This last controller definition allows simulation control on the basis of more realistic sensor characteristics or, if resultant temperature is used, on the basis of comfort considerations rather than (say) air temperature alone as is often done.

(b) The controller type: proportional, derivative or integral, or some mix.

(c) The location of the plant interaction point. This defines the point within a zone where plant energy is input/extracted at the dictates of the zone control function plant schedule. Again a number of possibilities exist. A convective system will exchange energy directly with the zone air nodes, a radiant panel will, in addition, exchange energy directly with surface nodes and an under-floor system will interact with intra-constructional nodes.

(d) The plant schedule. This is specified by subdividing a typical weekday, Saturday and Sunday (one set for each week, month or season as required) into a number of distinct periods during which stated control conditions are in force. For each period the start and finish times are given, along with an index which defines the control objectives for the period. Five possibilities are offered.

A controlled period is one for which lower and upper temperature limits are specified. Then, during the subsequent simulation, an attempt is made to maintain the zone control point temperature between the limits, when the control period is in force, by the utilisation of cooling or heating power as permitted by the defined plant system (see B31 of figure 7.15). If more (or less) power is required than is available then the control point temperature is forced to depart from the desired conditions.

A floating temperature period is one for which no power is available and therefore no control can be attempted.

A pre-heating or pre-cooling period can precede only a control period. The objective of this period type is to raise or lower the control point temperature in such a manner that by the end of the period a value has been attained which corresponds to the upper or lower value (whichever is the nearer) of the range specified for the following control period.

A specified injection/extraction period is one during which a specified amount of energy is added or extracted from the plant interaction node.

An unspecified injection/extraction period is one during which an energy addition or extraction occurs according to the physical laws incorporated within the overall plant matrix equation set. A number of control loops are specified for the plant system and, for each loop, the sensor and actuator nodes are defined, along with the control law which relates the sensed and actuated region states. The plant components are assumed to be decoupled from the interaction node if the control point conditions fall within the specified range.

The foregoing facilities permit the creation of complex time varying control regimes to allow the examination of, for example, the intermittent operation of cooling/heating plant, optimum start times, night set-back, morning boost, and so on. Real and idealised plant interactions can be mixed to facilitate the search for an optimum control schedule.

B5: Once all control functions are defined, they are associated with the building zones as required and the simulation is ready to commence. Prior to this, a page of data is displayed which gives an indication of the computing time and disk demands as estimated for the problem, simulation period and time-stepping scheme currently defined. If the predictions are considered excessive then the terminate option can be selected and the job submitted, instead, for batch processing to obtain any discounts offered by the installation (see figure 7.16 and accompanying text).

B6: The simulation controller now initiates the time-stepping procedure.

B7: (and B8) The results database is established and all input data and simulation control information transferred as a header to the results to be entered.

B9: The climatic database is accessed and linear interpolation applied to the hourly data (but only for time-steps less than one hour) to establish the present and future time-row values for the current time-step.

B10: At the first time-step only, all nodal temperatures are initialised on the basis of a steady-state calculation procedure. All nodal energy injections are zeroised.

B11: All time-dependent computations are initiated.

B12: Surface natural and forced convection coefficients are determined by the theory of §5.5.

B13: Plant component heat transfer coefficients are determined as required. See chapter 6.

B14: The view factors database is accessed or the simplified view factors accepted and the 'grey' body radiation coefficients evaluated for all internal surface pairings according to the formulation of §5.4.1.

B15: Solar gains at all external and internal surfaces are evaluated on the basis of the shading and insolation data held in the shading/insolation database. If no such database exists, or if data are not available for the time or surface under consideration, then the default data are utilised to give a standard or zero shading profile and a defined insolation plane or planes—perhaps the floor plane(s) in the case of an open plan office, the major pick-up plane(s) in the case of a solar collector or a passive solar feature such as a Trombe wall, or all internal surfaces in the case of a window device promoting diffusing direct gain. Solar computations adhere to the theory given in §§5.3.2 and 5.3.4 and will require the data determined by elements B16 to B19 inclusive.

B16: Solar angles are a function of year and day time as described in §5.3.1.

B17: Initial self-shading tests will identify those surfaces which can be excluded from the direct radiation calculations since they cannot 'see' the sun (see §5.2.1).

B18: The window properties database is accessed to establish the transmission, absorption and reflection factors for prevailing incidence angles (see §5.3.3).

B19: These window properties are replaced if a blind/shutter period is in force.

B20: External longwave radiation exchanges are determined by the techniques of §5.4.3 and will require the estimation of sky, ground and surrounding building temperatures.

B21: The casual heat injection is extracted from the list compiled at stage A10 of figure 7.13. Apportioning of the radiant energy between surface nodes and the convective energy between air nodes is performed on the basis of information describing the location of the heat sources.

B22: The default data of stage A9 are accepted or the air flows database (if one exists) accessed and the current zone-time data located.

B23: Windows and doors can be treated in the same way as multi-layered constructions, that is by a network of connecting nodes. Alternatively, since they usually have negligible thermal capacity, they can be treated on a zero node basis to minimise the size of the whole system matrix scheme. At each time-row an effective U-value is determined on the basis of the prevailing surface conditions as evaluated at stages B12, B14 and B20. The conductive flowpath is then added to the nearest air node simulation equation to effectively couple this air node directly to ambient conditions. Window and door surface temperatures are evaluated for use in the longwave radiation computations and window longwave radiation modified by the surface resistance to allow application at the air point.

B24: The basic fabric thermophysical properties can now be modified on the basis of previous time-step temperature or moisture (independently assessed) conditions.

B25: The nodal equation coefficients are finalised and transferred to the appropriate partitioned matrix (see chapter 4) according to the connectivity information held in the system configuration file created by the input module.

B26: The matrix solution controller is invoked for the current time-step.

B27: (to B50) See figure 7.15 and related text.

B51: Once the simultaneous solution is achieved for the time-step, the future time-row results are transferred to the results database. A number of save levels are available ranging from control temperature and plant interaction data only, through to full energy balance and a complete nodal conditions history. Appendix F describes the organisation of the results database.

B52: Results sets can be saved or destroyed to facilitate database management.

Figure 7.15 shows the matrix solution stages for each control node location possibility. The strategy is to tailor the partitioned matrix operations to the controller type imposed on the zone. For air point simulation control steps B28–B30 operate to produce the zone characteristic equation (see figure 4.14) which relates the control point temperature to the plant input at the defined interaction node(s). The process of B31 is delayed until all zones have been processed. Steps B31 through B35 are then performed to determine the ideal and then actual plant injections. Note that many building-side design studies can be performed without the need for steps B32–B35. The assumption is then made that the required power is always available without delay (as with direct electrical equipment). Lastly, steps B36–B37 are performed to complete the simultaneous solution.

The procedure for the other control point types is similar, deviating only in the treatment of the building-side partitioned matrices as dictated by the change in controller location. One exception occurs in the case of fictitious point (or multinode) control. It is not possible to inject plant energy at the fictitious point and the plant interaction point must be defined in the usual way. The B49 and B50 steps must therefore iterate to establish the plant injection (or extract) at the interaction node(s) required to give the desired fictitious point (weighted multinode) temperature. Considering each step in turn:

B27: Controls the matrix processing for the air point control case.

B28: Each multi-layered construction matrix is forward reduced to produce a construction characteristic equation (see figure 4.9).

B29: The construction CEs are then applied to the corresponding zone surface equations to eliminate the next-to-surface coefficients (see figure 4.12(a)).

B30: The adjusted zone matrix is then forward reduced to produce the whole zone CE which relates the control point to the plant interaction point (see figure 4.14).

Figure 7.15 Expansion of section D of figure 7.14.

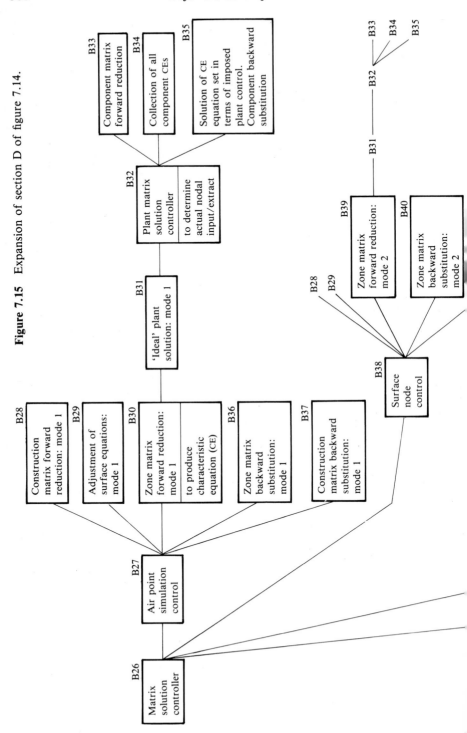

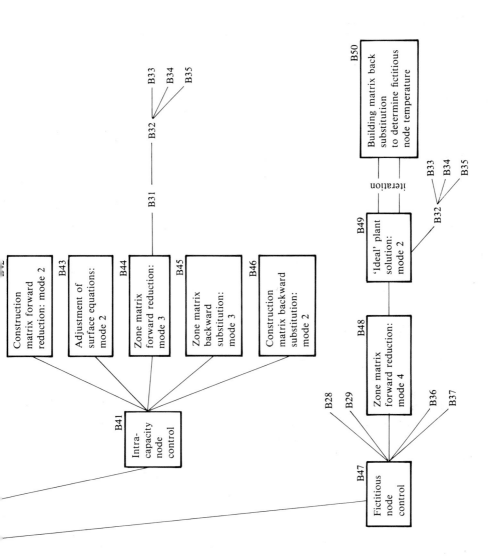

B31: The multi-zone CE set is solved (as described in §4.2.2), in terms of the user-defined plant schedule, to give the heating or cooling power to achieve the desired air point temperature.

B32: The plant matrix controller is now invoked to find the simultaneous solution of the plant matrix and so determine the actual nodal input or extract achievable by the plant system.

B33: Each component matrix is forward reduced to give the component CE.

B34: All component CEs are gathered together.

B35: And solved in terms of the imposed system regulation. The backward substitution operation is then completed for each component matrix.

B36: Now that the actual zone plant injection/extraction is established, zone matrix backward substitution gives the zone nodal conditions.

B37: And construction matrix backward substitution gives the construction nodal conditions to complete the time-step solution.

B38: Controls the matrix processing for the surface node control case.

B39: As B30 but the control node is carried through into the zone CE as shown in figure 4.14(*b*).

B40: As B36, but the backward substitution is associated with the scheme of B39.

B41: Controls the matrix processing for the intra-capacity node control case.

B42: As B28 but the control node is carried through into the construction CE as shown in figure 4.10.

B43: As B29 but with the introduction of the intra-capacity control node as shown in figure 4.12(*b*).

B44: As B30 but the control node is carried through into the zone CE as shown in figure 4.14(*c*).

B45: As B36 but backward substitution is associated with the scheme of B44.

B46: As B37 but backward substitution is associated with the scheme of B42.

B47: Controls the matrix processing for the fictitious point control case.

B48: Is as B30, B39 or B44 depending on the location of the nodes comprising the fictitious point. B30 is used for an air node mix, B39 for an air/surface node mix and B44 for an air/surface/capacity node mix.

B49: As B31 but with an iterative search for the CE solution. This is done by iterative adjustment of the plant input (to the plant interaction nodes) until the fictitious node condition (at the future time-row) is attained.

B50: Back substitution in the entire building-side matrix is required to determine the nodal conditions to permit fictitious node temperature evaluation for the current iterative step.

Trace and batch facility

Figure 7.16 details the elements of the trace and batch feature.

The trace facility allows output (to terminal or disk file) of any specified parameter not normally available at results recovery time. This includes all intermediate computations, available only during subsystem processing to determine nodal heat injections and the various heat transfer coefficients which are introduced to the matrix structure at each time-step. Thus each system subroutine has a trace feature to allow examination of performance as the simulation proceeds. This is an essential aid to software validation. Also, the recovery of such items as sky and ground temperatures, surface convection coefficients and the percentage of zone admitted solar radiation that is lost again to outside (or another zone) by window retransmission is an important enhancement to system output potential.

The batch facility allows module operation on the basis of an instruction set established prior to simulation. A number of simulation requests can be defined with simple conditional statements imposed between each. This mode of operation is useful in large scale parametric studies with the possibility of run termination should some criterion be met. The elimination of all interactive features (menuing, terminal dialogue, etc) will require some sophistication on the part of the routines of the utility and graphics libraries. It is also helpful if all terminal interaction is centralised.

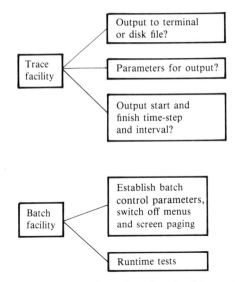

Figure 7.16   Expansion of section C of figure 7.12.

7.2.5 *Results recovery*

Figure 7.17 summarises the elements of the results recovery module. Since it is impossible to anticipate the output requirements for each designer/analysis combination, the recovery module should have extensive display potential. The output period can be any subset of the simulation period, with the output interval equal to, or possibly greater than, the simulation time-step, to allow small simulation time-stepping schemes for accuracy but larger recovery increments to promote cost effective results analysis. Note that since a large proportion of the

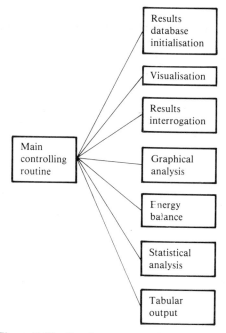

Figure 7.17   Results recovery: logic structure.

simulation cost is related to disk transfers and since multiple access to the results database will generally be required, the cost of results recovery will often exceed the simulation cost. Six output categories are on offer:

Visualisations
As with other system modules, the ability to produce perspective views will greatly aid results interpretation by showing the relative position of all building components. Also, if colour terminal facilities are available it is possible to produce pseudo thermographs by applying surface colouring on the basis of the computed surface temperatures or heat transfer.

Results interrogation
This allows a rapid investigation of the result set by extracting summary data relating to a range of typical design questions.

Graphical analysis
Allows the display of any predicted or boundary condition profile over time. Various other display techniques can be used (pie charts, 3-D plots, etc) to enhance the presentation.

Energy balance

The output of energy balance data, which show the gains and losses at any node, will demonstrate the causal relationships at any building or plant location and so identify areas for closer inspection.

Statistical analysis

Will include curve fitting, principal components analysis and occurrence search procedures.

Tabular output

Allows the display of numerical values against time.

Figure 7.18 shows the available command options for the ESP recovery module and appendix G gives some example result displays.

7.3   Indicative performance data

This section gives size and performance data for the system of §7.2. Such data can be used as a target in the evaluation of any new or existing system or to aid decisions on development priorities or hardware selection. Information is given on run timings and on core and disk requirements.

To illustrate the demands of a modelling system in operation, consider the ESP system as described in §7.2. Module source code (including comments) and object code disk file sizes are as in table 7.6.

Thus at least some 4 Mb of disk (source plus object) will be required if the entire system is to be kept on-line during the development and testing phase. In addition the permanent system and temporary databases will impose the requirements listed in tables 7.7 and 7.8, respectively. The minimum disk requirement to support a fully working system (but excluding system source code) is therefore some 3.8 Mb.

The core demand at runtime is dictated by the energy simulation module, which is the largest module in the system. This requires some 250 kb of primary memory for the program instructions with an additional 350 kb for the data. An average run, say 10 zones for 1 week (at a one hour time-step and with a moderate results save), requires approximately 95 seconds of CPU time on a DEC System 10 supporting (say) 30 substantial users (80 and 155 seconds for the minimum and maximum save options respectively; see the results database contents as described in appendix F). For a PRIME 750 under moderate loading slightly more CPU seconds are required. For this 'typical' problem the simulation module will perform between 3 and 20 k disk accesses to produce a results database of between 200 and 1550 kbytes depending on the requested save option. Other system modules, such as the input management and results recovery modules, will each consume approximately the same CPU time but do substantially more disk and terminal I/O.

STATSO

AFREQY		frequency analysis
BREGRS		regression and curve fitting
CMN&SD		means and standard deviation
–END		

REGRES

1RSETS	KLSURC	regression data location/surface convection
2RGPER	LISOLE	regression period/external surface solar absorption
3ZONDF	MISOLI	zones for consideration/internal surface solar
	NCSOLD	/direct normal solar radiation
4DPEND	OCSOLF	dependent variable definition/diffuse horizontal solar
5IPEND	PCAIRE	independent variable definition/external air temp.
	QCWINS	/wind speed
6MAX.V	RCWIND	max. value of variable used/wind direction
7MIN.V	SCRELH	min. value of variable used/relative humidity
8MEANV	TINSUL	mean value of variable used/insulation level
9DEFMV	UWINDO	user defined variable test used/window area
	VRATIO	/glazing or wall ratio used
ATAIRI	WVOLUM	internal air temp./internal volume
BTCONP	XFAREA	controller temp./floor area
CMNRAD	YENERG	mean radiant temp./energy consumption
DTWIND		wind direction
ELPLNT	ZSDREG	plant capacity/initiate standard regression
FLINFL	:GO	infiltration/commence regression
GLVENT		zone coupled air flow
HLGCON	?INFO	window conduction/regression details
ILSOLA	/CLEAR	window solar absorption/clear defined regression
JLCASG		casual gains
	–END	

ESPOUT

LIBRES	library definition
RESSET	result set definition
PEROUT	recovery period
ZONDFN	recovery zones
1VIEW	perspective views
2INTER	interrogation output
3GRAPH	graphical output
4TABUL	tabular output
5STATS	statistical analysis
6INTEG	result set integration
7INFO	status information
FINISH	

INTERO

RESSET	
PEROUT	
ZONDFN	
ALHEAT	heating load summary
BLCOOL	cooling load summary
CTAIRI	air temperature
DTRESL	resultant temperature
ETCONP	control (or sensed) temperature
FENERG	summary of energy demand
GLAIR	air loadings
HLGCON	window loads
ILDCON	door loads
JLCASG	casual loads
KLSURC	surface convection
LISOLG	solar pick-up
MCAUSL	causal load analysis
NCAUSE	causal energy analysis
–END	

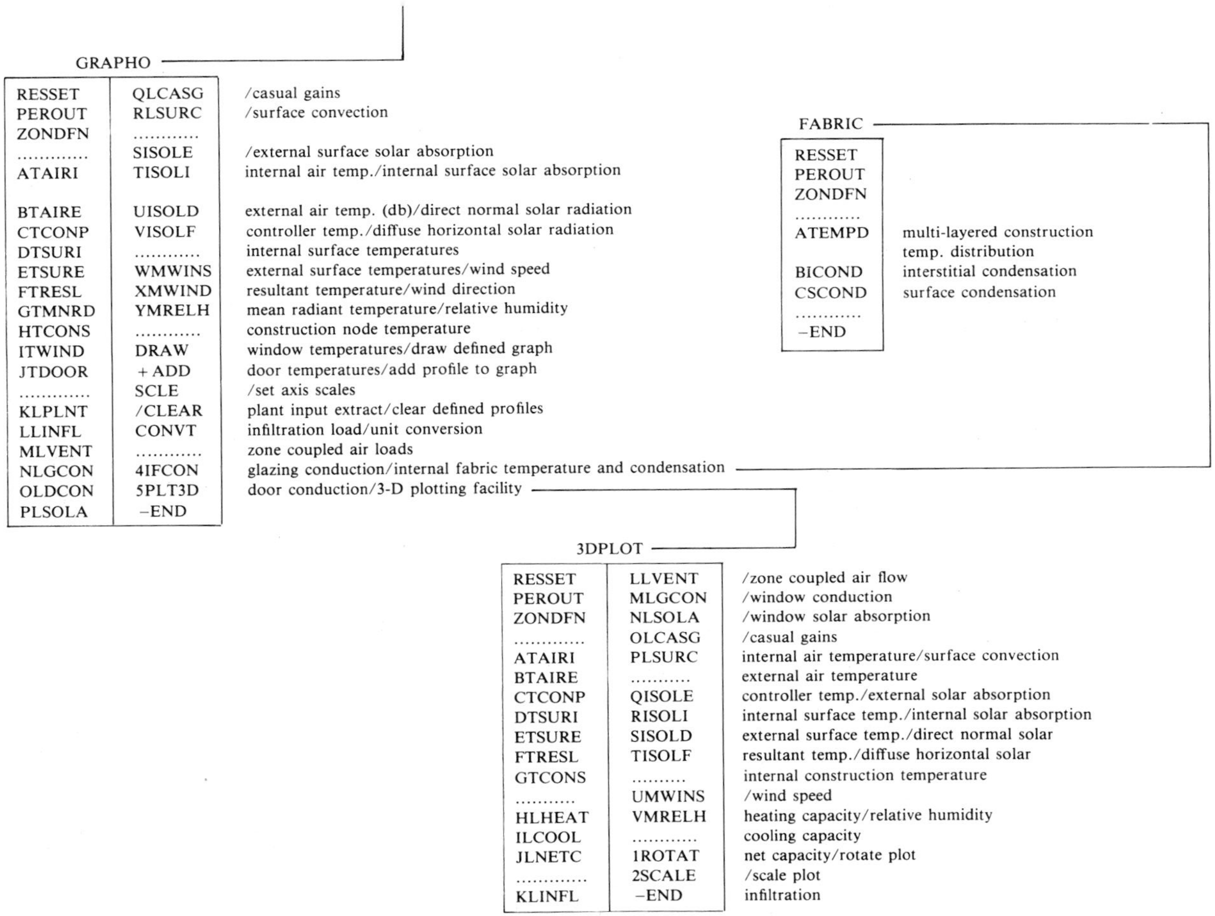

GRAPHO

RESSET		
PEROUT	QLCASG	/casual gains
ZONDFN	RLSURC	/surface convection
....		
ATAIRI	SISOLE	/external surface solar absorption
	TISOLI	internal air temp./internal surface solar absorption
BTAIRE	UISOLD	external air temp. (db)/direct normal solar radiation
CTCONP	VISOLF	controller temp./diffuse horizontal solar radiation
DTSURI		internal surface temperatures
ETSURE	WMWINS	external surface temperatures/wind speed
FTRESL	XMWIND	resultant temperature/wind direction
GTMNRD	YMRELH	mean radiant temperature/relative humidity
HTCONS		construction node temperature
ITWIND	DRAW	window temperatures/draw defined graph
JTDOOR	+ADD	door temperatures/add profile to graph
	SCLE	/set axis scales
KLPLNT	/CLEAR	plant input extract/clear defined profiles
LLINFL	CONVT	infiltration load/unit conversion
MLVENT		zone coupled air loads
NLGCON	4IFCON	glazing conduction/internal fabric temperature and condensation
OLDCON	5PLT3D	door conduction/3-D plotting facility
PLSOLA	-END	

FABRIC

RESSET		
PEROUT		
ZONDFN		
....		
ATEMPD		multi-layered construction
		temp. distribution
BICOND		interstitial condensation
CSCOND		surface condensation
-END		

3DPLOT

RESSET	LLVENT	/zone coupled air flow
PEROUT	MLGCON	/window conduction
ZONDFN	NLSOLA	/window solar absorption
	OLCASG	/casual gains
ATAIRI	PLSURC	internal air temperature/surface convection
BTAIRE		external air temperature
CTCONP	QISOLE	controller temp./external solar absorption
DTSURI	RISOLI	internal surface temp./internal solar absorption
ETSURE	SISOLD	external surface temp./direct normal solar
FTRESL	TISOLF	resultant temp./diffuse horizontal solar
GTCONS		internal construction temperature
	UMWINS	/wind speed
HLHEAT	VMRELH	heating capacity/relative humidity
ILCOOL		cooling capacity
JLNETC	IROTAT	net capacity/rotate plot
	2SCALE	/scale plot
KLINFL		infiltration
	-END	

Figure 7.18 The building-side results recovery menu driver for the ESP system.

Table 7.6　Module source code and object code disk file sizes.

Module	Source lines	Source kbytes	Object kbytes
Input data management	15971	488	248
Constructions management	823	24	58
Profiles management	551	17	46
Climate management	1032	30	98
Plant components management	693	22	52
Shading prediction	3180	89	130
Insolation prediction	3382	93	121
View factor prediction	4564	159	132
Air flow prediction	1520	30	72
Window properties prediction	3259	93	107
Energy simulation	20906	558	311
Results recovery	14758	384	210
Utilities library	2913	75	106
Graphics library	3954	75	106
Total	77506	2137	1797

Table 7.7　Requirements for permanent system databases.

Database	Contents	kbytes
Construction primitives	100 homogeneous elements	32
Construction composites	10 composites	4
Profiles	10 profiles	4
Plant components	100 components	120
Climate	1 year hourly	284
Total		444

Table 7.8　Requirements for temporary databases.

Database	Contents	kbytes
Shading/insolation	1 zone-facade-day	9
View factor	10 average zones	10
Air flow	10 average zones, 1 day	1
Window properties	10 windows	1
Energy results (1 week)	10 zones, full save	1547
	10 zones, minimum save	201
Total		1568 (222)

The performance of the utility modules will depend generally on the problem considered and, specifically, on the surface grid applied in the case of shading, insolation and view factor prediction, on the number and relative sizes of the leakage paths with air flow prediction, and on the number of wavebands considered with window property prediction. For example, the shading module will require some 1.4 CPU seconds, do 400 disk accesses for a 1 day, 1 facade, 1 window, 1 obstruction body, 10×10 surface grid problem; the insolation module will require some 30 CPU seconds and do 300 disk accesses for a 1 day, 1 window, 6 internal surface, 10×10 surface grid problem; the view factor module will require some 55 CPU seconds and do 10 disk accesses for a rectangular zone problem with 1 window, a 10×10 surface grid and 100 grid point hemispherical patches; the air flow module will require some 20 CPU seconds and do 120 disk accesses for a 1 day simulation of a 10 leakage path problem; and the window properties prediction module will require some 1.2 CPU seconds and do 20 disk accesses for a 3 element window problem with the solar spectrum subdivided into 10 wavebands.

7.4 Development environment

At the present time, hardware technology is changing rapidly, with powerful, but low cost, processors and terminals appearing in the marketplace. This section briefly considers these options (but without recommending specific manufacturers) and the computer languages appropriate to energy model applications software. And lastly, based on the statistics of the preceding section, a minimum processor/terminal specification is given which is capable of supporting interactive energy simulation work.

Processors
There are basically two processor environments which offer sufficient power for the development and/or use of advanced energy models. These are mainframes (and large minicomputers) capable of servicing (on a time-sharing basis) the needs of a large number of simultaneous users (up to 50 say) and small single- or multiple-user minis (or super micros) capable of handling less than (say) 5 simultaneous users.

In a mainframe environment the user is generally forced to accept the manufacturer's operating system (whether friendly or not), and may have restrictions imposed with respect to disk storage and available core. On a commercial mainframe, computing time and disk storage can be expensive so that, in view of the high demands made by energy models for both, operational costs can be prohibitive. On the other hand, many back-up facilities are available and software libraries such as the statistical packages NAG or SPSS, if on offer, can greatly reduce the programming burden.

In recent years a number of small, but powerful, systems have emerged which cater for single person or small group use. Based on 16 and 32 bit virtual memory technology, these 'workstations' are able to meet the demands of the large modelling systems without imposing the capital and running costs associated with the larger machines. Examples of such systems include APOLLO, SUN and WHITECHAPEL which come complete with integral high resolution graphics terminal (monochrome or colour) with multi-windowing, multi-tasking facilities. Most of these systems utilise the UNIX† operating system (Bourne 1982) (or some 'lookalike') which is one of the most powerful small machine operating systems available. Over 100 standard utilities are available with UNIX to perform such tasks as terminal independent graphics, communication networking and text processing. One attractive feature is that UNIX completely hides the hardware from the applications software so that all program I/O, whether to disk, terminal or another applications program, is treated in the same manner. In practice this means that the various modules of an energy system can be developed and tested independently, by disk file input and output, say, and then integrated into a single runtime system with each module receiving its data directly from another module as required. It is likely that UNIX will form the template on which future small machine operating systems will be based.

Software language

The most commonly used language, in terms of compiler availability and user experience, is FORTRAN. This is the traditional language for numerical type problems (such as energy modelling) and in its latest version, FORTRAN77, is rapidly gaining acceptance as a standard. PASCAL, although better suited for logical organisation, is, perhaps, not sufficiently numerically oriented to cater for the demands of numerical simulation. The C language is a powerful language designed for the implementation of large systems and tailored for UNIX. It has many facilities lacking in FORTRAN and, as UNIX continues to proliferate, C-type languages, will, perhaps, become more widely used.

Terminals

It is clear that the design community, with growing experience of CAD systems, will demand sophisticated graphics facilities. On the other hand the graphics demands of energy modelling systems (line graphs, simple perspectives, etc) are low when compared with the many other CAD application areas, especially visual impact. It is therefore possible to operate, in the first instance, with inexpensive, moderate resolution, raster graphics equipment. Capabilities can then be enhanced at some later date with the introduction of higher resolution and/or colour capabilities. The software centralisation of all terminal traffic will greatly reduce the difficulty of decoupling one terminal to introduce another.

An outline specification

Section 7.3 gave summary statistics for the multi-module ESP system described in §7.2. Based on these data, the minumum specification for a suitable

---

† UNIX is a trade mark of Bell Laboratories.

processor/terminal is: medium scale machine with capacity for upward expansion; high speed floating point processor; micro code or, preferably, hardware based; minimum of 740 kb user-accessible primary memory (to accommodate the simulation module); minimum of 80 Mb high speed secondary memory; a high quality editor, timesharing operating system and FORTRAN and perhaps C compilers; text and graphics terminal (low cost, in the first instance ?); terminal hard-copy device; and printer capable of text and simple graphics.

References

ABACUS 1981 User Manual for the Tektronix and Utilities Libraries *ABACUS Publication* (Glasgow: University of Strathclyde)

Bernard Y 1983 VOLUME: An Interactive Tool for Geometry Definition *Technical Rep., Laboratoire de Physique du Batîment* (Liège: University of Liège)

Bourne J R 1982 *The UNIX System* (New York: Addison-Wesley)

CADC 1976 *GINO-F User Manual* (Cambridge: Computer Aided Design Centre)

Clarke J A 1982 ESP Documentation, Section 12: The Output Program *ABACUS Publication* (Glasgow: University of Strathclyde)

Foley J D and Van Dam A 1982 *Fundamentals of Interactive Computer Graphics* (New York: Addison-Wesley)

GKS 1982 *Graphical Kernel System (GKS)—Functional Description* (International Standards Organisation)

Hopgood F R A, Duce D A, Gallop J R and Sutcliffe D C 1983 *Introduction to the Graphics Kernel System (GKS)* (London: Academic)

Mackey C O and Wright L T 1946 Periodic Heat Flow—Composite Walls or Roofs *Heating, Piping and Air Conditioning* **18** 107–110

Maver T W and Ellis J 1982 Implementation of an Energy Model within a Multidisciplinary Practice *Proc. CAD82*

Parkins R P 1978 3-D Input User Manual *ABACUS Publ. M17* (Glasgow: University of Strathclyde)

Stearn D D 1983 BIBMAK: Geometry Definition for the ABACUS Perspective Program BIBLE *ABACUS Publ. M35* (Glasgow: University of Strathclyde)

8

Validation and implementation in practice

There can be little doubt that existing theoretical and numerical techniques can permit an accurate modelling of a building and its associated plant system. However, notwithstanding the worldwide research effort over the past decade—an effort which has seen the emergence of such systems as BLAST (Hittle 1979), DOE (1980), DEROB (Arumi 1979) and TRNSYS (1979) in the United States, BUNYIP (Wooldridge 1983a) in Australia, LPB (Delorme and Trokay 1979) in Belgium, JULLOTA (Kallblad and Higgs 1981) in Sweden and ESP (Clarke 1982) and Tas° (1984) in the UK (to name but a few)—there remains a credibility gap which acts in conjunction with the financial problems to restrict the rate of model uptake and resulting improvement through use. This gap, between what exists and the design profession's ability to comprehend and make use of what exists, can only be bridged by a concerted effort on two complementary fronts: the establishment of a model validation methodology designed to test the predictive accuracy of existing and future models and the application of the methodology to selected systems to permit the declaration of 'valid' codes and so encourage uptake; and the demonstration of modelling potential by trial implementation in practice, in the form of demonstration projects.

Thus, the complementary tasks of validation and trial implementation in practice operate to reveal, over time, the potential of computer systems for the predictive modelling of building performance during the design phase.

8.1 Model validation

Ultimately, it is only by comparing model predictions with the corresponding results from actual buildings in use that a model's usefulness as a predictive agent can be ascertained. Unfortunately, this task is fraught with difficulties, due to the lack of comprehensive data relating to the performance of real buildings and the shortcomings inherent in even the most sophisticated technique, which make it impossible to model reality exactly. Commonly encountered problems include:

The monitoring of air movement and specification of the results in a form meaningful to modellers.

The on-line validation, organisation and management of large data sets in a manner which allows estimation of the data reliability.

The as yet ill-understood actions of occupants with regard to window opening, blind operation and control manipulation.

The accurate determination of thermophysical properties as required by the models.

The cost associated with the data logging system, sensor installation and quality technical staff.

The availability of simultaneously recorded climatic data relating to the test site. This is especially problematic with many historical performance data sets.

And the necessity to ensure, by constant checking, that all instruments remain as calibrated for the duration of the monitoring scheme.

In consequence, validation based on empirical data is expensive and time consuming to achieve and so is normally pursued at national or international level, although many individual research groups will undertake small monitoring schemes which focus on specific areas of building performance.

In 1974 the International Energy Agency (IEA) was established to administer a programme of energy-related projects agreed by a number of member countries. One project, undertaken between March 1979 and December 1983, undertook an empirical data gathering programme aimed at a large, occupied commercial office building. The captured data were then used in a validation study of a number of prediction models from Australia, Belgium, Holland, Switzerland and the UK. Table 8.1 gives summary details for these models. Detailed reports exist on the monitoring scheme (IEA1 1983) and the utilisation of the data for model testing purposes (IEA2 1983). From each validation project

Table 8.1   IEA Annex 4: participant programs.

Program	Source
AMBER	Oscar Faber Consulting Engineers, UK
ATKOOL	Atkins Research & Development, UK
DOE2	EMPA, Switzerland
ENCO2	Pilkington Flat Glass Ltd, UK
ENERGY	Ove Arup, UK
ESP	ABACUS, University of Strathclyde, UK
LPB1	LPB, University of Liège, Belgium
TEMPER	CSIRO, Australia
THERM	British Gas, UK
WTEO1	TNO, Holland

of this kind, one recurring conclusion emerges: it is likely that many compensating factors exist within models which allow different models to arrive at similar results, each by a slightly different route. For this reason there is now general consensus that additional testing procedures are required to complement empirical testing by forming a validation methodology. These are: comparative evaluation to allow the direct comparison of different models when applied to the same problem, perhaps hypothetical, perhaps simplified; and analytical verification, in which a model is tested under special boundary conditions for which an exact solution is known.

The US Solar Energy Research Institute (SERI), under contract to the US Department of Energy, has established such a three part methodology (Judkoff

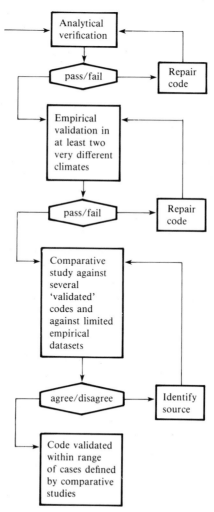

Figure 8.1 The SERI validation method (from Judkoff *et al* 1983).

et al 1983). The objective of their project, now complete, was to develop a validation methodology and to collect quality empirical data for use with the methodology when applied to four US systems possessing passive solar capabilities. The systems include DOE(2.1), BLAST(2MRT and 3.0), DEROB(3 and 4) and SUNCAT(2.4) (Palmiter and Wheeling 1979). Figure 8.1 shows the SERI three-part method which includes:

Analytical verification

This consists of two tests. In the first, the response of a homogeneous slab, adiabatic on one side, is exposed to a monthly step function in ambient temperature on the other side. The ambient coupling is considered convective only, the remaining climatic parameters are set to zero and all other heat transfer properties are held constant. In the second test the slab is considered to be insulated at both sides and receiving a constant diffuse shortwave flux at one side, which is, in addition, convectively coupled to a constant 26.7 °C. For both tests the analytical solution is given and compared with the corresponding code predictions.

Empirical validation

Data are established for an unoccupied house and a two-zone test cell. The empirical approach is outlined, the necessary data capture—to allow careful specification of model input and study of model output—is discussed, the solution to common monitoring problems outlined, and the comparisons to be applied between measurements and predictions are discussed.

Comparative studies

This is used in conjunction with the analytical and empirical methods to gain an insight into code operation and to test any existing claims of validity accompanying any candidate model. The test problem should be kept simple to ensure inter-code input equivalence. This method can precede the other methods to prove the need for, and then identify the target area of, any subsequent analytical or empirical study. It can also be used, after a model has been processed through analytical and empirical tests, as a final check against other models already validated.

At the present time (September 1984) a similar study is under way in the UK. The Science and Engineering Research Council (SERC) are funding university groups to research the problem of model validation at its fundamental level. Analytical test procedures will be developed, existing data sets will be examined and packaged for use and supplementary data sets proposed if necessary. The SERC funded work is intended to complement the other UK validation activities. For example, the Energy Modelling Group at the Building Research Establishment are active in model validation, especially analytical studies (Gough 1984), and the Energy Technology Support Unit, as part of its passive solar studies, is concerned in part with empirical validation in relation to test cells.

To demonstrate empirical validation in relation to a real building case, recall the IEA project mentioned earlier in this section. In many ways this project was a unique study: model representatives, the project participants, were given an opportunity to influence the empirical data collection through frequent meetings with the monitoring team; and throughout the model testing phase additional performance data could also be obtained by access to a working data logging system.

The temporal proximity of model testing and data collection also caused some difficulties. Model testing was expected to proceed in tandem with data collection and so often relied on incomplete data sets. Inevitably, logging errors were introduced to the modelling exercises before detection and, in the time available, it proved impossible to associate reliable accuracy estimates with the collected data. For these reasons—and there is little evidence to suggest that similar problems will not occur with existing and future data sets—monitored data had to be regarded with the same suspicion as model predictions. Indeed, on a number of occasions predictions highlighted areas of uncertainty which were subsequently traced to equipment malfunction. Also, the chosen building was a large, multi-storey, open plan office complex, making the specification of model input data (thermophysical properties, occupant action, air leakage paths, etc) a difficult task. Because of these problems, empirical validation was treated as a three-stage process in the case of the ESP system described in chapter 7. The technique applied to other participating models is described in the project report referenced previously.

First, a detailed building description was established which adhered as closely as possible to the consensus interpretation of the building specification as assessed from an input questionnaire completed by each participant. It is unlikely that any model will achieve immediate agreement with the monitored data because of input data inadequacy in a number of important areas. In the IEA project problem areas included zone coupling air flow between core and perimeter zones, the apportioning of electrical activity (which was centrally measured) between the various zones, and the presence of different energy supply equipment (heating and cooling) in the same zone. Because of such difficulties it is likely that any input data set will require some preliminary investigation in those areas that attract the greatest concern. In this way the relationship between predicted and monitored profiles can be adjusted (some would suggest that this is cheating!) until some 'best' agreement is obtained which corresponds to the most likely combination of system properties. Of course the number of simulations required to achieve some 'best' result is reduced as the problem is simplified or the input data specification tightened. Figure 8.2 shows one example profile pair which demonstrates the consensus obtained (for the ESP system) for an office core zone over warm July conditions. The departure of predictions from monitored values during the weekend period with no occupants and no cooling plant operation (30/31 July) is due to the use of non-integrated solarimeter readings as average time-step flux injections.

The second stage is concerned with the plausibility of these, and other, profiles since it is possible so to adjust parameters that profile consensus is achieved, for

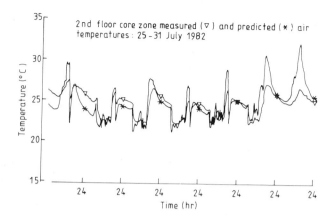

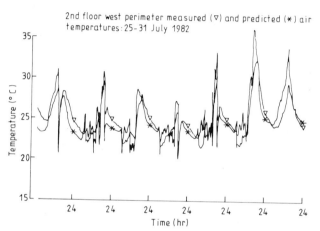

Figure 8.2 IEA Annex 4: prediction versus measurement for the ESP system.

each isolated zone, but at the expense of whole system energy balance. Figure 8.3 gives the energy balance for the air of the zone considered in figure 8.2. This shows the eight flowpaths deemed to cancel at the air point:

E1   is the convective energy transfer between all opaque surfaces (including zone contents) and the air point(s). Some dynamic processes (such as window direct solar penetration, transient conduction within the building fabric, internal longwave radiation and the radiant portion of casual gains) are included in this term since they each conspire to raise or lower surface temperatures to influence heat transfer by the convective flowpath. These processes can themselves be studied by constructing additional energy balances for surface nodes and for constructional nodes.

E2   is the convective casual gain from lights, occupants and equipment.

E3   is the portion of window absorbed solar energy which is transmitted to the zone air.

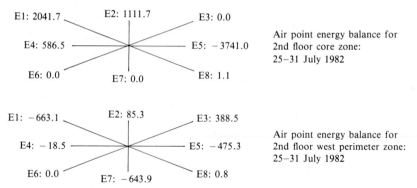

Figure 8.3 Energy balance information (kW h) for the predicted profiles of figure 8.2.

E4   is the energy transfer by inter-zone air flow.
E5   is the VAV system cooling injection to the zone air.
E6   is the infiltration energy exchange.
E7   is the window conduction if the air nodes are connected directly to ambient nodes by a window resistance. If windows are treated as multi-node systems (as walls are) then this term will be included in E1.
E8   is the air stored energy (or residual capacity) brought about by the fact that a nodal temperature difference exists between the beginning and end of the period for which the energy balance data were requested. The term is small in the case of air but will increase for energy balance applied to capacity regions.

Study of such energy balance data allows an assessment of the cause and effect relationships implicit in any model prediction and aids in the identification of dominant flowpaths in each of the models examined.

The third stage is then concerned to test the sensitivity of these energy balance flowpaths—and hence profile fit—by parametric study of the various design parameters. If sensitivity is low then some confidence can be placed in model predictions for the problem studied. If sensitivity is high, then further empirical or analytical study is indicated. Table 8.2 summarises the sensitivity of the overall predicted profile (of figure 8.2) to a number of parametric changes.

Many other monitoring and validation activities are currently under way worldwide. These include the IEA Annex 10 project (IEA3 1983) concerned with the validity of plant simulation techniques and the work of the EEC Passive Solar Modelling Group (EEC1 1984) to test and recommend models for passive solar architectural design applications. In some countries, organisations have implemented databases which hold information on the many monitoring exercises conducted to date. For example Lawrence Berkeley Laboratory (US) hold such a database in the energy field and the Air Infiltration Centre (UK) in the related air flow field.

Table 8.2 Summary of parametric investigation of building sensitivity (see Clarke *et al* 1984) relative to the profiles of figure 8.2.

1 *Solar radiation:* a 10% increase gives 1.8% increase in occupied space mean weekly temperature (relative to the standard ESP result reported in the above reference) or 3.4% increase in plenum value. Core profile shape unaffected; plenum shape modified during sun-up hours.

2 *Electrical activity:* a 10% increase gives 1% increase in occupied space mean weekly temperature or 1.5% increase in plenum value. Core and plenum profile shapes largely unaltered.

3 *Duct losses increased:* duct modulus β† = 0.35 gives 1.2% increase in occupied space mean weekly temperature relative to the $\beta = 0.25$ case with a corresponding 0.7% plenum reduction.

4 *Duct losses reduced:* duct modulus $\beta = 0.15$ gives 0.9% reduction in occupied space mean weekly temperature relative to the $\beta = 0.25$ case with a corresponding 0.8% plenum increase.

5 *No convective coupling between zones:* core mean weekly temperatures fall by 0.8% rising to 5.2% reduction in the north perimeter. The plenum falls by 3%. Core undercooled, perimeters suffer from lack of damping.

6 *Ventilated lighting fittings modified:* (10% rad. down, 39% rad. up, 51% con. up) insignificant effect on occupied space, plenum temperatures rise by 0.4%.

7 *Perfect mixing in occupied space:* 0.8% increase in core mean temperature, 0.7% increase in plenum.

8 *Additional plenum heat source:* 3% increase in mean plenum temperature results from an additional 5 kW (unspecified) constant power source.

9 *Removal of contents capacity:* insignificant effect on occupied space mean temperatures but large core temperature swings result. Plenum suffers 2.6% temperature drop.

10 *No thermal capacity:* 3.4% reduction in weekly mean temperature in occupied space, 1.5% reduction in plenum. Exaggerated temperature swings in all zones.

11 *Fixed surface convection coefficients:* 1% increase in mean weekly temperatures in occupied space, 0.4% reduction in plenum.

12 *Window U-value reduced:* 13% reduction in window U-value produces 2.8% increase in occupied space mean weekly temperatures.

13 *Single versus multiple zone treatment:* combining core and perimeter zones lowers whole zone and plenum predictions by 1.2% with core temperature profiles showing an improved fit at the expense of plenum values.

14 *Controller sensed temperature:* relative to a simulation controlled on the basis of pure air temperature, simulation control assuming a more likely sensed temperature (say 85% air and 15% mean radiant) will result in 10% cooling energy increase accompanied by a small rise in mean weekly air temperatures.

15 *Conductivity values increased by 5%:* small effect in mean values especially in core and plenum, 0.4% variation in perimeter zones.

---

† The factor β is defined in the IEA Annex 4 topic paper by Wooldridge (1983b).

8.2　Implementation in practice

Even when a sophisticated modelling package is nearing the commercial exploitation stage, effective uptake by practice can still be highly problematic. In addition to the usual cost barriers, a number of factors will contribute to this situation.

Firstly, the formulation of a design problem, in a manner conducive to simulation processing, is a skill often difficult to acquire in the hustle and bustle of professional life. At the present time the average designer is given little incentive to master the art of formulating objective functions or extracting the causal relationships inherent in the simulation predictions. As a consequence, only a small number of designers acquire the necessary skills to use models effectively. Unfortunately this is often achieved 'out of hours' and unfunded.

Secondly, there is little doubt that energy models (in particular) will expose misconceptions and misunderstandings on the part of the user. Such exposure can create barriers in attitudes which may persist thereafter.

Thirdly, even when simulation skills have been acquired, the designer is often frustrated by the lack of time and financial resources acting to limit the scope of a modelling exercise and prevent, completely, any fundamental evaluation.

Lastly, if insufficient attention has been paid to user interface matters, it is likely that model use will diminish with time.

Implementation (as placed in the development sequence of chapter 7) is concerned with the risk-underwritten trial of model robustness, relevance and ease of use in the real-world, real-time context of design practice. Whereas validation studies address the efficacy of the model's theoretical basis, implementation studies perform the complementary role of testing operational efficacy. Implementation trials therefore strive to improve, over time, model usefulness and, by demonstrating cost-effective application to a range of typical design problems, encourage effective proliferation. Two basic strategies exist to promote implementation trials:

By establishing demonstration project grants, practice can be encouraged to apply existing models to ongoing design projects in order to generate case study material from which model usefulness can be judged. Any financial risk— associated with terminal equipment, software access and additional staff time—is lessened in the first instance as user experience grows. Two projects of this kind have already been undertaken in the UK. In one, the ESP system was applied to six live projects undertaken by a large multidisciplinary design practice (Maver and Ellis 1982) and, over an eighteen month period, model performance and impact observed. As a result of this exercise, model interface improvements were made and the practice went on to become a sophisticated user of advanced energy modelling software. In another project the SUNCODE model (now SERI-RES in its public domain form) was used to evaluate a number of house design submissions as part of the Energy Technology Support Unit's (UK) passive solar demonstration programme (ETSU 1984).

An alternative implementation approach is to establish a professional service in the form of a network of energy modelling advisory centres. Operating advanced simulation software, such centres would offer training, technical advice and, perhaps, initial simulation services to those members of the design profession keen to assess the potential of the emerging technology. Initially the service would be subsidised by the professional institutions and/or Government, and so users would be protected from the hardware and software costs whilst investigating the new generation models. The viability of such a service has already been established by two independent market research exercises, commissioned by the Scottish Development Agency, addressing the need for such a service within the UK. A comprehensive report on the findings was produced (SDA1 1981) from which a cost development plan was drawn up (SDA2 1981).

If either or both of these low-risk (to practice) implementation scenarios are pursued—and backed by a national training programme and a timely open debate on the role of energy modelling, in conjunction with energy targeting, within building regulations—then (and only then, in the author's opinion) will the true potential of design appraisal by simulation be realised.

The short case study descriptions which follow are intended to demonstrate the range of building types, job sizes and design problems which will benefit from the introduction of simulation models within the design process. These descriptions relate to the application of the ESP system to projects now complete.

Case study 1: resource allocation
Many Government agencies own housing which dates from the early 1950s. In recent times much of this housing stock has fallen due for upgrading. ESP has been used by a number of architectural practices to help establish the most cost-beneficial upgrading strategy.

A sample of houses in any estate are analysed, firstly in their original form, and subsequently with a range of design modifications formulated on the basis of the flowpath rank ordering to emerge from the initial simulations. The analyses focus predominantly on the issues relating to construction, fenestration and, in particular, ventilation and condensation.

In one instance the major heat flowpath was identified as being through a suspended timber floor. This occurred in a building for which substantial wall insulation was proposed. As a result the upgrading proposal was reformulated and the client's investment put to better use.

Case study 2: building conversion
ESP was used by a large regional council in Scotland to perform a detailed analysis of design possibilities in a proposed conversion of school premises to new office headquarters. The study involved an in-depth investigation of diversity of heating demand as affected by a variety of proposed zoning strategies, plant operating schedules and building modifications. The resulting report, in addition to presenting the technical findings, raised the following general points:

The total cost incurred in the modelling exercise (including data preparation, computing time and consultancy charges) was estimated at half the cost incurred in a parallel manual exercise which involved only the straightforward calculation of individual zone peak heating loads under steady-state conditions.

The results from the simulation exercise allowed a more detailed analysis of both the building and plant performance than would otherwise have been possible. In particular, the ability to include building and plant features simultaneously and to vary plant operational regimes was considered an invaluable facility.

The graphical presentation of results was also considered invaluable as a means of conveying information to the design team.

Case study 3: passive solar architecture
In an attempt to pursue energy savings attention is being turned, worldwide, to the harnessing of solar energy as a supplementary and renewable fuel source. The technical problems inherent in such passive solar design are often considerable and require sophisticated modelling methods. In a number of projects ESP has been used to investigate the performance of buildings incorporating passive solar design features.

The studies focus on such issues as the performance of direct gain, attached sunspaces or Trombe walls, or attempt to optimise blind control, movable insulation, shading devices or selective coatings applied to window surfaces.

In one particular case an architectural practice was commissioned by a large house builder to investigate the feasibility of attached sunspaces (passive solar collectors) applied to houses in the British climate. A simulation scheme was established in which thermostatically controlled vents, connecting the sunspace to occupied spaces, operate to transmit excess energy from the sunspace. If the occupied spaces overheated, sunspace venting to ambient would occur. The passive solar scheme was simulated over an annual period and results compared with an identical simulation applied to the original house. By this means it was established that the passive solar feature would reduce annual energy requirements by approximately 30%. The simulation results were compared with corresponding data obtained from a parallel monitoring exercise undertaken by the house builder and found to compare favourably.

Case study 4: innovatory design
Dynamic modelling systems are also well tailored to the analysis of innovatory design schemes. In one case ESP was used to model a proposed solar wall construction which formed a major part of a multi-million pound laboratory complex.

The movement of large quantities of air (air demanded by the laboratory processes) had suggested to the design team a scheme in which, in an attempt to capture solar energy, this air was passed over the entire south-facing building facade and contained within an outer glass skin. The model was used to simulate this 'solar wall' in order to predict the potential annual pick-up of solar energy and the corresponding reduction of the south facade heat loss due to the insulating effect of the additional glass and air space combination.

Modelling of the proposed scheme was complicated by the inclusion of ducts within the air space which caused wall shading and convective heat pick-up. However, this could be incorporated within the model without difficulty.

Case study 5: load diversity

The design of a building to house a computer and ancillary activities for a nationalised industry was already well advanced when the opportunity arose to use ESP. The fully air conditioned building (of approximately 6000 m^2) was to be built on four floors, each 30 m wide by 50 m long; office accommodation would be housed on the perimeter with rooms of more occasional occupancy in the core.

The stimulus for use of a dynamic energy model came from the decision to alter the building envelope from a lightweight cladding system to brickwork, with an associated increase in glazing from 25% to 40% differentially arranged in the four floors. The effect of these changes on the variable air volume (VAV) system, on the distribution ductwork and on the central air handling plant needed, as a matter of urgency, to be determined.

The analysis was applied to modules sited on all four corners of the building and halfway along each facade. For each module on each floor the peak load across its VAV terminals was computed and the accumulative effect on the central plant determined. In relation to the climatic data used the peak load on all space modules was seen to occur on a high air temperature July day and not, as had been previously assumed, within the month of September, when solar angles are lower. In addition the results showed that the loading in different spaces occurred at different times throughout the critical day; as a consequence, although individual VAV terminal duties had to be increased, no significant increase in load would be experienced by the central plant.

Case study 6: atria air flow

Accompanying improved living standards is an increased demand for leisure and shopping facilities. Many of the proposed designs incorporate extensively glazed roofs, especially over pedestrian malls with domed atria features. Because of the pressure and stack effects, the prediction of environmental conditions will depend largely on the temperature regime and on the distributed leakage. The key to accurate modelling is then the ability to handle energy and mass balance simultaneously.

In one particular application the intention was to create a large shopping, office and recreation complex within an outer glass construction forming a central atrium with high level walkways adjacent to the glass skin. To assess performance, the following information was required for 'typical' and 'design' climatic periods during summer and winter conditions: the internal air flow patterns and velocities and the effect of varying louvre and door openings; the effects of solar penetration and the optimum operation of a movable reflective blind arrangement distributed over the entire south facing glass skin; and the maximum and minimum temperatures in the offices and walkways and vertical gradients within the atrium. Based on detailed, and simultaneous, air flow and energy simulation the above issues were addressed to allow decisions to be taken on venting strategy, solar control and mall heating by office exhaust.

Case study 7: electrical storage simulation
In many modelling applications it is essential to treat building and heating systems simultaneously to preserve the dynamic interaction. In one project, it was required to simulate the performance of fan-assisted electrical storage units to be incorporated within a building's conversion to small factory workshop units. A computer description of the building was established, which included a detailed model of the electrical equipment, and a number of simulations performed to investigate such issues as comfort, unit control, charge/discharge ratios, weather anticipator efficiency and running costs. It was shown that the current off-peak electrical storage technology is technically capable of achieving satisfactory comfort conditions, provided a reasonable level of automatic output control is incorporated within each unit. On the other hand, unit charge control is often simplistic, resulting in inefficient charge/discharge ratios.

Case study 8: parametric studies
Parametric analysis of the energy performance of buildings is of fundamental importance to a more complete understanding of the issues relating to energy efficient building design, in order to increase the corpus of knowledge upon which subsequent designs may be conceived. ESP has been used as the technical 'engine' in a number of such studies; in one the model was used to study the cost benefit associated with different window designs in terms of their performance in both Britain and Scandinavia. Annual simulations were performed for a number of combinations of orientation, window size, window type and structural mass, sampled from the many thousands of combinatorial possibilities. Internal gains, solar penetration and window curtain operation was included and window performance judged against energy and comfort criteria.

Case study 9: plant simulation
Simultaneous building/plant modelling is a new development field which offers much promise for the study of advanced systems design and control. In one such study the design team wished to compare a conventional air conditioning design—VAV cooling, perimeter zone heating and with air extract to a ceiling plenum across zone lighting fittings to encourage heat recovery—with a variational approach in which VAV injection is to a subfloor plenum with subsequent exchange between this plenum and the occupied space (the KRANTZ system).

Issues to be studied included the effects on energy consumption and comfort conditions of a number of control options applied to both systems. Explicit multi-component plant models were established for both systems and combined with a multi-zone building model comprised only of the zones of interest. Both configurations were then subjected to various control regimes and subjected to short- and long-term simulation. Regimes studied included: intermittent versus continuous plant operation; solar reset of perimeter heating; and the use of night-time charging of the floor plenum with outside air in anticipation of next day cooling demand.

In each case the impact on central plant consumption was determined and assessed relative to the level of comfort achieved in the occupied zones.

Case study 10: future possibilities
It is not difficult to imagine a number of futuristic application possibilities. For example, with the emergence of powerful but low-cost computing systems it is possible to incorporate prediction software within the overall building/plant control system. Climatic data and spatial requirements are continuously fed to the model to allow, in the context of building and plant descriptions already established, an anticipation of the consequences of any proposed control inter-action. Thus, the control system has inbuilt predictive capability and so can anticipate future reality and take any corrective action in advance. Such heuristic controllers are at present in their infancy.

There is little doubt then, that model applications of the kind cited will form the cornerstone of future design practice. I would like to conclude this book by quoting Tom Maver, Professor of CAD at the University of Strathclyde:

> Today's justifiable concern with the energy-conscious design of buildings is a recurring echo of a fundamental theme dating back to the first human settlements: the need for *shelter* from an inclement environment. Of the factors which determine the degree of effective shelter provided by a building, none are more important, nor more worthy of the designer's consideration, than its *form*, *fabric* and *operation*; but these are also the determinants, for better or worse, of the entire range of attributes which make up the quality of the built environment—from the life-cycle cost of the building to its visual impact on the site.
>
> The exciting prospect which emerges, then, is of an integrated computer-based system in which the exploration of form, fabric and operation leads to a design solution embodying a balance between cost and performance, between investment and return, between need and aspiration.
>
> The new generation of dynamic energy models exemplifies the benefits which a sustained intellectual commitment can bring to the quality of the built environment. It is to be hoped that this endeavour will itself be a model, and an inspiration, for future developments in computer-aided design.

References

Arumi F N 1979 *The DEROB System: User's Manual* vols **1**, **2** (Austin, Texas)

Clarke J A 1982 ESP Documentation Set *ABACUS Publication* (Glasgow: University of Strathclyde)

Clarke J A and McLean D J 1984 Results from the Analysis of the Collins Building Second Floor and Plenum by the ESP System *Appendix C8, IEA Annex 4 Final Report, October 1983*

DOE 1980 DOE-2 User's Guide (Version 2.1) *Rep. no LBL-8689, Rev. 1* (Berkeley, Calif.: Lawrence Berkeley Laboratory)

Delorme Y and Trokay P 1979 Application des Facteurs de Reponse à l'étude Dynamique d'ensemble de Locaux; Description du Program LPB-1 *Rep. 78, LPB* (Liège: University of Liège)

EEC1 1984 Commission of the European Communities DG XII, *Final Report of Project A: Modelling of Passive Solar Architecture*, prepared by the Passive Solar Modelling Sub-Group

ETSU 1984 Various Reports (AERE Harwell, England; Energy Technology Support Unit)

Gough M 1984 Proposals for the Analytical Testing of Energy Models *Internal Rep.* (Building Research Establishment)

Hittle D C 1979 The BLAST Program (Version 2.0) User's Manual *Report (TR) E-153/ADA072272 (vol 1) and ADA0722730 (vol 2)* (US Army Construction Engineering Research Laboratory)

IEA1 1983 *Specification of the Collins Building Monitoring Scheme: IEA Annex 4* (Glasgow: University of Glasgow, Building Services Research Unit)

IEA2 1983 *IEA Annex 4: Glasgow Commercial Building Monitoring Project, Final Rep.* (Building Research Establishment)

IEA3 1983 *IEA Annex 10:* various publications, Laboratoire de Physique du Batîment (Liège: University of Liège)

Judkoff, Wortman, O'Doherty and Burch 1983 A Methodology for Validating Building Energy Analysis Simulations *SERI Rep.*

Kallblad K and Higgs F 1981 Building Energy Use Modelling in Sweden by Jullotta *Proc. 3rd Int. Conf. on Energy Use Management, October 1981*

Maver T W and Ellis J 1982 Implementation of an Energy Model within a Multidisciplinary Practice *Proc. CAD82, March 1982*

Palmiter L and Wheeling T T 1979 *SUNCAT 2.4 User Notes* (BUTTEMT Centre for Appropriate Technology, April 1979)

SDA1 1981 *Market Analysis Study for a Proposed Energy Advisory Service, Scottish Development Agency Report* (Glasgow)

SDA2 1981 *Cost Development Plan for a Proposed Energy Advisory Service, Scottish Development Agency Report* (Glasgow)

Tas° 1984 *Tas° User Manual* (Milton Keynes: Amazon Energy)

TRNSYS 1979 *TRNSYS Users Manual* (University of Wisconsin: Solar Energy Laboratory)

Wooldridge M 1983a *BUNYIP Users Manual* (CSIRO, Australia)

—— 1983b Duct Gains in the Variable Air Volume System of the Collins Building *Topic Paper 6, IEA Annex 4 Final Report, October 1983*

Appendix A

Fourier heat equation

Consider an elemental volume of some homogeneous material as shown in figure A.1. The heat flow into the elemental volume in the x direction is given by

$$dq_x = -k \, dy \, dz \, \delta\theta/\delta x$$

where k is the material conductivity (W m^{-1} $^{\circ}$C^{-1}) and θ is the temperature ($^{\circ}$C).

The corresponding heat flow out of the elemental volume in the x direction is given by

$$dq_{x+dx} = -k \, dy \, dz \, \frac{\delta[\theta + (\delta\theta/\delta x) \, dx]}{\delta x}.$$

Therefore the net rate of heat flow into the element in the x direction is:

$$dq_x - dq_{x+dx} = k \, dx \, dy \, dz \, \delta^2\theta/\delta x^2$$

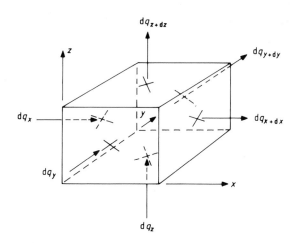

Figure A.1 An elemental homogeneous finite volume.

and in the limit this becomes

$$k \, dx \, dy \, dz \, \partial^2\theta/\partial x^2.$$

Similarly for the y and z directions:

$$dq_y - dq_{y+dy} = k \, dx \, dy \, dz \, \partial^2\theta/\partial y^2$$
$$dq_z - dq_{z+dz} = k \, dx \, dy \, dz \, \partial^2\theta/\partial z^2.$$

Thus the total net rate of heat flow into the element is

$$k \, dx \, dy \, dz \left(\frac{\partial^2\theta}{\partial x^2} + \frac{\partial^2\theta}{\partial y^2} + \frac{\partial^2\theta}{\partial z^2} \right).$$

It is also possible to have heat generation within the elemental volume and further, with unsteady conduction where temperature is dependent on time, heat will be stored within the elemental volume.

If q is the rate of heat generation per unit volume of the element, then the rate of heat generation is given by $q \, dx \, dy \, dz$. The rate of heat storage is governed by the temperature gradient and is given by

$$dx \, dy \, dz \, \varrho C \frac{\partial\theta}{\partial t}$$

where ϱ is the material density (kg m^{-3}), C the material specific heat (J kg^{-1} °C^{-1}) and t the time (s).

Now the rate at which heat is being stored (or released) within the elemental volume is equal to the net rate of heat flow plus the rate of heat generation (or extract). That is:

$$\varrho C \frac{\partial\theta}{\partial t} = k \left(\frac{\partial^2\theta}{\partial x^2} + \frac{\partial^2\theta}{\partial y^2} + \frac{\partial^2\theta}{\partial z^2} \right) + q$$

$$\Rightarrow \quad \frac{\partial\theta}{\partial t} = \alpha \left(\frac{\partial^2\theta}{\partial x^2} + \frac{\partial^2\theta}{\partial y^2} + \frac{\partial^2\theta}{\partial z^2} \right) + \frac{q}{\varrho C} \tag{A.1}$$

where α is the thermal diffusivity which is a property of the material; the larger its numerical value, the more rapidly is a temperature change propagated.

Equation (A.1) is known as the Fourier field equation in three space dimensions and with heat generation. For a steady state system with heat generation, equation (A.1) reduces to the Poisson equation:

$$\frac{\partial^2\theta}{\partial x^2} + \frac{\partial^2\theta}{\partial y^2} + \frac{\partial^2\theta}{\partial z^2} + \frac{q}{k} = 0$$

and for the case with no heat generation, the resulting formulation is the Laplace equation:

$$\frac{\partial^2\theta}{\partial x^2} + \frac{\partial^2\theta}{\partial y^2} + \frac{\partial^2\theta}{\partial z^2} = 0.$$

The Fourier equation in one space dimension and with heat generation is:

$$\frac{\partial \theta}{\partial t} = \alpha \frac{\partial^2 \theta}{\partial x^2} + \frac{q}{\varrho C}. \qquad (A.2)$$

The Fourier equation is the basis of the transient conduction element of most building energy prediction techniques and a variety of methods exist to achieve solution against appropriate initial and boundary conditions. In a few simple cases straightforward analytical solutions are possible but these tend to be unrealistic (due to the restrictions imposed by the special boundary conditions assumed in order to permit the analytical solution in the first place) and so difficult to apply in practice.

Appendix B

Thermophysical properties of building materials

ESP Construction database : Units - Conductivity W/mdeg.c
Density Kg/m**3
Specific Heat J/kgdeg.C

	CON.	DEN.	SHT.	EMIS.	SABS.
Classification description - ASBESTOS					
CEMENT	0.36	1500.	1000.	0.90	0.60
CEMENT SHEET	0.36	700.	1050.	0.96	0.60
ASBESTOS SHEET	0.16	2500.	1050.	0.90	0.96
INSULATION	0.16	577.	840.	0.90	0.60
Classification description - ASPHALT & BITUMEN					
BITUMEN FELT	0.50	1700.	1000.	0.90	0.90
ROOFING FELT	0.19	960.	837.	0.90	0.90
ASPHALT MASTIC ROOFING	1.15	2325.	837.	0.90	0.90
ASPHALT	1.20	2300.	1700.	0.90	0.90
BITUMEN COMPOSITION (FLOORS)	0.85	2400.	1000.	0.90	0.90
BITUMEN IMPREGNATED PAPER	0.06	1090.	1000.	0.90	0.90
Classification description - BRICK					
PAVIOUR	0.96	2000.	840.	0.93	0.70
BREEZE BLOCK	0.44	1500.	650.	0.90	0.65
INNER LEAF	0.62	1800.	840.	0.93	0.70
OUTER LEAF	0.96	2000.	650.	0.90	0.93
VERMICULITE INSULATING BRICK	0.27	700.	840.	0.90	0.65
Classification description - CARPET					
WILTON	0.06	186.	1360.	0.90	0.60
SIMULATED SHEEPS WOOL	0.06	198.	1360.	0.90	0.60
WOOL FELT UNDERLAY	0.04	160.	1360.	0.90	0.65
CELLULAR RUBBER UNDERLAY	0.10	400.	1360.	0.90	0.65
SYNTHETIC(DUNLOP TYPE)CARPET	0.06	160.	2500.	0.90	0.65

	CON.	DEN.	SHT.	EMIS.	SABS.

Classification description - CONCRETE

	CON.	DEN.	SHT.	EMIS.	SABS.
LIGHT MIX	0.38	1200.	653.	0.90	0.65
AERATED CONCRETE BLOCK	0.24	750.	1000.	0.90	0.65
AERATED CONCRETE	0.16	500.	840.	0.90	0.65
REFRACTORY INSULATING	0.25	10.	837.	0.90	0.65
VERMICULITE AGGREGATE	0.17	450.	837.	0.90	0.65
NO FINES CONCRETE	0.96	1800.	840.	0.90	0.65
FOAMED SLAG CONCRETE	0.25	1040.	960.	0.90	0.65
CONCRETE BLOCK INNER (3%MC)	0.51	1400.	1000.	0.90	0.65
FOAMED BLOCK INNER (3%MC)	0.16	600.	1000.	0.90	0.65
FOAMED BLOCK OUTER (5%MC)	0.17	600.	1000.	0.90	0.65
GLASS REINFORCED CONCRETE	0.90	1950.	840.	0.90	0.65
HEAVY MIX	1.40	2100.	653.	0.90	0.65

Classification description - EARTH

	CON.	DEN.	SHT.	EMIS.	SABS.
INFUSORIAL 9% M.C.	0.09	480.	180.	0.90	0.85
GRAVEL BASED	0.52	2050.	184.	0.90	0.85
COMMON EARTH	1.28	1460.	879.	0.90	0.85

Classification description - GLASS

	CON.	DEN.	SHT.	EMIS.	SABS.
GLASS BLOCK	0.70	3500.	837.	0.25	0.05
PLATE GLASS	0.76	2710.	837.	0.25	0.05
4MM CLEAR FLOAT	1.05	2500.	750.	0.25	0.05
6MM ANTI-SUN	1.05	2500.	750.	0.59	0.06

Classification description - INSULATION MATERIAL

	CON.	DEN.	SHT.	EMIS.	SABS.
FIBREBOARD	0.06	300.	1000.	0.90	0.50
WOODWOOL	0.10	500.	1000.	0.90	0.50
UREA FORMALDEHYDE FOAM	0.03	30.	1674.	0.90	0.50
THERMALITE	0.19	753.	837.	0.90	0.70
POLYURETHANE FOAM BOARD	0.03	30.	837.	0.90	0.50
SIPOREX	0.12	550.	1004.	0.90	0.40
P.V.C.	0.16	1379.	1004.	0.90	0.60
HARD RUBBER	0.15	1200.	1000.	0.94	0.92
CRATHERM BOARD	0.05	176.	837.	0.90	0.50
SILCON	0.18	700.	1004.	0.90	0.60
GLASSWOOL	0.04	250.	840.	0.90	0.30
ROOF INSULATION BOARD	0.19	960.	950.	0.90	0.55
FELT SHEATHING	0.19	960.	950.	0.90	0.90
EXPANDED POLYSTYRENE (EPS)	0.03	25.	1000.	0.90	0.30
EXPANDED PVC	0.04	55.	1000.	0.90	0.60
MINERAL FIBRES	0.04	100.	750.	0.90	0.60
CORK	0.04	105.	1800.	0.90	0.60
THATCH (STRAW)	0.07	240.	180.	0.90	0.50

Classification description - METAL

	CON.	DEN.	SHT.	EMIS.	SABS.
COPPER	200.00	8900.	418.	0.72	0.65
STEEL	50.00	7800.	502.	0.12	0.20
ALUMINIUM	210.00	2700.	880.	0.22	0.20

	CON.	DEN.	SHT.	EMIS.	SABS.

Classification description - PLASTER

DENSE	0.50	1300.	1000.	0.91	0.50
LIGHT	0.16	600.	1000.	0.91	0.50
PERLITE PLASTERBOARD	0.18	800.	837.	0.91	0.60
GYPSUM PLASTERING	0.42	1200.	837.	0.91	0.50
PERLITE PLASTERING	0.08	400.	837.	0.91	0.50
VERMICULITE PLASTERING	0.20	720.	837.	0.91	0.50
GYPSUM PLASTERBOARD	0.19	950.	840.	0.91	0.50

Classification description - SCREEDS & RENDERS

LIGHTWEIGHT CONCRETE	0.41	1200.	840.	0.90	0.80
CAST CONCRETE	1.28	2100.	1007.	0.90	0.65
GRANOLITHIC	0.87	2085.	837.	0.90	0.65
CEMENT SCREED	1.40	2100.	650.	0.91	0.65
WHITE RENDER DRY	0.50	1300.	1000.	0.91	0.50
RENDERING (1%MC)	1.13	1431.	1000.	0.91	0.50
RENDERING (8%MC)	0.79	1329.	1000.	0.91	0.50

Classification description - STONE

SANDSTONE	1.83	2200.	712.	0.90	0.60
GRANITE (RED)	2.90	2650.	900.	0.90	0.55
MARBLE (WHITE)	2.00	2500.	880.	0.90	0.46
LIMESTONE	1.50	2180.	720.	0.90	0.60
SLATE	2.00	2700.	753.	0.90	0.60
GRAVEL (GENERAL)	0.36	1840.	840.	0.90	0.60

Classification description - TILES

CLAY	0.85	1900.	837.	0.90	0.60
CONCRETE	1.10	2100.	837.	0.90	0.65
SLATE	2.00	2700.	753.	0.95	0.85
PLASTIC	0.50	1050.	837.	0.90	0.40
RUBBER	0.30	1600.	2000.	0.94	0.82
CORK	0.08	530.	1800.	0.90	0.60
ASPHALT/ASBESTOS	0.55	1900.	837.	0.90	0.70
P.V.C./ASBESTOS	0.85	2000.	837.	0.90	0.60
TILE BEDDING	1.40	2100.	650.	0.90	0.60
CEILING (MINERAL)	0.03	290.	2000.	0.90	0.60
CEILING (PLASTER)	0.38	1120.	840.	0.90	0.60

Classification description - WOOD

BLOCK	0.16	800.	2093.	0.90	0.65
HARDBOARD (MEDIUM)	0.08	600.	2000.	0.91	0.70
HARDBOARD (STANDARD)	0.13	900.	2000.	0.91	0.70
FIR (20% MOIST)	0.14	419.	2720.	0.90	0.65
FLOORING	0.14	600.	1210.	0.91	0.65
CORK BOARD	0.04	160.	1888.	0.90	0.60
CHIP BOARD	0.15	800.	2093.	0.91	0.65
WEATHERBOARD	0.14	650.	2000.	0.91	0.65
OAK (RADIAL)	0.19	700.	2390.	0.90	0.65
PLYWOOD	0.15	560.	2500.	0.90	0.65
SOFTWOOD	0.13	630.	2760.	0.90	0.65
PLYWOOD	0.15	700.	1420.	0.90	0.65
SOFTBOARD	0.55	350.	1000.	0.90	0.65

Appendix C

Alternative discretisation schemes

There are a number of alternative construction discretisation techniques which rely on thermal rather than geometrical criteria to achieve nodal placement. In §3.1 nodes were placed at the centre and boundary of each homogeneous element within a multi-layered construction. This means that the construction will be replaced by $2n + 1$ homogeneous and mixed property finite volumes, where n is the total number of homogeneous layers. Such a scheme is simple to implement, can achieve accuracy by simple layer subdivision, and ensures that interfacial temperatures and heat fluxes are available for output. Two alternative methods are outlined here. The first attempts to reduce the processing burden by combining two or more layers into a single equivalent layer. Thus, a wall of 6 homogeneous layers could be reduced to an m layer equivalent, where m can vary between 1 and 5. In the second method, the multi-layered construction is restructured to give a scheme in which each layer has identical thermal characteristics.

Layer combination
Mackey and Wright (1946) give a method for combining two or more homogeneous elements into an equivalent homogeneous layer:

$$(k\varrho c)_e = \frac{1.1 R_i (k\varrho C)_i + \sum_m 1.1 R_m (k\varrho C)_m}{R_e} + \frac{(k\varrho C)_o}{R_e} \left(R_o - 0.1 R_i - \sum_m 0.1 R_m\right)$$

where k is the conductivity ($\mathrm{W\,m^{-1}\,^\circ C^{-1}}$), ϱ is density ($\mathrm{kg\,m^{-3}}$), C is specific heat ($\mathrm{J\,kg^{-1}\,^\circ C^{-1}}$), R is resistance ($\mathrm{m^2\,^\circ C\,W^{-1}}$) and subscripts e, i, m and o refer to the equivalent element, and the innermost, intermediate and outermost elements for concatenation. If two elements are being combined, then obviously $m = 0$. If an intermediate element is an air gap then the corresponding $(k\varrho C)$ term is set to zero. Further, if the resistance of the outermost layer is small in comparison with the other layers, then the second term of this equation may become negative.

In this event the term is neglected and so the equivalent layer is defined by

$$(k\varrho C)_e = \frac{1.1R_i(k\varrho C)_i + \sum\limits_m 1.1R_m(k\varrho C)_m}{R_e}$$

The equivalent resistance is found from

$$R_e = \frac{x_o}{k_o} + \sum_m \frac{x_m}{k_m} + \frac{x_i}{k_i} \tag{C.1}$$

where x is the layer thickness (m) and x/k is R_o, the combined convective/ radiative resistance, in the case of an air gap.

The equivalent (ϱC) value is then obtained from

$$(\varrho C)_e = \frac{(k\varrho C)_e R_e}{(x_o + \sum\limits_m x_m + x_i)}. \tag{C.2}$$

Equations (C.1) and (C.2) can now be used to establish some equivalent layering scheme (1 layer, 2 layers, etc), which is then made discrete by any preferred nodal placement technique.

Dwell time restructuring

The dwell time of a homogeneous layer is defined as

$$t_d = (x^2 \varrho C)/k.$$

One procedure is to restructure a multi-layered construction such that the square root of the dwell time across each new layer is the same. This is the procedure adopted by Merriam *et al* (1982) in their EMPS 2 program.

The average value for the square root of the dwell time is determined from

$$\langle t_d^{1/2} \rangle = \left(\sum_{i=1}^{N} \frac{x_i(\varrho C)_i^{1/2}}{k_i^{1/2}} \right) \bigg/ N$$

where N is the total number of homogeneous layers. Commencing at layer 1, the individual dwell time square roots are summed until

$$\sum_i \frac{x_i(\varrho C)_i^{1/2}}{k_i^{1/2}} \geq \langle t_d^{1/2} \rangle.$$

When this condition occurs, the dwell time square root, for the last layer to be included in the summation, is subtracted and the layer subdivided until the average dwell time square root value is obtained. The technique then proceeds until the entire construction has been processed. It should be noted that after restructuring there may be more layers present than existed in the original construction. As before, any favoured nodal scheme can be applied, but now the nodes will more closely match the distribution of construction capacity.

References

Mackey C O and Wright L T 1946 Periodic Heat Flow — Composite Walls or Roofs *Heating, Piping and Air Conditioning* **18** 107–10

Merriam R L, Rancatore R J and Purcell G P 1982 EMPS 2: A Computer Program for Residential Building Energy Analysis *Proc. Int. Conf. Systems Simulation in Buildings* University of Liége Dec 1982

Appendix D

Nomenclature

Upper case

A	area, total absorption, finite difference equation coefficients
$A(p)$ etc	multi-layered construction transfer functions in the Laplace domain
C	finite difference equation coefficients, pressure coefficient, specific heat, ventilation conductance
CC	cloud cover factor
CRR	capacity rate ratio
E	coil effectiveness, illumination, z transfer function
F	Fourier number, light flux
Gr	Grashof number
I	a node of interest, radiation intensity
IRC	internally reflected component of daylight factor
K	absorption extinction coefficient, heat flow conductance, proportional gain factor
L	Laplace operator, thickness of multi-layered construction, convective load correction factor, luminance
L^{-1}	inverse Laplace transform
M	conduction transmission matrix
NTU	number of heat transfer units
Nu	Nusselt number
P	pressure, shading factor
Pr	Prandtl number
Q	total heat transfer
R	combined convective/radiative resistance, radiosity, sky radiation
RF	response factor
S	solar gain factor
SC	sky component of daylight factor
SHR	sensible heat ratio
T	action time
U	overall thermal transmittance

V	velocity, volume
W	fuel supply rate, precipitable water content, weighting factor
X, Y, Z	heat flux unit response functions
Y	year day number
Z	overall impedance

Lower case

a	absorptivity, admittance, altitude, finite difference equation coefficients at $t + \delta t$
b	finite difference equation coefficients at t
d	characteristic dimension, declination, decrement
d/dx	differential coefficient with respect to space dimension
d/dt	differential coefficient with respect to time dimension
$d\omega$	infinitesimal element of solid angle
e_t	equation of time
f	a function, diffuse view factor
f'	first derivative
g	gravitational constant, humidity ratio
h	enthalpy, heat transfer coefficient, height
i	incidence angle
i, j	complex operator
k	conductivity
l	fin length, thickness of homogeneous element
m	air mass, homogeneous element transfer function, mass flowrate
n	surface normal
p	complex number associated with Laplace transform
q	fuel heat content, heat flux per unit area
r	distance, ramp function, reflectivity
s	surface factor
$t + \delta t$	future time-row
v	volume flowrate

Greek symbols

α	shortwave absorptivity, diffusivity, a weighting factor
β	coefficient of expansion, bypass factor, surface elevation angle
γ	a large number, an arbitrary variable
δ	a small difference, change of phase
δt	time-step
$\nabla^2\theta$	div(grad θ)
ϵ	emissivity factor, error
θ	angle, temperature, present time
λ	wavelength

μ	fluid viscosity, index of refraction
ξ	an arbitrary point in time, pseudo solar angles
ϱ	density, reflectivity
σ	Stefan–Boltzmann constant, surface tension
τ	transmissivity, turbidity coefficient
$\boldsymbol{\phi}$	state variable vector
ϕ	time lag or lead
ω	angular frequency, propagation direction, surface–solar azimuth

Subscripts

C	convection
I	infiltration
K	conduction
L	convective casual gains
M	radiant casual gains
R	longwave radiation
S	shortwave radiation
ao	outside air temperature
b	surrounding buildings, boiling convection
c	capacity node, construction cavity state, control condition, convection condition, cooler condition
$d\beta$	direct solar on surface of inclination β
dh	direct horizontal solar
dt	direct solar beam transmission
ei	internal environmental temperature
f	free stream
f?	fluid node ?
fc	opaque material conduction
fg	of vaporisation
fs	opaque surface solar gain
g	ground condition
gc	transparent material conduction
h	a harmonic, humidifier state
mrt	mean radiant temperature
n	neutral height
r	angle of refraction, radiant component, return condition
$r\beta$	ground reflected to surface of inclination β
res	resultant temperature
rv	ground reflected to vertical
s	saturation state, sensible component
$s\beta$	sky diffuse solar to surface of inclination β
sc	screen temperature, solar constant
se	outside surface
si	inside surface

v	vapour state
w, l	liquid
z	zenith angle
∥	parallel component
⊥	perpendicular component

Superscripts

−	mean condition
d	dry air state
~	cyclic condition

Appendix E

Description of the ESP system

INTRODUCTION

This appendix summarises the structure of the ESP (Environmental Systems Performance) building/plant energy simulation system. The program modules which comprise the system are described and the application potential of ESP outlined. The machine environment required by ESP is then discussed and the various documents, which detail each facet of the system, listed.

E1.1 Purpose of ESP

ESP is a transient energy simulation system capable of modelling the energy and mass flows within combined building and plant systems. Any building, defined as a collection of interlocking polyhedral zones, specified in terms of geometry, construction and usage profiles, can be associated with a plant system consisting of a distributed network of plant components. The combined system can then be subjected to simulation processing under dynamic control. ESP is as applicable to existing buildings as it is to proposed new designs incorporating traditional or advanced technological features. The system has been designed to operate in 'interactive graphics' and 'automatic processing' modes and has sophisticated input/output facilities to enable the designer to address such design questions as:

-   What, and when, are the peak building or plant loads and what are the various causal contributions ?

-   What will be the effect of some design change, such as increasing wall insulation, changing the glazing type or distribution, re-zoning the building, re-configuring the plant system or changing the building/plant control schedules ?

-   What is the plant optimum start time ?

-   How will comfort levels vary throughout the building ?

-   How will temperature stratification, in terms of sensor location and zone terminal unit position, affect energy consumption ?

- What are the building infiltration and zone-coupled air flows and how can they be minimised ?

- What is the contribution to energy saving of a range of passive solar features ?

- What is the optimum arrangement of constructional elements to encourage load levelling and hence efficient plant operation ?

- What are the energy consequences of non-compliance with prescriptive energy regulations or, conversely, how should a design be modified to come within some deemed-to-satisfy performance target ?

and so on. This allows the designer to identify potential problem areas, appropriate building and plant modifications, energy saving operational strategies, comfort levels, fabric performance, condensation risks and, most importantly, the interrelation between design and performance parameters, which can be carried forward to influence future projects.

E1.2 Structure of ESP

Figure E1.1 shows the connections between the various program and database modules which comprise ESP. These include:

ESP imp A program which allows the interactive definition of the combined building/plant network to be subjected to some weather influence and simulated over time. All items, as input, are processed through legality and 'within acceptable range' checks before being located in an output file for transfer to the simulation engine ESPsim. Other facilities include a range of interactive editting commands and automatic access to construction, casual gain profile and plant component databases.

ESP con A program to manage (create, modify, delete, list) a primitive constructions database containing the thermophysical properties (conductivity, density, specific heat, solar absorptivity and emissivity) of a number of miscellaneous homogeneous elements. This program will also allow the creation and editting of a second project related database containing composite multilayered constructions formed from elements extracted from the primitives database.

ESP pro A program to manage a casual gain profiles database containing any number of standard or project specific profiles defining the time variation in occupancy, lighting levels, etc.

ESP plt A program to manage the plant components database containing component descriptions, component models (in matrix differential equation form), and essential manufacturers data.

ESP shd A program to predict the time-series shading of external opaque and transparent surfaces as caused by surrounding site and facade obstructions.

ESP ins A program to predict the time-series insolation of internal opaque and transparent surfaces as caused by solar penetration through windows.

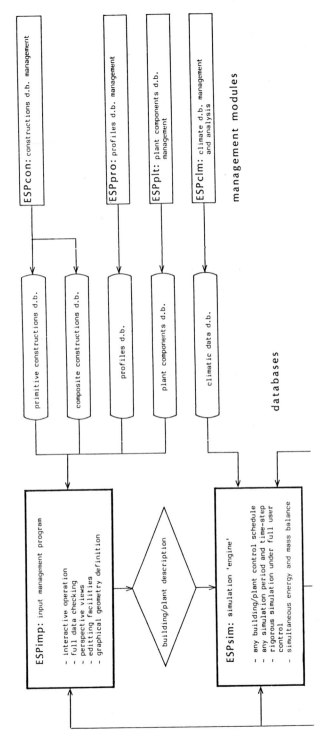

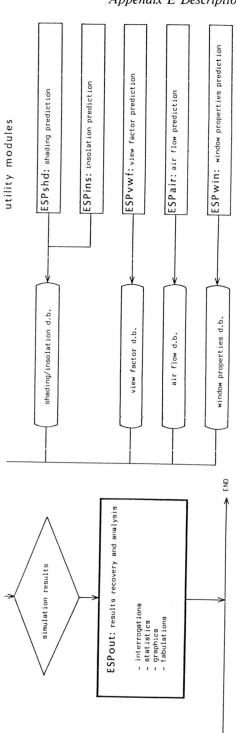

Figure E.1 The ESP system.

ESPvwf   A program to compute view factors between the surfaces bounding any building zone.

ESPair   A program which simulates the flow of air throughout a multizone system, connected by flow resistances representing windows, doors, cracks, etc, subjected to boundary wind loading and under the action of buoyancy effects.

ESPwin   A program to perform a spectral analysis of a multilayered window system to determine spectral transmittance, reflectance and absorptance.

ESPclm   A program to manage the climatological database allowing creation (by prediction), modification and analysis of the hourly time-series values of the climatic parameters retained.

ESPsim   Is the central simulation 'engine' which predicts building and plant energy flows and building air flows by a rigorous first principle modelling technique. As the program steps through small time increments, it continuously subdivides the building/plant network into a number of finite volumes. By applying energy balance techniques, unconditionally stable equation structures are produced, which can be solved by special matrix processing software. The techniques used for equation formulation and solution are described in detail elsewhere [1].

ESPout   Is the output program which operates on the simulation results located in a results database by ESPsim. A variety of output options are available: visualisations, result interrogation, graphical display, statistical interpretation and tabulations.

Thus the three main modules imp, sim and out can be used to investigate performance and, by iteration, to assess the consequence of any change to the building or plant design or control. The various databases are used to reduce the input task and the utility modules exist to allow the subsystems they address to br treated with greater rigour, if this can be justified by the design objectives in hand.

E1.3 Machine Environment

The ESP system requires the following hardware:

- A graphics terminal, preferably with Tektronix emulation. Nongraphical terminals can be used but are not recommended.

- A micro/mini-computer or mainframe, capable of supporting programs using up to 500Kbytes of memory.

- Hard disks (not floppy) with capacity greater than 3.8Mb to allow the storage of:

a   Module code (object only), 1.8Mb

b   Input data (including climate data), 0.5Mb

c   Results (at least) 1.5Mb

Table E.1 The ESP user documentation set.

Section	Contents
1	This section: describes the ESP system, hardware and software environment, and details the associated documentation.
2	General system operation: outlines a strategy for system application to example design problems.
3	ESP data structures: describes the creation and management of the required data sets via the input management module ESPimp.
4	Constructions database: describes the operation of ESPcon, to manage a primitives construction database and create multilayered constructions.
5	Casual gain profiles database: describes the creation and management of representative profiles via the module ESPpro.
6	Plant components database: management of existing plant components and insertion of new components via module ESPplt.
7	Climatological database: management, interrogation and severity analysis of climate data via module ESPclm.
8	Shading prediction: describes time-series analysis if external surface shading via module ESPshd.
9	Insolation prediction: describes time-series analysis of internal surface insolation via module ESPins.
10	Blackbody view factors: describes prediction of geometrical view factors via module ESPvwf.
11	Air flow simulation: describes simulation of air flow due to pressure and buoyancy effects via module ESPair.
12	Window spectral analysis: describes the prediction of spectral properties of window systems via module ESPwin.
13	Simulation analysis: describes the energy simulation module ESPsim; theory, validity and operation.
14	Results recovery: describes the recovery of the simulation results in various forms (synoptic, graphical, statistical, etc.) via module ESPout.
15	Software structures: implementation procedure, logic diagrams, subroutine trees, memory maps, variable definitions, etc.
16	Test sets and training examples: tests for implementation verification and user training.
17	Example applications: short case studies of a range of typical ESP applications in practice.
18	Essential libraries: details the UTILITY and GRAPHIC libraries required by ESP.

- Operating system supporting Fortran (preferably 77) and providing access to low level I/O for the graphics library.

  For advanced use of ESP, it is recommended that

- Some method of graphical hardcopy is available, preferably screen copy from the terminal.

- A fairly powerful processor with hardware floating point is used.

- A large amount of disk storage is available, possibly as much as 75Mb, to accomodate large simulations.

- A powerful operating system, preferably UNIX, to minimise implementation difficulties (see following section).

E1.4 Software

The ESP system consists of 10 modules and 2 libraries. The source code (heavily commented) is supplied and is described in detail in section 13. The ESP modules and 1 library (UTIL) are written in Fortran. The UTILITY library contains, along with general utility routines, all the file management routines, which will, of course, have to be tailored to the operating system being used. The other library (GRAPHIC) is mainly written in Fortran, but some low level I/O routines are written in 'c' or assembler. These low level routines will also have to be tailored to the operating system.

E1.5 Available Documentation

The user documentation set is extensive and divided into a number of distinct sections corresponding to separate program modules (covering theory, operation and validity) and other related issues such as software implementation, software proving and training, and graphic library requirements. Table E1.1 summarises this documentation set, all or part of which is available from ABACUS on request. A number of other papers and technical reports are also available. These describe the ESP system in general terms, describe a number of commercial applications of the system, and summarise current development plans.

Appendix F

Data structures of the ESP system

The reasons for inclusion of this appendix are twofold:

- It provides a record of the data required (files marked *) by advanced energy modelling systems and so is useful for those contemplating use in practice.

- And by acting as a reference, the information on internal data organisation may be helpful to those planning the development of future systems.

The contents and arrangement of each data-set is described in turn. Binary random access files are considered first, followed by ASCII sequential access files.

CLIMATE FILE: random access, 144 word records

Record	Contents
1	6 climate parameters for each hour of 1 day
	ie diffuse horizontal solar radiation
	dry bulb temperature
	direct normal solar radiation
	wind speed
	wind direction
	relative humidity
2	REPEAT for each day in year
.	
365	
366	Year to which climate relates eg 1984
367	Name of site at which climate was collected
368	Latitude and longitude of site

CASUAL GAINS PROFILE DATABASE: random access, 10 word records

Records		Contents
1	NITMS	- number of profiles in database
2	NCG	- number of casual gains in this profile
3	DESC	- description of this profile
4	ICGS1	- start hour of 1st casual gain of this profile
	ICGF1	- finish hour of 1st casual gain
	CGS1	- sensible percentage of casual gain
	CGL1	- latent percentage of casual gain
4	REPEAT	record 4 for each casual gain in this profile
. n		
n+1	REPEAT	records 2 to n for each profile in this database
. m		

NOTES
n	depends on the number of casual gains in the profile
m	depends on the number of profiles in the database

PRIMITIVE CONSTRUCTION DATABASE: random access, 35 word records

Record		Contents
1	NE	- classification code
	IDESC()	- classification description
2	VAL()	- construction thermophysical data
		ie conductivity
		density
		specific heat
		emissivity
		absorpttivity
	IDESC()	- element description
3	REPEAT	record 2 for each element in the classification
.		(max 19)
20		
21	REPEAT	records 1 - 20 for each classification (max 30)
.		
600		
601	NCLASS	- number of primitive constructions in this database
	ICLN()	- index number of each classification entry
		in order of entry to database

NOTES
The primitive construction database is always 601 records

MULTILAYER CONSTRUCTION DATABASE: random access, 10 word records

Record		Contents
1	NITEMS	- number of multilayered constructions in this database
	I1	- constant = 601
2	NE	- number of elements in this construction

```
3          DESC()  - description of this construction
4          IREF    - element code (index into primitive database)
           ITHK    - thickness of element
5          REPEAT record 4 for each element & air gap in construction
. n
n+1        REPEAT record 2 to n for each construction in database
. m
```

NOTES
```
The reference code for an airgap is zero
n          depends on the number of elements in the construction
m          depends on the number of constructions in the database
```

SHADING: random access, 24 word records

```
Record              Contents

   1       ISHD(12)        - information index for each month
                                = 0      no data
                                = 1      shading data only
                                = 2      insolation only
                                = 3      shading & insolation data

           ISADD(12)       - start record for each months data
   2       IRECX           - number of the next record after the end
                               of the file

           NSUR            - number of surfaces in body being shaded
           NGL(MS)         - number of windows in each surface of body
 3 to      PO(MS,MT)       - % shading, hourly, for each surface
 3+NSUR


                           THEN, IF ISHD = 1 OR 3
                                    (shading exists for this month)

   i       ISC(MS)         - if surface is considered, contains offset from
   .                           here to record containing data about its
                               windows - else 0
     j     PI(MGT,MT)      - % shading hourly for each window

                           THEN, IF ISHD = 2 OR 3
                                    (insolation exists for this month)

   k       ISC(MS)         - if surface is considered, contains offset from
   .                           here to record containing data about its
                               windows - else 0
   .       IRS(MGT,MT,3)   - 2 surfaces insolated by window and flag if one
                               of these contains a window
     1     PI(MGT,MT,2)    - % insolation falling hourly on each of the
                               surfaces in IRS
   n       REPEAT records 2 - n for each month of shading data that exists
```

NOTES
```
   i       = 3+NSUR
   j       = i+total number windows considered for shading
   k       = if no shading = 3+NSUR, otherwise = j+1
   1       = k+1+total number of windows considered for insolation
```

SIMULATION RESULTS FILE: random access, 20 word records

Records		Contents

1	NSIM	- number of simulation result-sets in file
2	NST	- record number for start of simulation result-sets
. 32		
32	NCOMP	- number of zones in scheme
	NCON	- number of constructions in scheme
	NFP(NCOMP)	- default shading plane for each zone
	NPCOMP	- number of plant components in scheme
33-37	LSNAM	- scheme name
38	NCODE(NCOMP)	- code number for each zone
	NPCODE(NPCOMP)	- database code number for each plant component
39-43	LCLIM(NCOMP)	- climate file name
44	LPROJ(NCOMP)	- project file names for all zones
. 5i+44		
5i+45	LGEOM(NCOMP)	- geometry file names for all zones
. 10i+44		
10i+45	LSHAD(NCOMP)	- thermal properties file names for all zones
. 15i+44		
1	IC1(NCON)	- start point of each connection
1+j	IE1(NCON)	- construction (component) number of start point
1+2j	ICT(NCON)	- connection type

$$\begin{aligned} &= 0 \quad \text{to external conditions} \\ &= 1 \quad \text{to environment as current zone} \\ &= 2 \quad \text{to constant, known conditions} \\ &= 3 \quad \text{to another zone (component) given} \\ &\qquad \text{in IC2, IE2} \end{aligned}$$

1+3j	IC2(NCON)	- number of connecting zone (component)
1+4j	IE2(NCON)	- connecting zone (component) construction (node)
1+5j+i	ZNAME(NCOMP)	- zone name for all zones
1+5j+2i	NCONST(NCOMP)	- number of constructions in each zone
1+5j+3i	NELTS(NCOMP,NCONST)	- number of elements in each construction
m+k	NGAPS(NCOMP,NCONST)	- number of air gaps in each construction
m+2k	NPGAP(NCOMP,NCONST,3)	- position of air gaps in each construction
m+5k	NWINS(NCOMP,NCONST)	- number of windows in each construction
m+6k	NDOORS(NCOMP,NCONST)	- number of doors in each construction
m+7k	SNA(NCOMP,NCONST)	- opaque surface area of each construction
m+8k	ZOA(1)	- total opaque surface area bounding zone
	ZGA(1)	- total glazing area
	ZDA(1)	- total door area
m+9k	REPEAT record 1+5j+3i+9k for each zone	
m+i+9k	CHARDM(NCOMP,NCONST)	- plane charactistic dimension
m+i+10k	PAZI(NCOMP,NCONST)	- azimuth of each surface
m+i+11k	PELV(NCOMP,NCONST)	- azimuth of each surface
m+i+12k	ZGAE	- sum of external glazing areas
	ZGAI	- sum of internal glazing areas
	ZDAE	- sum of external door areas
	ZDAI	- sum of internal door areas
m+i+13k	REPEAT record 1+5j+4i+12k for each zone	
nst		

NOTES

i	= NCOMP, ie number of zones
j	= NCON, ie number of connections

```
k        = NCOMP * NCONST, ie number of zones * number of surfaces per zone
l        = 15i+45
m        = 1+5j+3i
nst      = m+2i+13k  =  START OF 1st RESULT SET

nst      ISD1                    - simulation start day
         ISD2                    - simulation start month
         ISD2                    - simulation finish day
         ISM2                    - simulation finish month
         ISDS                    - simulation start year-day
         ISDF                    - simulation finish year-day
         ISTEP                   - number of simulation timesteps per hour
         ISAVE                   - save option level
nst+1    NCF                     - number of control functions
         ICASCF(NCOMP)           - control function for each zone
         ISATS                   - Saturday control function flag
         ISUNS                   - Sunday control function flag
nst+2    NCT                     - control point node type
         NCS                     - surface number of control point
         NCE                     - element number of control node
         NCP                     - position in element of control node
         NIT                     - plant interaction point node type
         NIS                     - surface number of plant interaction point
         NIE                     - element number of plant interaction node
         NIP                     - position in element of plant interaction node
         NOCP(3)                 - number of control periods for Weekdays,
                                   Saturdays and Sundays
nst+3    CPS(NOCP)               - start hour of each of the above periods
         CPF(NOCP)               - finish hour of each of the above period
nst+4    IPDI(NOCP)              - control flag for each period
nst+5    QHMX(NOCP)              - available heating capacity
         QCMX(NOCP)              - available cooling capacity
nst+6    TL(NOCP)                - lower control point temperature
         TU(NOCP)                - upper control point temperature

                    THEN, IF ISATS = 1 OR ISUNS = 1

nst+7    REPEAT records nst+3 to nst+6 for Saturday and Sunday
nst+8    REPEAT records nst+2 to nst+7 for each control function
nst+9    PCI                     - plant control loop function index

             THEN FOR EVERY BUILDING ZONE AND PLANT NETWORK,
             DEPENDING ON SAVE OPTION, EITHER         (save option 1)

n        TA                      - air node temperature
         T                       - value of control node temperature
         Q                       - value of plant input/extract at interaction node
         E                       - value of plant network energy consumption
         GG                      - zone humidity level
. y
```

NOTES

    nst   = NST from record 1, depends on the problem specification
    n   depends on the number and type of control periods

                        OR   (save option 2)

n	TA	- air node temperature
	Q	- value of plant input/extract at interaction node
	TS(NSUR)	- internal surface temperatures
	QV1	- infiltration
	QV2	- zone-coupled ventilation
n+1	QWE	- external window conduction
	QWI	- internal window conduction
	QDE	- external door conduction
	QDI	- internal window conduction
	QSA	- air point solar energy absorbed
	QSI(NSUR)	- solar energy adsorbed at opaque surfaces
n+3	QSE(NSUR)	- solar energy adsorbed at external opaque surfaces
	QCASR	- total radiant injection
	QCASC	- total convective injection
n+4	H(NPCOMP)	- plant component enthalpy or temperature
	G(NPCOMP)	- plant component humidity ratios
n+5	GG	- zone humidity level
	I	- zone infiltration
	ZCA(NCOMP)	- zone-coupled air flow
	ILR	- internal longwave flux exchange
	ELR	- external longwave flux exchange
. y		

OR   (save option 3)

n	TA	- air node temperature
	Q	- value of plant interaction node input/extract
	TS(NSUR)	- internal surface temperatures
	QV1	- infiltration
	QV2	- zone-coupled ventilation
n+1	QWE	- external window conduction
	QWI	- internal window conduction
	QDE	- external door conduction
	QDI	- internal window conduction
	QSA	- air point solar energy adsorbed
	QSI(NSUR)	- solar energy adsorbed at opaque surfaces
n+3	QSE(NSUR)	- solar energy adsorbed at external opaque surfaces
	QCASR	- total radiant injection
	QCASC	- total convective injection
n+4	TC(NELTS)	- temperature of construction node
n+5	H(NPCOMP)	- component enthalpy
	G(NPCOMP)	- component humidity ratio
n+6	GG	- zone humidity level
	I	- zone infiltration rate
	ZCA(NCOMP)	- zone coupled air flow
	P	- zone pressure
n+7	CC	- control condition
	AS	- actuator status
	FR	- plant component flowrates
. y		
y+5	REPEAT records n to y for each simulation timestep and building zone	
z	REPEAT records NST to here for each result-set in the file	

NOTES

y	depends on the number of constructions in the zone and components in the plant network
z	depends on the number of result sets in the file

TRANSITIONAL SHADING FILE: <u>random access</u>, <u>20 word records</u>

Record		Contents
1	IRECE(MS)	– record number containing start of shading data for each surface
2	MON	– month number
	ISC(NSUR)	– flags if surface is considered
	IYD	– year day number
	XLAT,XLONG	– latitude and longitude of site
3	ETYPE	– type of body
		= REC rectangle
		= REG regular
		= GEN general
	NSUR	– number of surfaces in body
	NTV	– number of vertices defining body
	NVER(NSUR)	– number of vertices defining each surface
4	JVN(1,NVER(1))	– list of vertices defining surface 1
5	REPEAT i for each surface
. i		
i+1	NGL(NSUR)	– number of windows in each surface
i+2	XGL(1,1),ZGL(1,1)-	coord of bottom left corner of window 1
	DXGL(1,1),DZGL(1,1)-	width and height of window 1
i+3	REPEAT i+2 for each window in body
. j		
j+1	NB	– number of site obstructions
j+2	XO(1),Y(1),Z(1)	– coordinate of corner of obstruction 1
	DX(1),DY(1),DZ(1)-	length, breadth and height of obstruction 1
	ANG(1)	– orientation of obstruction 1 (relative to north)
j+3	REPEAT j+2 for each obstruction
. k		
k+1	NOX,NOZ	– number of mesh points on surfaces
	NGX,NGZ	– number of mesh points on windows

NOTES
i	depends on the number of surfaces considered
j	depends on the number of windows considered
k	depends on the number of site obstructions

J	XFT(NVER)	– transformed x coordinates of surface 1 vertices
J+1	ZFT(NVER)	– transformed z coordinates of surface 1 vertices
J+3	ISUNUP	– sun flag
		= 1 sun up

THEN, IF ISUNUP = 1

J+4	IS	– shading flag for surface 1
		= -2 not shaded
		= -1 totally shaded
		= 0 mixed shading/insolation

THEN, IF IS = 0

J+5	IOSHD(1,20)	– shading flag for each mesh point across surface 1
J+6	REPEAT J+5 for each mesh point down surface 1
. K		

K+1	IGSHD(1,1,NGZ),IGSHD(1,2,NGZ)
	- shading flag for each mesh point across window 1
	of surface 1
K+2	REPEAT K+1 for each mesh point down window 1
L	
L+1	REPEAT K+1 to L for each window in surface 1
M	
M+1	REPEAT L to M for each surface considered

NOTES

J	depends on the size of the geometry specification
K	depends on the number of mesh points per surface
L	depends on the number of mesh points per window
M	depends on the number of windows per surface
N	depends on the number of surfaces considered

SHADING PICTURE FILE: random access, 24 word records

Record	Contents	
1	MON	– month
	MV	– number of vertices
	NG	– number of windows
2	NS(24)	– number of shadow polygons on face for each hour
3	XFT(NV)	– x coordinate of surface vertices (local to face)
4	ZFT(NV)	– z coordinate of surface vertices (local to face)
5	XGL(NG)	– x coordinates of window corner (local to face)
6	ZGL(NG)	– z coordinates of window corner (local to face)
7	DXGL(NG)	– width of windows
8	DZGL(NG)	– height of windows
9		
10	ISUNUP	– sun up flag for hour 1
		= 0 sun not up

THEN, IF ISUNUP = 1

11	XS(1),ZS(1) – shadow polygon 1 coordinates
	XS(2),ZS(2), etc
	XS(6),ZS(6)
12	REPEAT record 11 for each shaddow polygon
i	REPEAT records 11 to i-1 for each hour

NOTES
  i   depends on NS, ie number of shadow polygons on surface

TRANSITIONAL INSOLATION FILE: <u>random access</u>, <u>20 word records</u>

Records		Contents
1	IRECS(NSUR)	- record number of start of data for each surface
2	MON	- month number
	ISC(NSUR)	- Surface considered
		= 1 yes
	IYD	- year day number
	XLAT,XLONG	- site position
3	ETYPE	- body shape
		= REC rectangular
		= REG regular
		= GEN general
	NSUR	- number of surfaces defining body
	NTV	- number of vertices defining body
	NVER(NSUR)	- number of vertices defining each surface
4	JVN(NVER)	- vertices defining surface 1 (anticlockwise)
5	REPEAT record 4 for each surface	
. i		
i+1	X(NTV)	- x coords of vertices defining body
. j	Y(NTV)	- y coords of vertices defining body
	Z(NTV)	- z coords of vertices defining body
j+1	NGL(NSUR)	- number of windows in each surface
j+2	XGL,ZGL	- coordinates local to surface of window corner
	DXGL,DZGL	- width and height of window
j+3	REPEAT record j+2 for each window in each surfaces	
. k		
k+1	NGL(NSUR)	- number of doors in each surface
k+2	XGL,ZGL	- coordinates local to surface of door corner
	DXGL,DZGL	- width and height of door
k+3	REPEAT record k+2 for each door in each surfaces	
. l		
l+1	NGX,NGZ	- number of mesh points on a window

NOTES

i	depends on the number of surfaces in the body
j	depends on the number of vertices defining the body
k	depends on the number of windows in the body
l	depends on the number of doors in the body

I	IHR	hour in day
	IS	- surface number
	IG	- window number
	PINS(2)	- percentage insolation on insolated surfaces
I+2	IGINS(NGX,NGZ)	- insolation index for each mesh point
		= -2 insolation falls on internal window
		= -1 mesh point shaded
		= 0 sun not up
		> 0 insolated surface number
I+3	REPEAT record I+1 to I+2 for each window in each surface	
J	REPEAT record I to J for each hour in day	

NOTES

I depends on problem size
J depends on number of mesh points on windows

3D PLOT DATA FILE: random access, 10 word records

```
    Records            Contents

       1      ITITLE(10)      - title of plot
       2      XTITLE(2)       - title of x axis
              YTITLE(2)       - title of y axis
              ZTITLE(2)       - title of z axis
       3      NXPNTS          - number of divisions along x axis
              NZPNTS          - number of divisions along z axis
       4      XAXIS(NXPNTS)   - x value for each x axis division (x ordinate)
     . i
      i+1     ZAXIS(NXPNTS)   - z value for each z axis division (z ordinate)
     . j
      j+1     DATA(NXPNTS)    - data for surface heights at all x ordinates
     . k                               for 1st z ordinate
      k+1     REPEAT record j+1 ti k for each z ordinate
```

NOTES

$$i = 4 + NXPNTS/20$$
$$j = i + 1 + NZPNTS/20$$
$$k = j + 1 + NXPNTS/20$$

PLANT COMPONENT LIBRARY: random access, 20 word records

```
    Records            Contents

       1      DESC            - file identified = "Plant component library"
       2      NPC             - number of components in library
              NXTREC          - next available record after end of library
       3      IRECS(140)      - start record of each component
       .
       9
      10      INDXPC          - Internal library index
              ICODE           - user specified code (0-32000)
              ZCDATE          - date of insertion ( A16)
              NNODE           - no. nodes in nodal scheme ( <20)
              NMATX           - no. non-zero coefficients ( < 80)
              NMISC           - no. misc data items ( < 80)
              ZGTYPE          - component classification ( < 40 chars)
      11      ZCDESC          - component description ( < 80 chars)
      12      NDCON           - connectivity of each node
      13      NDPOS()         - coeff positions
       .
       i
      i+1     DATAMS()        - any numeric data associated with component
       .
       j
      j+1     REPEAT records 10 to k for each component in library
       .
       k
```

NOTE

$$i \text{ is } 13 + NMATX / 20$$
$$j \text{ is } k + NMISC / 20$$
k   depends on total number of components

PROJECT FILE: ascii file, free format (*)

Items		Description
1	LPROJ( )	- name of zone
2	IPJ	- thermostatic control on:
		= -2 incoming air to control ventilation only
		= -1 zone temperature to control ventilation only
		= 0 none
		= 1 input air temperature to control ventilation and infiltration
		= 2 zone temperature to control ventilation and infiltration
	TLO	- lower temperature for thermostat
	TUP	- upper temperature for thermostat
3	NAC1	- number of air change periods during weekday
4	IACS1	- start hour for 1st air change period
	IACF1	- finish hour for 1st air change period
	ACI1	- infiltration air change rate
	ACV1	- ventilation air change rate
	IPT1	- ventilation index
	TA1	- source temperature
5	REPEAT record 3 for each weekday air change period	
. n		
n+1	REPEAT record 2 to n for Saturday air change periods	
. m		
m+1	REPEAT record 2 to n for Sunday air change periods	
. l		
l+1	NCG1	- number of casual gains periods during weekday
l+2	ICGS1	- start hour for 1st casual gain
	ICGF1	- finish hour for 1st casual gain
	CGS1	- sensible magnitude of casual gain
	CGl1	- latent magnitude of casual gain
	RADC1	- radiant portion of casual gain
	CONC1	- convective portion of casual gain
l+3	REPEAT record l+2 for each weekday casual gains period	
. k		
k+1	REPEAT record l+1 to k for Saturday casual gains periods	
. j		
j+1	REPEAT record l+1 to k for Sunday casual gains periods	
. l		
l+1	IAIRFL	- air file flag
		= 0 use data from project file
		= 1 24 hourly data values in file
		= 2 data values for each time-increment in file
l+2	ICASFL	- casual gains file flag
		= 0 use data from project file
		= 1 24 hourly data values in file
		= 2 data values for each time-increment in file

NOTES

n	depends on NAC1, ie number of weekday air change periods
m	depends on NAC1, ie number of Saturday air change periods
l	depends on NAC1, ie number of Sunday air change periods
k	depends on NCG1, ie number of weekday casual gains periods
j	depends on NCG1, ie number of Saturday casual gains periods

GEOMETRY FILE: ascii file, free format (*)

Items		Description
1	ETYPE	- zone shape
		= REC rectangular shape
		= REG regular shape
		(vertical walls of constant height)
		= GEN general shape

THEN, DEPENDING ON ZONE SHAPE
                        EITHER (rectangular zone)

2	X	- x coordinate of bottom left hand corner
	Y	- y coordinate of bottom left hand corner
	Z	- z coordinate of bottom left hand corner
	DX	- length
	DY	- breath
	DZ	- height
	ANG	- angle of orientation

OR (regular shape)

2	N	- number of walls
	Z1	- height of floor
	Z2	- height of ceiling
	ANG	- angle of rotation of body coordinates from north
3	X,Y	- coordinate of base vertex of wall
4	REPEAT item 3 for each wall in zone (anticlockwise)
. n		

OR (general shape)

2	NTV	- number of vertices in zone
	NSUR	- number of surfaces in shape
	ANG	- angle of rotation of body coordinates from north
3	X,Y,Z	- coordinate of vertex
4	REPEAT item 3 for each vertex in zone
. m		
m+1	NVER	- number of vertices in 1st surface
	JVN(NVER)	- ordered (anticlockwise) list of vertices of 1st surface
m+1	REPEAT item n+1 for each surface in zone
. n		

NOTES

m	= 4+NTV, ie number of vertices in zone	
n	depends on complexity of zone	
n+1	NGL(NSUR)	- number of windows in each surface
n+2	XGL,ZGL	- coordinate (local to surface) of bottom left hand corner of window
	DXGL,DZGL	- width and height of window
n+3	REPEAT item n+2 for each window in surface	
. k		
k+1	REPEAT items n+2 to k for each surface with windows	
. l		
l+1	NDO(NSUR)	- number of doors in each surface
l+2	XDO,ZDO	- coordinate (local to surface) of bottom left hand corner of door
	DXDO,DZDO	- width and height of door
l+3	REPEAT item l+2 for each door in surface	
. j		
j+1	REPEAT items l+2 to i for each surface with doors	
. i		
i+1	NDP	- number of default planes for shading
	IDPN(3)	- 2 default insolation planes and window flag
i+2	ISHAD	- shading data type
		= 0 use default
		= 1 shading/insolation file provided
i+3	IVFI	- view factor data type
		= 0 use default
		= 1 view factor file provided

NOTES

l	depends on the total number of windows in the zone
k	depends on the total number of windows in the zone
j	depends on the total number of doors in the zone
i	depends on the total number of doors in the zone

ZONE CONSTRUCTION FILE: ascii file, free format (*)

Items		Description
1	NE	- number of elements in 1st surface
	NAIRG	- number of air gaps in 1st surface
	.	
	.	REPEAT for each aurface in zone
2	IPAIRG	- position of 1st air gap in 1st surface
	RAIRG()	- resistance of 1st air gap in 1st surface
	.	
	.	REPEAT for each air gap in 1st surface
3	REPEAT record 2 for each surface with air gaps	
. n		
n+1	CON	- conductivity of 1st element in 1st surface
	DEN	- density of 1st element in 1st surface
	SHT	- specific heat of 1st element in 1st surface
	THK	- thickness of 1st element in 1st surface
	REPEAT for each element in 1st surface	

n+3	REPEAT record n+1 for each surface in zone	
. m		
m+1	GTR(5)	- window solar transmissivity for 1st window in 1st surface with windows (5 angles of incidence)
	GHGF(5)	- 5 window solar heat gain factor for 1st window
	GU	- window U value for 1st window
	REPEAT for each window in 1st surface with windows	
m+2	REPEAT record m+1 for each surface with windows	
. k		
k+1	DU	- door U value for 1st door in 1st surface with doors
	REPEAT for each door in 1st surface with doors	
k+2	REPEAT record m+1 for each surface with doors	
. l		
l+1	EMISI()	- internal surface emmissivity for each surface
	EMISE()	- external surface emmissivity for each surface
l+2	ABSI()	- internal surface solar absorptivity for each surface
	ABSE()	- external surface solar absorptivity for each surface
l+3	IBCI	- blind/shutter control index

Notes

n	depends on the number of surfaces with air gaps
m	depends on the number of surfaces defining the zone
k	depends on the number of surfaces with windows
l	depends on the number of surfaces with doors

CONFIGURATION FILE: ascii file, free format (*)

Items		Description
1	INDCFG	- configuration file index
		= 1 building only
		= 2 plant only
		= 3 building and plant

THEN, IF INDCFG =1 or 3

2	SNAME(72)	- building system name
3	XLAT	- site latitude
	XLONG	- site longitude
4	NZONES	- number of zones in scheme
5	NCODE	- code number (arbitary) for all zones
. n+4		
n+5	PFN(72)	- project file names for all zones
. 2n+4		
2n+5	GFN(72)	- geometry file names for all zones
. 3n+4		
3n+5	CFN(72)	- construction file names for all zones
. 4n+4		
m	NCON	- total number of opaque surfaces in scheme
m+1	IC1	- start point of inter-zone connection, zone no.
	IE1	- start point of inter-zone connection, surface no.
	ICT	- type of connection
		= 0 connect to external conditions
		= 1 connects to zone identical to zone IC1
		= 2 connects to constant conditions
		= 3 connects to another zone IC2

	IC2	- finish zone of connection OR constant temperature
	IE2	- finish surface of connection OR constant incident radiation
m+2 .1		REPEAT item m+1 for each connection

THEN, IF INDCFG = 2 or 3

I	PNAME(72)	- plant system name
I+1	NPCOMP	- number of plant components
I+3	NPCODE	- component code (located in matrix in ascending code)
I+4	NPLOC	- component location in plant component library
	NPNAME	- generic name of component (as in library)
	PDMISC(4)	- associated miscellaneous data for component

I+5 . J		REPEAT I+3 to I+4 for all plant components

J+1	NPCON	- number of component connections
	NCCON1	- component upstream
	NCNOD1	- node upstream
	NCCON2	- component downstream
	NCNOD2	- node downstream

J+3 . K		REPEAT J+2 for each connection

THEN, IF INDCFG = 3

K+1	NPENV	- component location in local environment

K+2 . L		REPEAT record K+1 for each component

NOTES

n	= NZONES, ie the number of zones in the scheme
m	= 4n+5, ie depends on the number of zones
I	depends on wheather building is considered
J	= I+6+NPCOMP, ie depends on number of components
K	= J+NPCON, ie depends on number of connections
L	= K+NPCON, ie depends on number of components

BLIND CONTROL FILE: ascii file, free format (*)

Item		Description
1	NPER	- number of blind control periods per day
2	IST(NPER)	- start hour of each period
3	IEN(NPER)	- finish hour of each period
4	DST(NPER)	- direct solar transmittance for 5 angles of incidence
5	THGF(NPER,5)	- total solar transmittance for 5 angles of incidence
6	TTR(NPER)	- U-value of window during each period

ZONE CASUAL GAINS FILE: ascii file, free format (* optional)

Item		Description
1	IHDR	- Contains string "* CASUAL GAINS FILE"
2	QCASC	- convective casual gains
3	QCASR	- radiant casual gains
4	REPEAT record 2 to 3 for each simulation timestep, including startup	

ZONE AIR FLOW FILE: ascii file, free format (* optional)

Item		Description
1	IHDR	- Contains string "* AIR FLOWS FILE"
2	ACI	- infiltration air changes
3	ACC(NCOMP)	- coupled air changes, from other zones
4	REPEAT record 2 to 3 for each simulation timestep, including startup	

PLANT CONTROL FILE: ascii file, free format (*)

Items		Description
1	NCT	- control point node type where:

$$
\begin{array}{ll}
= 1 & \text{zone air point node} \\
= 2 & \text{inside opaque surface node} \\
= 3 & \text{intra-constructional node} \\
= 4 & \text{resultant temperature node} \\
= 5 & \text{plant component node} \\
= 6 & \text{outside temperature/radiation}
\end{array}
$$

THEN, DEPENDING ON NCT, EITHER
(NCT = 2)

2	NCS	- surface number to which control node belongs

OR (NCT = 3)

2	NCS	- surface number to which control node belongs
3	NCE	- element number within which control node is situated
4	NCP	- defines control node position within element where

$$
\begin{array}{ll}
= 1 & \text{outermost element boundary} \\
= 2 & \text{element centre plane} \\
= 3 & \text{innermost element boundary}
\end{array}
$$

OR (NCT = 4)

2	TWAIR	- temperature weighting of air component for mixed temperature control

```
 3          TWRAD            - temperature weighting of radiant component, mixed
                              temperature control

                              THEN, IF TWRAD # 0 ( AND NCT = 4)

 4          IMRTT            - type of mixed temperature sensor
                                   = 1      selected surfaces only
                                   = 2      all surfaces considered
 5          ITSC             - transparent surface flag
                                   = 1      consider transparent surfaces

                              THEN, IF IMRTT = 1 ( AND NCT = 4)

 6          NSMART           - number of surfaces considered
 7          IMRTS            - surface number of surfaces considered

                              OR (NCT = 5)

 2          NCC              - defines plant component
 3          NCP              - defines component node

NOTES
 i               = 2,3,4,5,6 or 8, depending on NCT and TWRAD and IMRTT,
                                  ie control point type
 j               = i+1 or i+4, depending on NIT, ie plant interaction node type
 k               j+4,j+6 or j+8, depending on IPDI, ie type of control periods
 l               depends on ISATS, ie whether Saturday is same as weekdays or not

 i          NIT    - defines actuator node
                                   = 1      air point node
                                   = 2      inside opaque surface node
                                   = 3      intra-constructional node
                                   = 4      plant network node

                              THEN, IF NIT = 2 OR 3

 i+1        NIS    - surface number to which plant interaction node belongs

                              THEN, IF NIT = 3

 i+2        NIE    - element number within which plant interaction node is
 i+3        NIP    - defines plant interaction node position within element wher
                                   = 1      outermost element boundary
                                   = 2      element centre plane
                                   = 3      innermost element boundary

                              THEN, IF NIT = 4

 i+1        NIC    - defines component
 i+2        NIP    - defines node
 i+3        TYPE   - defines controlled variable:
                                   = 1      temperature
                                   = 2      mass flow
                                   = 3      heat input/extract

                              CONTROL LOOP DETAILS

 j          NOCP   - number of distinct control periods during
                     typical weekdays
```

j+1	CPS	– start hours of each of the above periods
j+2	CPF	– finish hours of each of the above periods
j+3	IPDI	– the control flag for each period where

= 1	ideal control
= 2	no control
= 3	boost
= 4	fixed action
= 5	defined control law

THEN, IF IPDI = 1, 3, 4 OR 5

j+4	QHMX	– upper controlled variable capacity

IPDI = 1,3 maximum heating capacity available
IPDI = 4   fixed heat injection

j+5	QCMX	– lower capacity

IPDI = 1,3 maximum cooling capacity available
IPDI = 4   fixed heat extraction

THEN, IF IPDI = 1, 4 OR 5

j+6	TL	– lower control point temperatures for any period
j+7	TU	– upper control point temperatures for any period
k	ISATS	– saturday control function flag

= 1   same as weekday

THEN, IF ISATS = 0

k+1		REPEAT items j to k-1 for saturday control data
l	ISUNS	– sunday control function flag

= 1   same as weekday
= 2   same as saturday

THEN, IF ISUNS = 0

l+1	REPEAT items j to k-1 for saturday control data

WINDOW FILE: ascii file, free format (* optional)

Items		Description
1	NTE	– number of transparent elements
	KFORTS	– type of transparent element

= 0	air or vaccum
= 1	transparent substrate
= 3	opaque substrate

2	NOCODE	– optical code (in window database)
	THICK	– element thickness

3		REPEAT item 2 for each element
.		
i		
	NOSPRO	- spectral property index
	NWAVES	- number of monochromatic wavebands considered
	NEEWB	- energy spacing between wavebands
		= 0 defined wavebands
		= 1 equal energy weightings

THEN, IF NEEWB = 0

i+1	LAMDA	- waveband boundaries
j	RIWL	- waveband dependancy for each element
j+1	RN	- mean refractive index of each waveband for element 1
	RK	- mean absorption coefficient for each waveband, element 1
j+2		REPEAT item j+1 for each element
.		
k		
k+1	ICMEM	- optical code information (=0)

NOTES	i	depends on number of elements
	j	depends on energy waveband spacing type
	k	depends on number of elements

SITE OBSTRUCTION FILE: ascii file, free format (* optional)

Items		Description
1	XLAT	- site latitude
	XLONG	- site longitude
	NB	- number of obstruction blocks
2	XO,YO,ZO	- location of origin of block 1
	DX,DY,DZ	- length, width and height of block 1
	ANG	- angle of rotation of block 1
3		REPEAT item 2 for all obstruction blocks
i	NOX,NOZ	- number of opaque grid points in x,z direction
	NGX,NGZ	- number of glazing grid points in x,z direction

NOTES	i	depends on number of site obstruction blocks

<u>AIR</u> <u>FLOW</u> <u>PROBLEM</u> <u>FILE</u>: <u>ascii</u>, <u>free</u> <u>format</u> (<u>* optional</u>)

Items		Description
1	NSP	- number of spaces
	NCON	- number of air flow connections
2	KDSP	- name of space
	KSPT	- space type
		1 = internal
		2 = external
	STEM	- space temperature
	SVOL	- space volume
	SUPV	- ventilation supply
	SAZ	- Azimuth
	IPDN	- Pressure Distribution Index
		(pointer to database)
	HREF	- Space height reference
3		REPEAT item 2 for all spaces
. n		
n+1	KCNT	- connection type
		1 = area
		2 = crack
		3 = door
		4 = other (user defined)
	KDPS	- positive space reference
	KDNE	- negative space reference
	FCOF	- coefficient for crack & door
	FEXP	- exponent for crack & door
	AFAR	- connection area

NOTE   n   depends on the number of spaces

Appendix G

ESP system output examples

Figure G.1 A perspective view of the proposed St Enoch Development in Glasgow. ESP was used to predict air movement internally and the effects of solar penetration through the glass envelope. Display by BIBLE via ESPsim.

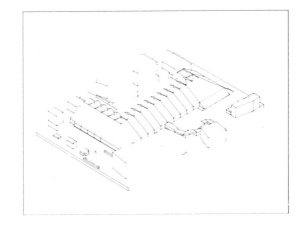

Figure G.2 Internal air temperature for a zone controlled to 21°C during the occupied period. Also shown — the broken line — is the plant power required. Display by ESPout.

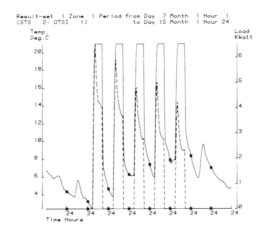

Figure G.3 Internal surface temperatures (∨), air dew point temperature (×), mean radiant temperature (+) and resultant temperature (∆) for a zone controlled to 21°C during the occupied period only. Display by ESPout.

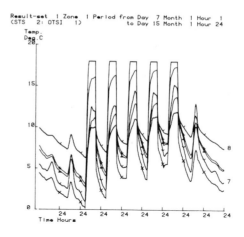

Figure G.4 Infiltration load versus ambient temperature (solid line). Display by ESPout.

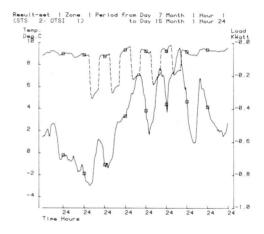

Figure G.5 Solar absorption (the chain line) and surface temperature (the ∧ mark) displayed against ambient air temperature for a solar pick-up wall. Display by ESPout.

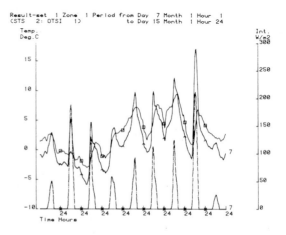

Figure G.6 Zone internal air temperature displayed as a three-dimensional surface. Note the plant shutdown at weekends. Display by ESPout.

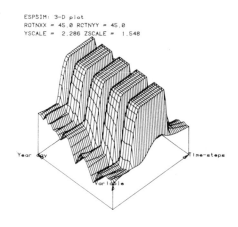

Figure G.7 Variation of multi-layered construction temperature distribution with time. This is useful for the study of insulation and capacity positioning. Display by ESPout.

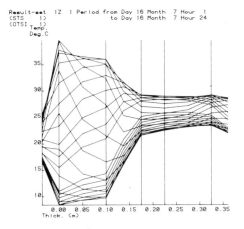

Figure G.8 A zone energy balance showing the gains and losses during the different plant control states. Display by ESPout.

```
Causal energy breakdown (KWhrs) for zone 1
Period from Day  1 Month  1 Hour  1
            to Day 31 Month  1 Hour 24

Simulation time-step =    2/hour
Output time-step increment =   1 (results not averaged)
```

	Heat.period		Cool.period		OFF period	
	G.	L.	G.	L.	G.	L.
Infiltration air load	0.	-50.	0.	0. ,	0.	-24.
Ventilation air load	0.	0.	0.	0.	0.	0.
Window conduction: external	0.	-192.	0.	0.	0.	-92.
Window conduction: internal	0.	0.	0.	0.	0.	0.
Door conduction: external	0.	-50.	0.	0.	0.	-24.
Door conduction: internal	0.	0.	0.	0.	0.	0.
Air point solar load	7.	0.	0.	0.	5.	0.
Convective casual load	109.	0.	0.	0.	4.	0.
Opaque surface convection	0.	-564.	0.	0.	124.	-2.
Plant capacity	745.	0.	0.	0.	0.	0.
Totals	862.	-856.	0.	0.	133.	-141.

Figure G.9 Output obtained from the simulation trace facility giving the results of solar calculations. Display by ESPsim.

```
Subroutine MZSOLG      Trace output   2
                       Zone      1

Day No. = 196 (i.e. 15th of July),  Time 15:00 Hours

Future time-row is day 196 hour  15
All output W/m**2

Azimuth =  244.3 Altitude =   45.1
Direct normal intensity        =       11.0
Diffuse horizontal intensity   =      300.0

Surface External   Internal
        Solar      Solar
        Absorption Absorption

    1    183.37      4.07
    2    180.24      6.75
    3    180.24      8.09
    4    186.74      8.09
    5    180.24      8.09
    6    186.74      8.09
    7    281.76      7.47
    8      0.00      8.27

Solar energy gain at the air point =      98.36
```

>>

Figure G.10 Boiler supply water temperature (the upper curve) against fuel supply rate. The cycle time is one minute. Display by ESPout.

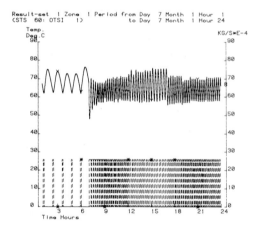

Figure G.11 Room load and radiator heat output (top and second top curves respectively) and radiator inlet and outlet temperatures (third and bottom curves respectively) during steady state conditions.

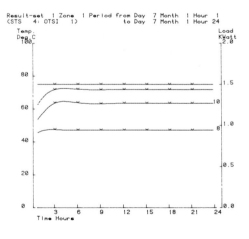

Figure G.12 The saturated and vapour pressure gradients through a multi-layered construction indicating the occurrence of interstitial condensation. Display by ESPout.

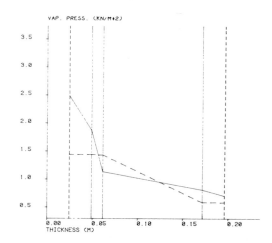

```
VAP. PRESS. (KN/M+2)

3.5

3.0

2.5

2.0

1.5

1.0

0.5

     0.00        0.05        0.10        0.15        0.20
     THICKNESS (M)
```

Figure G.13 A listing of some entries in the construction primitives database. Display by ESPcon.

```
ESP Construction database : Units - Conductivity W/mdeg.c
                                    Density Kg/m**3
                                    Specific Heat J/kgdeg.C

Element Element                              CON.  DEN.  SHT.  EMIS. SABS.
Code    Description

Classification number       -   1
No.of elements filed        -   5
Classification description - BRICK

    1    PAVIOUR                            0.96 2000.  840.  0.93  0.70
    2    BREEZE BLOCK                       0.44 1500.  650.  0.90  0.65
    3    INNER LEAF                         0.62 1800.  840.  0.93  0.70
    4    OUTER LEAF                         0.96 2000.  650.  0.90  0.93
    5    VERMICULITE INSULATING BRICK       0.27  700.  840.  0.90  0.65

Classification number       -   2
No.of elements filed        -  12
Classification description - CONCRETE

   21    LIGHT MIX                          0.38 1200.  653.  0.90  0.65
   22    AERATED CONCRETE BLOCK             0.24  750. 1000.  0.90  0.65
   23    AERATED CONCRETE                   0.16  500.  840.  0.90  0.65
   24    REFRACTORY INSULATING              0.25   10.  837.  0.90  0.65
   25    VERMICULITE AGGREGATE              0.17  450.  837.  0.90  0.65
   26    NO FINES CONCRETE                  0.96 1800.  840.  0.90  0.65
   27    FOAMED SLAG CONCRETE               0.25 1040.  960.  0.90  0.65
   28    CONCRETE BLOCK INNER (3%MC)        0.51 1400. 1000.  0.90  0.65
   29    FOAMED BLOCK INNER (3%MC)          0.16  600. 1000.  0.90  0.65
   30    FOAMED BLOCK OUTER (5%MC)          0.17  600. 1000.  0.90  0.65
   31    GLASS REINFORCED CONCRETE          0.90 1950.  840.  0.90  0.65
   32    HEAVY MIX                          1.40 2100.  653.  0.90  0.65
```

Figure G.14 A listing of two entries in the multi-layered constructions database. Display by ESPcon.

```
         CONSTRUCTION    1      SANDSTONE HEAVY WALL
         DB CODE  THICKNESS     DESCRIPTION
             81      0.700             STONE
                                    SANDSTONE
              0      0.015      AIR GAP

            101      0.025            PLASTER
                                     DENSE


       Standardised "U" value =   1.27
         CONSTRUCTION    2      CAVITY BRICK
         DB CODE  THICKNESS     DESCRIPTION
              4      0.105             BRICK
                                   OUTER LEAF
              0      0.050      AIR GAP

              3      0.105             BRICK
                                   INNER LEAF
            102      0.012           PLASTER
                                     LIGHT


       Standardised "U" value =   1.41
```

>>

Figure G.15 A summary of a plant component as held in the ESP plant components database. Descriptive information, a component matrix template (defining energy balance) and manufacturers' data are held. Display by ESPplt.

```
PLANT COMPONENT                                          No :    2
──────────────

Component Description

    Generic type    : Water filled radiator/ WCH
    Description      : Single panel - inlet top, outlet bottom
    Insertion date   : 15-Aug-84 16:19        Component code :    40

Nodal Scheme Description

    Number of nodes  :   6      No of nonzero matrix elements :    14
    Matrix positions :   1,   2,   7,   8,  13,  15,  16,  21,  22,  27,
                       29,  30,  35,  36,
    Node connections :   1,   0,   0,   0,   0,   0,

Miscellaneous Data

    No. data items   :  12
    Misc data        :
            0.0001,        0.0002,      0.0584,      0.0884,      0.0003,
            0.0011,        0.1200,      0.1500,   7900.0000,    460.0000,
         1000.0000,     4200.0000,
```

>>

Figure G.16 Climate graphical display showing dry bulb temperature (solid curve), direct normal radiation (broken curve) and wind speed (chain curve). Display by ESPclm.

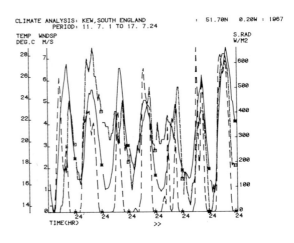

```
CLIMATE ANALYSIS: KEW,SOUTH ENGLAND           : 51.70N   0.20W : 1967
        PERIOD: 11. 7. 1 TO 17. 7.24
```

Figure G.17 A min/max analysis of dry bulb temperature occurring over a user-specified January week. Display by ESPclm.

```
CLIMATE ANALYSIS: KEW,SOUTH ENGLAND           : 51.70N   0.20W : 1967
        PERIOD: 9. 1. 1 TO 15. 1.24

DRY BULB TEMPERATURE     DEG C

D  M  T        MIN    T          MAX    T         MEAN

 9, 1  1-24   -3.0 @ 7         1.3 @17          -1.0
10, 1  1-24   -1.5 @ 4         3.3 @22           1.2
11, 1  1-24    3.4 @ 1         7.2 @15           5.6
12, 1  1-24    1.5 @ 4         7.7 @16           4.9
13, 1  1-24    3.3 @ 1         9.6 @16           7.3
14, 1  1-24    0.9 @ 8         6.8 @15           3.7
15, 1  1-24    1.5 @18         4.1 @ 1           2.5

MONTH          -3.0 @ 7, 9     9.6 @16,13        3.4

ALL PERIOD     -3.0@ 7, 9, 1   9.6@16,13, 1      3.4
```

>>

Figure G.18 Temperature statistics extracted for each hour in a period of interest. Display by ESPout.

```
CLIMATE ANALYSIS: KEW,SOUTH ENGLAND                    : 51.70N   0.20W : 1967
           PERIOD:  9. 1. 1 TO 15. 1.24

MONTH  1          : DRY BULB TEMPERATURE        DEG C

HR      MIN     MAX      MEAN    DEV.N       LODAY   HIDAY
 1     -2.1     4.3      2.1      2.4        -2.1     3.3
 2     -2.6     5.5      2.1      2.7        -2.6     5.5
 3     -2.6     6.5      2.2      2.8        -2.6     6.5
 4     -2.7     6.8      1.9      3.0        -2.7     6.8
 5     -2.7     6.8      2.1      3.0        -2.7     6.8
 6     -2.9     7.1      2.1      3.1        -2.9     7.1
 7     -3.0     7.3      2.2      3.2        -3.0     7.3
 8     -2.8     7.2      2.2      3.2        -2.8     7.2
 9     -2.8     7.3      2.3      3.2        -2.8     7.3
10     -2.6     7.7      2.6      3.2        -2.6     7.7
11     -1.8     8.1      3.0      3.1        -1.8     8.1
12     -0.7     8.7      3.8      3.0        -0.7     8.7
13     -0.3     8.9      4.3      3.0        -0.3     8.9
14     -0.1     9.3      4.7      3.1        -0.1     9.3
15      0.5     9.4      5.1      3.1         0.5     9.4
16      1.1     9.6      5.3      3.0         1.1     9.6
17      1.3     9.1      5.1      2.8         1.3     9.1
18      1.1     8.2      4.9      2.8         1.1     8.2
19      1.1     7.6      4.7      2.5         1.1     7.6
20      0.9     7.1      4.6      2.4         0.9     7.1
21      0.8     7.0      4.3      2.2         0.8     6.5
22      0.5     6.4      4.0      2.0         0.5     5.7
23     -0.5     6.1      3.8      2.1        -0.5     6.1
24     -1.0     4.8      3.3      1.9        -1.0     4.8

MIN    -3.0     4.3      1.9
MAX     1.3     9.6      5.3
MEAN   -1.0     7.4      3.4               >>   -1.0     7.3
```

Figure G.19 Degree days less than a 15.3°C base for a cold winter week. Display by ESPclm.

```
CLIMATE ANALYSIS: KEW,SOUTH ENGLAND                    : 51.70N   0.20W : 1967
           PERIOD:  9. 1. 1 TO 15. 1.24

DEGREE DAY ANALYSIS AT  15.3 DEG C

D   M   T        <        >
 9,  1  1-24   16.30    0.00
10,  1  1-24   14.10    0.00
11,  1  1-24    9.68    0.00
12,  1  1-24   10.40    0.00
13,  1  1-24    8.03    0.00
14,  1  1-24   11.60    0.00
15,  1  1-24   12.85    0.00

MONTH:AV/DAY   11.85    0.00

MONTH:TOTAL    82.95    0.00

TOTAL:AV/DAY   11.85    0.00

PERIOD TOTAL   82.95    0.00


                                >>
```

Figure G.20 Climatic severity index values and partial breakdowns for the Kew 1967 collection. Display by ESPclm.

```
Climatic Severity Index for :
KEW,SOUTH ENGLAND                 51.70N   0.20W : 1967
HOUSE TYPE  16

Month    CSIT    CSIRD    CSIRF    CSIW    CSI
  1      4.8      0.7      0.0      0.3     5.8
  2      4.3      0.6      0.0      0.5     5.4
  3      3.6      0.4     -0.0      0.6     4.7
  4      3.4      0.5     -0.0      0.5     4.4
  5      2.1      0.4     -0.0      0.5     2.9
  6      0.7      0.3     -0.0      0.3     1.3
  7     -0.6      0.2     -0.0      0.3    -0.2
  8      0.1      0.3     -0.0      0.2     0.6
  9      1.0      0.5     -0.0      0.2     1.8
 10      2.1      0.6      0.0      0.5     3.1
 11      4.2      0.6      0.0      0.2     5.0
 12      4.8      0.7      0.0      0.2     5.8


Annual


Total   30.6      5.8     -0.0      4.3    40.6

Mean     2.5      0.5     -0.0      0.4     3.4


                                >>
```

Figure G.21 A tabulation of one day's climatic data from the Kew 1967 collection. Display by ESPclm.

```
DAY 17 OF MONTH  7

HR D.B. TEMP. DR.N. RAD. DF.H. RAD. WD. VEL. WD. DIR. REL. H.
    DEG.C        W/M^2       W/M^2      M/S     DEG.F.N.    %
  1    16.30        0.          0.       0.00       0.    81.00
  2    16.20        0.          0.       0.00       0.    85.00
  3    15.20        0.          0.       0.30      45.    86.00
  4    15.90        0.          0.       0.80      85.    81.00
  5    15.20        1.         10.       1.00      95.    81.00
  6    15.90        7.         41.       1.30     110.    80.00
  7    18.20      140.         77.       2.20     130.    78.00
  8    20.60      405.         95.       4.40     155.    68.00
  9    22.30      575.        105.       5.40     165.    64.00
 10    23.60      622.        130.       5.70     170.    60.00
 11    25.00      634.        158.       6.40     165.    55.00
 12    26.20      605.        217.       7.00     160.    50.00
 13    26.70      557.        241.       7.20     165.    48.00
 14    27.10      588.        214.       7.20     170.    48.00
 15    28.00      610.        224.       7.20     170.    45.00
 16    28.70      585.        218.       7.00     160.    43.00
 17    28.20      475.        172.       6.70     155.    45.00
 18    27.50      390.        123.       6.20     155.    45.00
 19    26.70      235.         81.       4.90     150.    49.00
 20    25.80       49.         40.       3.10     150.    52.00
 21    23.80        4.          0.       2.10     160.    62.00
 22    23.60        0.          0.       2.20     190.    64.00
 23    22.50        0.          0.       2.10     210.    69.00
 24    21.80        0.          0.       2.10     225.    70.00

CONTINUE WITH ANOTHER DAY ?
>
```

Figure G.22 Predictions of external surface shading as caused by surrounding and facade obstructions. Display by ESPshd.

```
SHADING INFORMATION FOR MONTH JUL

OPAQUE SURFACE  2 NO.OF WINDOWS  4

HOUR OPAQUE  WINDOW SHADING FOR WINDOW NUMBER .......
      SHADING        1      2      3      4
  1 SUN NOT UP
  2 SUN NOT UP
  3 SUN NOT UP
  4    62           100    100      0      0
  5    72           100    100     60      0
  6    68            96    100    100      0
  7    60            40    100    100      0
  8    70            20    100    100     95
  9    30             0     10     10     10
 10 SURFACE NOT SHADED
 11 SURFACE NOT SHADED
 12 SURFACE NOT SHADED
 13 SURFACE FULLY SHADED
 14 SURFACE FULLY SHADED
 15 SURFACE FULLY SHADED
 16 SURFACE FULLY SHADED
 17 SURFACE FULLY SHADED
 18 SURFACE FULLY SHADED
 19 SURFACE FULLY SHADED
 20 SURFACE FULLY SHADED
 21 SUN NOT UP
 22 SUN NOT UP
 23 SUN NOT UP
 24 SUN NOT UP
MEAN OVERALL SURFACE SHADING DURING SUN-UP HOURS =    68
CONTINUE ?
>
```

Figure G.23 Predictions of internal surface insolation as caused by direct solar penetration through a specified window. Display by ESPins.

```
Insolation information for month No.  3

Surface  1     Window  1

                          Insolation
                         ------------
Hour   Shading    Surf  Value    Surf  Value    Surf
  1  Sun not up
  2  Sun not up
  3  Sun not up
  4  Sun not up
  5  Sun not up
  6  Sun not up
  7    0.0         6   100.0      0     0.0      6
  8    0.0         6    76.0      8    24.0      6
  9    0.0         6    36.0      8    64.0      6
 10    0.0         6    12.0      8    88.0      6
 11    0.0         8   100.0      0     0.0      8
 12    0.0         8   100.0      0     0.0      8
 13    0.0         8   100.0      0     0.0      8
 14    0.0         2     8.0      8    92.0      2
 15    0.0         2    28.0      8    72.0      2
 16    0.0         2    60.0      8    40.0      2
 17    0.0         2   100.0      0     0.0      2
 18  Sun not up
 19  Sun not up
 20  Sun not up
 21  Sun not up
 22  Sun not up
 23  Sun not up
 24  Sun not up

                              >>
```

Figure G.24 Solar azimuth versus altitude for a specified location and date. Display by ESPins.

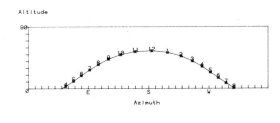

>>

Figure G.25 Inter-surface black body view factors as predicted. Display by ESPvwf.

```
VIEW FACTOR INFORMATION
-----------------------

        1        2        3        4        5        6        7        8
1    0.0000   0.1335   0.0190   0.0178   0.1117   0.1069   0.3056   0.305
2    0.1335   0.0000   0.1069   0.1117   0.0178   0.0190   0.3056   0.305
3    0.0402   0.2072   0.0000   0.1801   0.0000   0.0000   0.2863   0.286
4    0.0289   0.2208   0.1801   0.0000   0.0000   0.0000   0.2851   0.285
5    0.2208   0.0289   0.0000   0.0000   0.0000   0.1801   0.2851   0.285
6    0.2072   0.0402   0.0000   0.0000   0.1801   0.0000   0.2863   0.286
7    0.1566   0.1566   0.0691   0.0726   0.0726   0.0691   0.0000   0.403
8    0.1566   0.1566   0.0691   0.0726   0.0726   0.0691   0.4033   0.000
```

>>

Figure G.26 Lobby infiltration (solid curve) and air exchange with the living zone in a house. Predictions by ESPair.

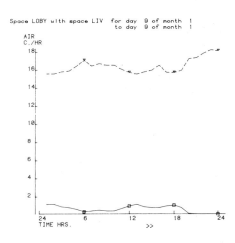

Figure G.27 A table of zone infiltration rates and zone coupled air flow as predicted by ESPair.

Tabular Output : Infiltration and total zone coupled air flow (AC/Hr)

Day 9 of month 1

Hr	LIV		KIT		LOBY		LAND		BED1		BED2		LOFT	
1	0.0	14.5	1.3	0.0	1.2	16.2	0.0	0.2	0.0	0.1	0.1	0.0	0.0	0.1
2	0.0	14.5	1.3	0.0	1.2	16.2	0.0	0.2	0.0	0.1	0.1	0.0	0.0	0.1
3	0.0	14.3	1.1	0.0	0.9	16.3	0.0	0.5	0.0	0.1	0.1	0.0	0.0	0.1
4	0.0	14.3	1.1	0.0	0.8	16.4	0.0	0.5	0.0	0.1	0.1	0.0	0.0	0.1
5	0.0	13.8	0.8	0.0	0.6	16.7	0.0	0.5	0.0	0.1	0.0	0.0	0.0	0.1
6	0.0	13.2	0.5	0.1	0.4	17.1	0.0	0.5	0.0	0.1	0.0	0.1	0.0	0.1
7	0.0	13.8	0.6	0.0	0.4	16.5	0.0	0.5	0.0	0.1	0.0	0.1	0.0	0.1
8	0.0	13.6	0.7	0.0	0.5	16.8	0.0	0.5	0.0	0.1	0.0	0.0	0.0	0.1
9	0.0	13.7	0.6	0.0	0.4	16.7	0.0	0.5	0.0	0.1	0.0	0.1	0.0	0.1
10	0.0	13.7	0.6	0.0	0.4	16.7	0.0	0.5	0.0	0.1	0.0	0.1	0.0	0.1
11	0.0	14.0	0.8	0.0	0.7	16.4	0.0	0.2	0.0	0.1	0.0	0.0	0.0	0.1
12	0.0	14.3	1.1	0.0	1.0	16.3	0.0	0.2	0.0	0.1	0.1	0.0	0.0	0.1
13	0.0	14.5	1.3	0.0	1.2	16.2	0.0	0.2	0.0	0.1	0.2	0.0	0.0	0.1
14	0.0	14.3	1.1	0.0	0.9	16.3	0.0	0.2	0.0	0.1	0.1	0.0	0.0	0.1
15	0.0	14.2	1.0	0.0	0.8	16.4	0.0	0.5	0.0	0.1	0.1	0.0	0.0	0.1
16	0.0	13.8	1.0	0.0	0.8	16.9	0.0	0.5	0.0	0.1	0.1	0.0	0.0	0.1
17	0.0	14.3	1.1	0.0	1.0	16.3	0.0	0.2	0.0	0.1	0.1	0.0	0.0	0.1
18	0.0	14.4	1.3	0.0	1.1	16.3	0.0	0.2	0.0	0.1	0.1	0.0	0.0	0.1
19	0.0	14.2	1.0	0.0	0.8	16.4	0.0	0.5	0.0	0.1	0.1	0.0	0.0	0.1
20	0.2	13.2	0.4	0.2	0.2	17.3	0.0	0.5	0.0	0.1	0.0	0.1	0.0	0.1
21	0.5	13.1	0.2	0.2	0.1	17.4	0.0	0.5	0.0	0.1	0.0	0.1	0.0	0.1
22	0.7	12.7	0.1	0.5	0.1	17.9	0.0	0.7	0.0	0.1	0.0	0.1	0.0	0.1
23	1.2	12.4	0.0	0.8	0.1	18.3	0.0	1.0	0.0	0.1	0.0	0.1	0.0	0.1
24	1.1	12.5	0.0	0.8	0.1	18.2	0.0	1.0	0.0	0.1	0.0	0.1	0.0	0.1

>>

Figure G.28 Shortwave transmittance (*T* axis and lower curve), absorptance (*A* axis and upper curve) and reflectance (difference between curves) for heat absorbing double glazing. Predictions by ESPwin.

T
1.0
0.8
0.6
0.4
0.2
0.0
0 10 20 30 40 50 60 70 80 90
ANGLE
0.0
0.2
0.4
0.6
0.8
A
1.0
>>

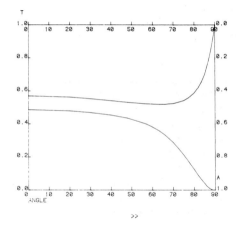

Figure G.29 Daylight factor levels for a regular grid placed at the working plane height.

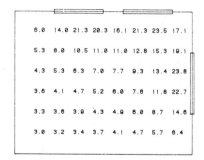

WORKING PLANE HEIGHT = 0.5 M
6.0 14.0 21.3 20.3 16.1 21.3 23.5 17.1
5.3 8.0 10.5 11.0 11.0 12.8 15.3 19.1
4.3 5.3 6.3 7.0 7.7 9.3 13.4 23.8
3.6 4.1 4.7 5.2 6.0 7.6 11.8 22.7
3.3 3.6 3.9 4.3 4.9 6.0 8.7 14.6
3.0 3.2 3.4 3.7 4.1 4.7 5.7 6.4

Index